Applied Many-Body Methods in Spectroscopy and Electronic Structure

Applied Many-Body Methods in Spectroscopy and Electronic Structure

Edited by
Debashis Mukherjee
Indian Association for the Cultivation of Science
Calcutta, India

PLENUM PRESS • NEW YORK AND LONDON

Library of Congress Cataloging in Publication Data

Applied many-body methods in spectroscopy and electronic structure / edited by Debashis Mukherjee.

p. cm.

Includes bibliographical references and index.

ISBN 0-306-44193-4

1. Spectrum analysis. 2. Electronic structure. 3. Many-body problem. 4. Chemistry, Physical and theoretical. I. Mukherjee, D. (Debashis), date.

QC451.A66 1992 92-3802

535.8′4—dc20 CIP

Proceedings of a symposium on Applied Many-Body Methods in Spectroscopy and Electronic Structure, held December 10–12, 1990, in Calcutta, India

ISBN 0-306-44193-4

Printed in the United States of America

FOREWORD

There has been a steady advance of the atomic and molecular many-body methodology over the last few years, with a concomitant development of versatile computer codes. Understanding and interpretation of electronic structural features and the associated spectroscopic properties via many-body techniques are becoming competitive with those obtained with the traditional formalisms. Since the many-body techniques are not yet a part of the repertoire of the "black-box tools" of electronic structure and spectroscopy, it seems worthwhile to take stock now of the recent progress in certain selected areas.

The present volume is more in the nature of proceedings of a "Paper Symposium," rather than of one which actually took place. We did organize in Calcutta, between December 10 and 12, 1990, a small meeting on Applied Many-Body Methods to Spectroscopy and Electronic Structure, jointly organized by the Indian Association for the Cultivation of Science and the S.N. Bose National Centre for Basic Sciences. Several leading practitioners were invited, among which some could not come for various reasons. We later decided to publish a volume of a more enhanced scope and invited other experts as well to contribute papers on topics which are related in spirit with the coverage of the meeting. The volume embodies such diverse topics as a modern pedagogical survey of the Bruckner-Goldstone theory, recent progress in the Propagator theories (Multiconfiguration Propagator theories, Algebraic Diagrammatic Construction and studies on photoionization cross-section), Coupled-Cluster approaches to the parity non-conserving processes, Complex Scaling methods for atomic and molecular resonances and other nonstationary states, and recent developments of Effective Hamiltonian formalisms in the framework of Coupled-Cluster theory (for treating geometry and spectroscopic constants and studies on quasi-Hilbert and quasi-Fock theories) and Many-Body Perturbation theories of Magnetic Hamiltonians.

I am deeply grateful to all the authors of the volume for their contributions and cooperation. Thanks are due to Professors U.R. Ghatak and C.K. Majumdar, the two Directors of the host Institutes for their unstinting help. My pre- and post-doctoral students, particularly Mr. G. Sanyal, Mr. S. Hannan, and Dr. B. Kundu, and my brother Dr. S. Mukherjee have ably assisted me in the editing process. Finally, I thank Ms. Janie Curtis of the Plenum Publishing Company for her kind patience during the period the articles were being collected and collated.

Calcutta, July, 1991 — Debashis Mukherjee

CONTENTS

THE MANY-BODY PERTURBATION THEORY OF BRUECKNER AND GOLDSTONE

Werner Kutzelnigg

Lehrstuhl für Theoretische Chemie
Ruhr-Universität Bochum
W-4630 Bochum, Germany

After a historical introduction to the many-body problem the landmark papers on the many-body perturbation theory by Brueckner (1955) and by Goldstone (1957) are reviewed in the light of the present knowledge of many fermion systems. Brueckner started from conventional Rayleigh-Schrödinger perturbation theory, and his derivation of what he called the linked-cluster theorem was pedestrian but very tedious and not sufficiently general. Goldstone used the apparatus of quantum electrodynamics for a very elegant (but to some extent criticizable) derivation of the linked cluster expansion, that showed at first glance little resemblance to that of Brueckner. It is shown that most ingredients of Goldstone's approach are not necessary for a simple and transparent derivation of the linked cluster expansion. It is only compulsory to work in Fock space. A very simple modern derivation is then given. Finally some insight into the time-dependent theory is gained by means of its regularization.

1. HISTORICAL BACKGROUND

In his well-known book[1] Nozière makes the statement that 'before 1950 there was practically no many-body problem'. This reflected a general feeling, but it is true only if one refers to many-body effects in extended systems, the many-electron problem in atomic theory is certainly older. That the electron interaction is the main bottleneck on the way to an accurate theory of atoms has been known since the early days of quantum mechanics. Hylleraas[2] succeeded around 1930 in finding a satisfactory numerical solution of the two-electron problem present in the He-atom, although he was not able to generalize his ansatz to the case of more than two electrons. A much cruder approach to atoms in general was found by Hartree[3] and improved by others[4] in the mean-field one-configuration approach, now known as Hartree-Fock, that has also turned out quite successful in molecular theory[5]. The multiconfiguration Hartree-Fock method was initiated by Jucys[6] in the early 50ies. That an n-electron wave function can be expanded in Slater determinants and that such an expansion can, in principle, be made exact,

Applied Many-Body Methods in Spectroscopy and Electronic Structure
Edited by D. Mukherjee, Plenum Press, New York, 1992

provided that the basis set from which the Slater determinants are built up, is complete, has been known in the 30ies. This can e.g. be seen from the classical textbooks of Kramers[7] and Frenkel[8], where the former used the configuration space, the latter the Fock space formulation.
So it was known, at least in principle, long before 1950 how the quantum mechanical many-body problem could be solved, although successful practical applications of the configuration interaction (CI) approach, as it is now called, had to wait until fast electronic computers became available, which happened in the late 50ies.

There are mainly four branches of physics, where one deals with many-particle systems. Complementary to the theory of atoms and molecules already mentioned, there is solid-state physics, nuclear physics and statistical mechanics. While in statistical mechanics a genuine many-body theory had been developed rather early, e.g. by Ursell[9] and Mayer[10] (which, by the way inspired later the 'coupled-cluster' expansion in many-fermion theory[11]) solid state physics was until quite recently rather successful in terms of an effective-one-particle theory and the extensive use of the quasiparticle concept without much worrying about many-electron effects. The need for a satisfactory many-particle theory (in fact a many-fermion theory) arose around 1955 independently in nuclear theory and in the theory of atoms and molecules.

The motivation was different in these two domains. In the theory of small electronic systems (atoms or molecules), that we shall henceforth refer to as 'quantum chemistry', the advent of computers made it possible to perform Hartree-Fock calculations even for medium-sized molecules. At the same time it became clear that the Hartree-Fock approximation is not accurate enough for many purposes. A series of papers by Löwdin in 1955 on the CI approach to the many-electron problem[12] and related work[13] appeared at a time when the importance of many-electron effects became generally appreciated - although it took still a long time before many-electron calculations on this level became really standard[14]. For a long while, ab initio quantum chemistry was dominated by SCF ('self-consistent field' i.e. Hartree-Fock) calculations.

In nuclear theory the point was that the Hartree-Fock method is not applicable, if - as it seemed obvious in the 50ies - the nuclear interaction potential is highly repulsive (keyword 'hard core') for short distances between the nucleons. The Hartree-Fock ansatz allows two interacting particles to occupy the same region of space. Since the Hartree-Fock expectation value diverges, and hence Hartree-Fock could not be used, one had to find a different starting point. An obvious possibility was to start from a system of non-interacting particles and to treat the interaction by means of perturbation theory. The use of perturbation theory to take care of the electron interaction had already been proposed in 1934 by Møller and Plesset [15] in a paper that became very popular in Quantum Chemistry much later[16]. The first question to answer in a systematic study of many-body perturbation theory, was whether to use the Rayleigh-Schrödinger (RS) or the Brillouin-Wigner (BW) variant. There were some a priori arguments in favour of BW, since it often appeared to converge faster. A main advantage of BW seemed to be that for some model systems (e.g. a 2x2 matrix) it is exact, while RS can never be exact. In many-body theory, however, BW failed. It could be shown that for an extended system the 2nd order BW perturbation energy is not proportional to the number n of particles. The 2nd order RS perturbation energy, is proportional to n. In an important paper [17] Brueckner analyzed the application of RS perturbation theory to a system of interacting fermions and he found that the higher orders $E^{(k)}$ of the energy contain contributions that have a wrong dependence on the number of particles. He was able, however, to show that 'wrong' contributions cancel with other terms. As an example take $E^{(3)}$ which consists of two terms (see sec. 2; g is the reduced resolvent of the unperturbed Hamiltonian).

$$E^{(3)} = <\Phi|VgVgV|\Phi> - <\Phi|V|\Phi><\Phi|Vg^2V|\Phi> \tag{1.1}$$

The contributions to the first term on the r.h.s. in (1.1) with the wrong dependence on n cancel with identical contributions to the second term. The remaining contributions are 'linked' (in a sense to be explained later) and are for an extended system proportional to n. To demonstrate this cancellation turned out to be rather tedious, especially for higher orders. Brueckner was not able to give a general proof valid to all order for his 'linked cluster theorem', and hence of extensivity of RS perturbation theory.

Other authors, including Hubbard[18], Hugenholtz[19] and Goldstone[20] found general proofs for the linked-cluster theorems, but these were unexpectedly far-fetched and not easy to understand. Only Goldstone's derivation [20], although certainly rather unconventional, was very elegant and not really hard to follow, provided than one was ready to learn a sort of a new language.

If one places Goldstone's derivation of the linked-cluster expansion into the historical context it appears extremely straightforward and by no means esoteric. In fact Goldstone made use of the formalism current in that branch of physics which was most 'in' in the 50ies, namely quantum electronic dynamics (QED), as it had shortly ago been developped by Tomonaga[21], Schwinger[22] and Feynman[23]. The application of the formalism of QED to many-fermion systems is particularly convincing if one realizes that QED is a theory of electrons interacting via a quantized radiation field.

The ingredients from QED that Goldstone took over to many-body theory were

a. The time-dependent formulation of the theory, even for a time-independent perturbation.
b. The 2nd quantization notation, i.e. the use of creation and annihilation operators.
c. The particle-hole formalism based on an arbitrary choice of a physical vacuum.
d. The 'interaction representation', which is intermediate between the Schrödinger and the Heisenberg representation.
e. The evaluation of the energy by the Gell-Mann-Low[24] formula.
f. Wick's theorem[25] for the evaluation of (time-ordered) products of creation and annihilation operators.
g. The use of Feynman-like diagrams, not only in order to illustrate the analytical expression, but as a real instrument in the derivation of the final results.

Brueckner's original[17] and Goldstone's later[20] derivation of the linked-cluster expansion were formulated in so different languages that it was hard to see how they are related, especially to see the clue for the linked-cluster theorem, that appeared almost trivially in Goldstone's approach and that was so hard to prove in the traditional perturbation theory used by Brueckner.

In QED divergencies that have to be removed by 'renormalization', play a central role. Fortunately the more serious divergencies that plague QED did not arise in Goldstone's application of the QED formalism. Only one divergency - that in the spirit of QED is rather harmless - survived, and is responsible for one somewhat problematic aspect of Goldstone's many-body theory (see sec.3). Looked at from a more rigorous point of view it is certainly a drawback of Goldstone's derivation that the adiabatic limit of an exponentially switched perturbation does not exist for the wave function, although this limit can be performed for the energy.

Goldstone's derivation was readily accepted and had a large impact on the mind of many-particle physics. Starting with Goldstone's paper[20] many-particle physics became to a large extent applied quantum field theory rather than a branch of conventional quantum mechanics. One can say that the field-theoretical formulation of the many-body problem tied this branch of physics to 'main-stream physics', which was at that time dominated by QED, and made it so more fashionable. At the same time the change of language involved in this process increased the gap between many-body physics and quantum chemistry, where there was no motivation to switch to field theoretical language, except for some outsiders[26]. In spite of some early applications of Goldstone's theory to atoms[27], it took until ~1970 until this formalism was applied to molecules. One reason for this reluctance was probably that Goldstone's theory was formulated for free-electron states and it was not obvious whether the main results could be generalized to molecular orbitals. On the other hand the developments of quantum chemistry were even more reluctantly taken care of by nuclear physicists. It is not too astonishing that the old 'separated-pair theory' proposed by Hurley, Lennard-Jones and Pople around 1953 [28] has recently been 'rediscovered' in nuclear physics [29].

The original motivation of Goldstone's paper[20], which is very obvious from its title, has been to prove the linked-cluster theorem conjectured by Brueckner[17]. But rather than to prove Brueckner's conjecture, Goldstone has derived an alternative expansion of its own right. The relation to Brueckner's conjecture was rather indirect. After Goldstone's paper some authors tried to prove or to derive his expansion in a simpler or more direct way. Here one can mention the work of Frantz and Mills[30], who showed in a time-independent way by means of their 'factorization theorem' that the Goldstone linked-diagram expansion is a solution of the Schrödinger equation for a many-fermion system. Brandow[31] started from the time-independent Brillouin-Wigner expansion in order to derive the linked-diagram expansion in a rather tedious way, making also use of the factorization theorem. This opened a straightforward way to a generalization of many-body perturbation theory to open-shell states.

An interesting contribution was the Lie algebraic formulation of perturbation theory by Primas[32] (see also Yaris [33]).The Lie algebraic structure, when done in Fock space (this was not so clearly specified in Primas' approach) does, in fact, guarantee that only connected diagrams contribute to the energy. An alternative simple but too general proof of the connectedness of MB-PT was given by Caianiello[34]. In fact, in order to rederive the Goldstone expansion, it is not sufficient to show that RS perturbation theory only contains connected diagrams, one must further show that there are no renormalization diagrams, but instead so-called exclusion-principle-violating diagrams.

In the meantime the interest in many-body theory had shifted to degenerate or quasidegenerate states. It soon turned out that the generalization of the Goldstone formalism to such states causes a lot of difficulties and that a time-independent formalism is advantageous. (The rich literature on quasidegenerate many-body perturbation theory starting with C. Bloch[35] can here not be reviewed nor can certain interesting controversies[36] and their final solution[37] be discussed). A rather straightforward modern version of quasidegenerate many-body theory is the formalism of 'Quantum Chemistry in Fock space'[37] that has recently been reviewed in a didactic form[38]. A central aspect of this theory is the 'separation theorem' which immediately implies that only connected diagrams can contribute to additively separable quantities like the energy or the logarithm of the wave operator. Combination of the separation theorem with a particle-hole picture leads to a very simple derivation of Goldstone's linked cluster expansion as a by-product. One can also rationalize the meaning of 'folded diagrams' that arose in Brandow's attempt [31] to generalize Goldstone's diagrammatic expansion to the quasidegenerate case.

Looking at the long history of the linked-diagram theorem it is really astonishing how fundamental and easy to prove it is, and that a simple proof had not been found earlier[37,39].

The aim of the present paper is not to follow the landmark papers of Brueckner[17] and Goldstone[20] as closely as possible but rather to revisit them in the light of the present understanding of many-body theory. We therefore stress essential aspects and skip details. It turns out useful to introduce the concept of diagrams for the interpretation of Brueckner's results, although he has actually not used diagrams. On the other hand it is not necesssary to refer extensively to diagrams to clarify the key features of Goldstone's derivation, although in Goldstone's original derivation diagrams play a central role. At the end we shall give a modern derivation of Goldstone's expansion to stress how simple this really is and to stress that the essential point is to formulate the wave operator in Fock space, i.e. withouth specifying the particle number. Separability and connectedness are Fock space properties. A nice feature of Fock space theory is that one need not worry about summation restrictions in sums over spinorbital labels. These are automatically taken care of by the anticommutation properties of the fermion operators. Consequently so called EPV (exclusion-principle-violating) diagrams arise naturally only in a Fock space theory.

Various didactic introductions to many-body theory, in particular to the use of diagrams are available in the literature. Among these we mention especially ref. 40-42. These have little overlap with the main concern of the present text. We hope to offer beginners a relatively easy access to many-body perturbation theory, but we also want to give initiated readers some new insight.

In this review we only consider closed-shell states. For open-shell states the reader is referred to ref. 43 and 44.

2. TRADITIONAL PERTURBATION THEORY AND BRUECKNER'S LINKED-CLUSTER THEOREM

The probably simplest way to derive RS as well as BW perturbation theory is as follows (Readers interested in a mathematically rigorous formulation of perturbation theory are referred to Kato's famous book[45], where among other things the conditions for convergence of the perturbation expansion are discussed). Let

$$H = H_o + \lambda V \tag{2.1}$$

with H_o the unperturbed Hamiltonian and λV the perturbation. The eigenfunctions of H and H_o are called Ψ and Φ respectively

$$H\Psi = E\Psi \tag{2.2a}$$

$$H_o\Phi = E_o\Phi \tag{2.2b}$$

We choose the intermediate normalization

$$< \Phi|\Phi >=< \Phi|\Psi >= 1 \tag{2.3}$$

and decompose Ψ as

$$\Psi = \Phi + \chi; \quad < \Phi|\chi >= 0 \tag{2.4}$$

Insertion of (2.4) into (2.2a) and use of (2.1 and 2.2b) leads to

$$(H - E)\chi = (\Delta E - \lambda V)\Phi; \quad \Delta E = E - E_o \tag{2.5}$$

We define the reduced resolvent $G_o(z)$ of H_o with the property

$$G_o(z) \cdot (z - H_o) = Q \equiv 1 - |\Phi >< \Phi| \tag{2.6}$$

Application of $G_o(E)$ on the left to (2.5) yields

$$[1 - \lambda G_o(E)V]\chi = \lambda G_o(E)V\Phi \tag{2.7}$$

After formal inversion of the operator in brackets one gets

$$\chi = \lambda[1 - \lambda G_o(E)V]^{-1}G_o(E)V\Phi \tag{2.8}$$

We abbreviate

$$\tilde{g} \equiv G_o(E) \tag{2.9}$$

and make a Taylor expansion in λ, which leads to the BW series

$$\chi = \left\{\lambda\tilde{g}V + \lambda^2\tilde{g}V\tilde{g}V + ...\right\}\Phi \tag{2.10}$$

$$\Delta E = \lambda < \Phi|V|\Psi >= \lambda < \Phi|V|\Phi > +\lambda^2 < \Phi|V\tilde{g}V|\Phi > +\lambda^3 < \Phi|V\tilde{g}V\tilde{g}V|\Phi > +... \tag{2.11}$$

The RS series is obtained if one applies $G_o(E_o)$ on the left to (2.5)

$$[1 - G_o(E_o)(\lambda V - \Delta E)]\chi = \lambda G_o(E_o)V\Phi \tag{2.12}$$

We abbreviate

$$g \equiv G_o(E_o) \tag{2.13}$$

invert the operator in brackets and expand

$$\chi = \lambda[1 - g(\lambda V - \Delta E)^{-1}]gV\Phi = \{\lambda gV + \lambda g(\lambda V - \Delta E)gV + \lambda g(\lambda V - \Delta E)g(\lambda V - \Delta E)gV + ...\}\Phi \tag{2.14}$$

$$\Delta E = \lambda < \Phi|V|\Psi >= \lambda < \Phi|V|\Phi > +\lambda^2 < \Phi|VgV|\Phi > +\lambda^2 < \Phi|Vg(\lambda V - \Delta E)gV|\Phi > +... \tag{2.15}$$

The RS series is obtained if one orders ΔE in powers of λ

$$\Delta E = \lambda E^{(1)} + \lambda^2 E^{(2)} + ... \tag{2.16}$$

and eliminates successively ΔE from (2.14) and (2.15). The result is

$$\begin{aligned}
&E^{(1)} =<\Phi|V|\Phi> \\
&E^{(2)} =<\Phi|VgV|\Phi> \\
&E^{(3)} =<\Phi|VgVgV|\Phi> - <\Phi|V|\Phi><\Phi|Vg^2V|\Phi> \\
&E^{(4)} =<\Phi|VgVgVgV|\Phi> - <\Phi|V|\Phi><\Phi|Vg^2VgV|\Phi> - <\Phi|V|\Phi><\Phi|VgVg^2V|\Phi> \\
&\quad + <\Phi|V|\Phi>^2<\Phi|Vg^3V|\Phi> - <\Phi|VgV|\Phi><\Phi|Vg^2V|\Phi>
\end{aligned} \tag{2.17}$$

The first contributions $<\Phi|Vg...gV|\Phi>$ are called 'direct' terms, the other ones, starting with $E^{(3)}$, like $<\Phi|V|\Phi><\Phi|Vg^2V|\Phi>$ 'renormalization' terms. They come from the expansion of ΔE in (2.14).

Obviously the BW series is formally simpler, since it only involves 'direct' terms

$$\begin{aligned}
&\tilde{E}^{(1)} =<\Phi|V|\Phi> \\
&\tilde{E}^{(2)} =<\Phi|V\tilde{g}V|\Phi> \\
&\tilde{E}^{(3)} =<\Phi|V\tilde{g}V\tilde{g}V|\Phi> \\
&\tilde{E}^{(4)} =<\Phi|V\tilde{g}V\tilde{g}V\tilde{g}V|\Phi>
\end{aligned} \tag{2.18}$$

but each $\tilde{E}^{(k)}$ depends via $\tilde{g}$ on the exact energy E, such that the BW series is not genuinely an expansion in powers of λ. The RS-series is an expansion in powers of λ, but contains also so-called 'renormalization terms' like $<\Phi|V|\Phi><\Phi|Vg^2V|\Phi>$ etc.

If we introduce the 'spectral representation' of the reduced resolvent (note that $\Phi \equiv \Phi_o$)

$$G_o(z) = \sum_{\mu=1}^{\infty} (z - E_\mu)^{-1} |\Phi_\mu><\Phi_\mu| \tag{2.19}$$

in terms of the eigenstates (Φ_μ) and eigenvalues (E_μ) of H_o, - assuming, for simplicity's sake, that all these are discrete - we can reformulate (2.17) and similarly (2.18) in the sum-over-states form

$$E^{(2)} = \sum_{\mu=1}^{\infty} <\Phi|V|\Phi_\mu> (E_o - E_\mu)^{-1} <\Phi_\mu|V|\Phi> \tag{2.20a}$$

$$\begin{aligned}
E^{(3)} = &\sum_{\mu=1}^{\infty}\sum_{\nu=1}^{\infty} <\Phi|V|\Phi_\mu> (E_o - E_\mu)^{-1} <\Phi_\mu|V|\Phi_\nu> (E_o - E_\nu)^{-1} <\Phi_\nu|V|\Phi> \\
&- <\Phi|V|\Phi> \sum_{\mu=1}^{\infty} <\Phi|V|\Phi_\mu> (E_o - E_\mu)^{-2} <\Phi_\mu|V|\Phi>
\end{aligned} \tag{2.20b}$$

If one applies perturbation theory to a many-body problem, H_o is in the form

$$H_o = \sum_k h(k) \tag{2.21}$$

i.e. a one-electron operator, and λV can be written as

$$\lambda V = \sum_{k<l} V(k,l) \tag{2.22}$$

The unperturbed wave function (we assume that the state that we consider is non-degenerate) is a Slater determinant. We designate the spin-orbitals occupied in Φ as ψ_i, those unoccupied (virtual) as ψ_a. The ψ_i together with the ψ_a are supposed to form a complete basis of one-particle Hilbert space.

Brueckner[17] was worried by the following observation. It is rather easy to see that for an extended system $E^{(1)}$ or $E^{(2)}$ are proportional to the number of particles, i.e. they are 'extensive' quantities, as the total energy should be. The second term to $E^{(3)}$ (see 2.17 or 2.20b) a so-called renormalization term, is however a product of two factors, each of which goes as $\sim N$. If there are not similar contributions $\sim N^2$ in the first term of $E^{(3)}$, $E^{(3)}$ would have the wrong dependence on the number of particles and hence be unphysical. (One of the arguments for the refutation of the Brillouin-Wigner series (2.18) has actually been that except $\tilde{E}^{(1)}$ none of the $\tilde{E}^{(k)}$ is extensive). Brueckner therefore tried to show that all terms $\sim N^2$ (or $\sim N^3$ etc.) cancel exactly, such that any $E^{(k)}$ of the Rayleigh-Schrödinger expansion is extensive. To follow Brueckner's argument we must evaluate the expressions (2.20) explicitly.

Since V is a two-electron operator only those matrix elements $< \Phi|V|\Phi_\mu >$ in (2.20) are non-vanishing for which Φ_μ is at most doubly excited with respect to Φ. Designating singly and doubly 'excited' state determinants as Φ_i^a and Φ_{ij}^{ab} respectively, we can rewrite (2.20) as

$$\begin{aligned} E^{(2)} = & \sum_{i,a} < \Phi|V|\Phi_i^a > (\varepsilon_i - \varepsilon_a)^{-1} < \Phi_i^a|V|\Phi > \\ & + \sum_{i<j} \sum_{a<b} < \Phi|V|\Phi_{ij}^{ab} > (\varepsilon_i + \varepsilon_j - \varepsilon_a - \varepsilon_b)^{-1} < \Phi_{ij}^{ab}|V|\Phi > \end{aligned} \tag{2.23a}$$

$$\begin{aligned} E^{(3)} = & \sum_{i,a} \sum_{j,b} < \Phi|V|\Phi_i^a > (\varepsilon_i - \varepsilon_a)^{-1} < \Phi_i^a|V|\Phi_j^b > (\varepsilon_j - \varepsilon_b)^{-1} < \Phi_j^b|V|\Phi > \\ & + \sum_{i<j} \sum_{a<b} \sum_{k<l} \sum_{c<d} < \Phi|V|\Phi_{ij}^{ab} > (\varepsilon_i + \varepsilon_j - \varepsilon_a + \varepsilon_b)^{-1} < \Phi_{ij}^{ab}|V|\Phi_{kl}^{cd} > (\varepsilon_k + \varepsilon_l - \varepsilon_c - \varepsilon_d)^{-1} < \Phi_{kl}^{cd}|V|\Phi > \\ & + \sum_{i,a} \sum_{k<l} \sum_{c<d} < \Phi|V|\Phi_i^a > (\varepsilon_i - \varepsilon_a)^{-1} < \Phi_i^a|V|\Phi_{kl}^{cd} > (\varepsilon_k + \varepsilon_l - \varepsilon_c - \varepsilon_d)^{-1} < \Phi_{kl}^{cd}|V|\Phi > \\ & + \sum_{i<j} \sum_{a<b} \sum_{k,c} < \Phi|V|\Phi_{ij}^{ab} > (\varepsilon_i + \varepsilon_j - \varepsilon_a - \varepsilon_b)^{-1} < \Phi_{ij}^{ab}|V|\Phi_k^c > (\varepsilon_k - \varepsilon_c)^{-1} < \Phi_k^c|V|\Phi > \\ & - < \Phi|V|\Phi > \sum_{i,a} < \Phi|V|\Phi_i^a > (\varepsilon_i - \varepsilon_a)^{-2} < \Phi_i^a|V|\Phi > \\ & - < \Phi|V|\Phi > \sum_{i<j} \sum_{a<b} < \Phi|V|\Phi_{ij}^{ab} > (\varepsilon_i + \varepsilon_j - \varepsilon_a - \varepsilon_b)^{-2} < \Phi_{ij}^{ab}|V|\Phi > \end{aligned} \tag{2.23b}$$

Here the ε_p are one electron energies.

The expressions (2.23) are already rather lengthy, but they are not yet final. We must next express the matrix elements such as $< \Phi|V|\Phi_{ij}^{ab} >$ between Slater determinants in terms of matrix elements over spin orbitals like

$$V_{rs}^{pq} =< \psi_r(1)\psi_s(2)|V(1,2)|\psi_p(1)\psi_q(2) > \tag{2.24}$$

The tensor notation for these matrix elements[37] is convenient, but not necessary.

Slater[46] has derived rules for the matrix elements of one and two-electron operators between Slater determinants constructed from an orthogonal set of spin-orbitals. A reader who is familiar with Slater's rules can skip the following lines until eqn. (2.33). For didactic reasons we choose here an alternative way towards the evaluation of the matrix elements between Slater determinants, namely via a Fock space (2nd quantization) formalism. If one wants to understand many-body theory one has to learn this formalism anyway. If one defines $a_p^\dagger$ (or a^p) as the 'creation operator' for an electron in the spin orbital ψ_p and a_q as the destruction operator in the spin orbital ψ_q, a_p^q as the excitation operator from spin orbital ψ_p to spin orbital ψ_q, a_{pq}^{rs} as the excitation operator from spin orbital pairs ψ_p, ψ_q into the pairs $\psi_r\psi_s$ etc., then the 'excited' wave function Φ_i^a as well as the electron interaction V can be written as

$$\Phi_i^a = a_i^a\Phi = a^a a_i\Phi = a_a^\dagger a_i\Phi \tag{2.25}$$

$$V = \frac{1}{2}V_{rs}^{pq}a_{pq}^{rs} = \frac{1}{2}V_{rs}^{pq}a^r a^s a_q a_p = \frac{1}{2}\sum_{p,q,r,s} V_{rs}^{pq}a_r^\dagger a_s^\dagger a_q a_p \tag{2.26}$$

In the first two expressions in (2.26) we have used a tensor notation together with the Einstein summation convention over repeated indices[37] which simplifies the notation considerably. One then gets

$$< \Phi|V|\Phi_i^a >= \frac{1}{2}V_{rs}^{pq} < \Phi|a_{pq}^{rs}a_i^a|\Phi > \tag{2.27}$$

where the summation over p, q, r, s is implied. In order to evaluate $< \Phi|...|\Phi >$ we first use the generalized Wick theorem[37] according to which a product of basis operators like a_q^p, a_{pq}^{rs} etc. (which are each in normal order, e.g. all annihilation operators right to all creation operators) is equal to the normal product (the first term in (2.28)) plus all contractions of an upper label of a right factor with a lower label of a left factor.

$$a_{pq}^{rs}a_i^a = a_{pqi}^{rsa} + \delta_p^a a_{iq}^{rs} + \delta_q^a a_{pi}^{rs} \tag{2.28}$$

We hence get

$$< \Phi|V|\Phi_i^a >= \frac{1}{2}V_{rs}^{pq} < \Phi|a_{pqi}^{rsa} + \delta_p^a a_{iq}^{rs} + \delta_q^a a_{pi}^{rs}|\Phi > \tag{2.29}$$

We next note that the first operator in the expectation value gives no contribution, because a spin-orbital ψ_a not occupied in Φ is created and $< \Phi|a^a = 0$. For the two other terms we note that the δ factors suppress one summation index, and further that non-vanishing contributions only arise if spin-orbitals ψ_k occupied in Φ are first annihilated, and then the same ones created, i.e. if all labels at the operators refer to occupied spin orbitals (label i, j, k) and if the same labels (possibly permuted) arise as upper and lower labels. Hence

$$\begin{aligned} < \Phi|V|\Phi_i^a > &= \frac{1}{2}V_{rs}^{aq} < \Phi|a_{ik}^{ik}\delta_i^r\delta_k^s\delta_q^k + a_{ik}^{ki}\delta_i^s\delta_k^r\delta_q^k|\Phi > \\ &+ \frac{1}{2}V_{rs}^{pa} < \Phi|a_{ki}^{ki}\delta_k^r\delta_p^k\delta_i^s + a_{ki}^{ik}\delta_i^r\delta_i^k\delta_p^k|\Phi > \\ &= \frac{1}{2}V_{ik}^{ak} < \Phi|a_{ik}^{ik}|\Phi > + \frac{1}{2}V_{ki}^{ak} < \Phi|a_{ik}^{ki}|\Phi > \\ &+ \frac{1}{2}V_{ki}^{ka} < \Phi|a_{ki}^{ki}|\Phi > + \frac{1}{2}V_{ik}^{ka} < \Phi|a_{ki}^{ik}|\Phi > \end{aligned} \tag{2.30}$$

where one sums over k. Finally we note that the matrix elements of operators like a_{ik}^{ik} are equal to 1 if the relative permutation of upper labels with respect to the lower labels is even, otherwise they are equal to -1, and that $V_{rs}^{pq} = V_{sr}^{qp}$. So

$$< \Phi|V|\Phi_i^a >= V_{ik}^{ak} - V_{ki}^{ak} \tag{2.31}$$

If we define the antisymmetrized interaction

$$\bar{V}_{rs}^{pq} = V_{rs}^{pq} - V_{rs}^{qp} \tag{2.32}$$

the final result for our matrix element and its adjoint is

$$< \Phi|V|\Phi_i^a >= \bar{V}_{ik}^{ak}; \quad < \Phi_i^a|V|\Phi >= \bar{V}_{ak}^{ik} = (\bar{V}_{ik}^{ak})^\dagger \tag{2.33}$$

Introducing the short hand notation

$$(\bar{V}_{rs}^{pq})_H = \bar{V}_{rs}^{pq}(\varepsilon_p + \varepsilon_q - \varepsilon_r - \varepsilon_s)^{-1} \tag{2.34}$$

we can finally rewrite the first term of (2.23a) as

$$(\bar{V}_{ik}^{ak})(\bar{V}_{al}^{il})_H \tag{2.35}$$

(with an implied summation over k, l, i and a). Using a similar argument for the second term in (2.23a) we finally get

$$E^{(2)} = \bar{V}_{ik}^{ak}(\bar{V}_{al}^{il})_H + \frac{1}{2}\bar{V}_{ij}^{ab}(\bar{V}_{ab}^{ij})_H \tag{2.36}$$

The result (2.36) is conveniently illustrated by diagrams, as in fig. 1 where (a) represents the first term on the r.h.s. of (2.36) and (b) the second term. In these diagrams a matrix element like $\bar{V}_{ik}^{ak}$ is represented by a bold face horizontal line (a vertex) with the upper labels symbolized by arrows that enter the vertex and the lower labels by arrows that leave the vertex. Any line that leaves one vertex must enter another vertex (or the same vertex) and vice versa. Lines that symbolize orbitals $(i, j, k...)$ always carry downgoing arrows, lines for virtual orbitals $(a, b, c...)$ upgoing arrows. Energy denominators are indicated by rectangular boxes, orbital energies corresponding to an ingoing line appear with a plus sign, those of an outgoing line with a minus sign.

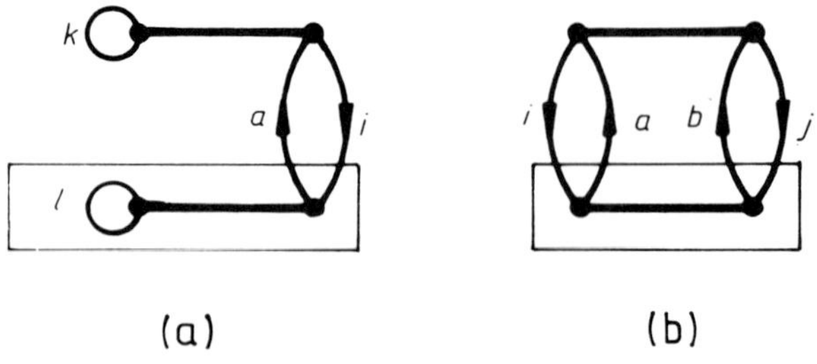

Figure 1. Diagrams to illustrate the contributions to the 2nd order perturbation energy $E^{(2)}$. Diagram (a) corresponds to the first, (b) to the second term in eqn. (2.36). Rectangular boxes indicate energy denominators. Vertices refer to the antisymmetrized electron interaction.

Brueckner[17] has not used diagrams. The diagrams of fig. 1 have been inspired by those of Goldstone[20], but they differ from these in two aspects: (a) our vertices represent antisymmetrized interactions (2.32), like in the diagrams of Hugenholtz[19] (where point vertices rather than line vertices are used) or Brandow[31]. (b) We explicitly indicate energy denominators.

The analogous reformulation of $E^{(3)}$ as given by (2.23b) is much more tedious. Let us first consider the first line of (2.23b). We know already how to evaluate the first and last factors in this sum, see (2.33). The middle factor is

$$< \Phi_i^a|V|\Phi_j^b >= \frac{1}{2} V_{rs}^{pq} < \Phi|a_a^i a_{pq}^{rs} a_j^b|\Phi > \tag{2.37}$$

The result depends on whether or not $a = b$ or $i = j$. We hence consider these cases separately.

(1) $i \neq j, a \neq b$. Since ψ_a and ψ_b are not occupied in Φ, the normal product gives a vanishing matrix element, and so do single contractions in which an a or b survives. Hence only the full contractions

$$\delta_p^b \delta_a^r a_{jq}^{is} + \delta_q^b \delta_a^r a_{pj}^{is} + \delta_p^b \delta_a^s a_{qs}^{ir} + \delta_q^b \delta_a^s a_{jp}^{ir} \tag{2.38}$$

give contributions and this only if the unspecified labels (s, q in the first term of (2.38)) are chosen such that operators like a_{ij}^{ij} or a_{ji}^{ij} result. It is then straightforward to see that

$$< \Phi_i^a|V|\Phi_j^b >= \bar{V}_{ja}^{ib} \tag{2.39}$$

The two Slater determinants Φ_i^a and Φ_j^b are 'doubly excited' with respect to each other and their interaction is an exchange-type matrix element.

(2) The next case is $i \neq j, a = b$. Now in addition to (2.38) also the operator contributes in which just a is contracted with b

$$a_{jpq}^{irs} \tag{2.40}$$

In order to get a non-vanishing matrix element, of the two labels (r, s) one must be equal to $j(\neq i, k)$ the other to some k, of (p, q) one is i, the other k. There is hence an extra contribution (in addition to (2.39)).

$$\frac{1}{2} V_{rs}^{pq} < \Phi|a_{jik}^{ijk} \delta_j^r \delta_p^i \delta_k^s \delta_q^k + a_{jik}^{ikj} \delta_k^r \delta_p^i \delta_j^s \delta_q^k + 2 \text{ more terms}|\Phi >= \bar{V}_{jk}^{ik} \tag{2.41}$$

$$< \Phi_i^a|V|\Phi_j^a >= \bar{V}_{ja}^{ia} + \sum_{\substack{k \\ (\neq i,j)}} \bar{V}_{jk}^{ik} \tag{2.42}$$

(3) $i = j, a \neq b$. If we put $i = j$ in (2.38), we see that the other two labels (s, q in the first term) must be equal to $k \neq i$, hence

$$< \Phi_i^a|V|\Phi_i^b >= \sum_{\substack{k \\ (\neq i)}} \bar{V}_{ka}^{kb} \tag{2.43}$$

(4) $i = j, a = b$. From (2.38) we get, as in case 3, (2.42) with $a = b$ while (2.40) now becomes

$$a_{ipq}^{irs} \tag{2.44}$$

such that one label of (r,s) and (p,q) must be $k(\neq i)$, the other $l(\neq i,k)$, hence instead of (2.41) we get $\sum_{\substack{k,l\\(\neq i)}} \bar{V}_{kl}^{kl}$ and the final result is

$$< \Phi_i^a|V|\Phi_i^a >= \sum_{\substack{k\\(\neq i)}} \bar{V}_{ka}^{ka} + \frac{1}{2}\sum_{\substack{k,l\\(\neq i)}} \bar{V}_{kl}^{kl} \tag{2.45}$$

We put all these results together and get for the first line in (2.23b)

$$\sum_{i\neq j}\sum_{a,b}\sum_{k,l} \bar{V}_{ik}^{ak}(\bar{V}_{ja}^{ib}(\bar{V}_{bl}^{jl})_H)_H + \sum_{i\neq j}\sum_{a}\sum_{k,l}\sum_{\substack{m\\(\neq i,j)}} \bar{V}_{ik}^{ak}(\bar{V}_{jm}^{im}(\bar{V}_{al}^{jl})_H)_H$$
$$+ \sum_{i}\sum_{a,b}\sum_{k,l}\sum_{\substack{m\\(\neq i)}} \bar{V}_{ik}^{ak}(\bar{V}_{ma}^{mb}(\bar{V}_{bl}^{il})_H)_H + \sum_{i,a}\sum_{k,l}\sum_{\substack{m,n\\(\neq i)}} \bar{V}_{ik}^{ak}(\bar{V}_{mn}^{mn}(\bar{V}_{al}^{il})_H)_H \tag{2.46}$$

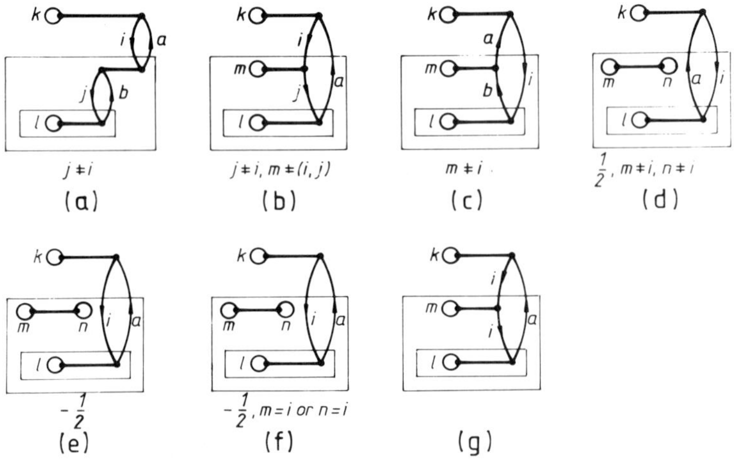

Figure 2. Diagrams for contributions to the 3rd order energy $E^{(3)}$. Diagrams (a) to (d) correspond to direct terms, eqn. (2.46), (e) to renormalization terms, eqn. (2.47), (f) and (g) are different representations of the sum of (d) and (e)

What makes things complicated is the need to consider different cases depending on whether or not certain labels agree, and the final summation restrictions.

Again we illustrate the result (2.46) in terms of diagrams. On fig. 2, (a) to (d) represent respectively the four sums in (2.46). The summation restrictions can be rationalized in terms of the Pauli principle. At the middle vertex in (a) ingoing particles (labels i,b) are first annihilated before the outgoing particles (labels j,a) are created. This implies, that $a = b$ is allowed, however $i = j$ is not, because would mean the spinorbital with label i is twice destroyed before it is created once, which is in conflict with the Pauli principle. (ψ_i can, of course, be annihilated once, since it is occupied in Φ). A diagram of type (a) with $(i = j)$ would be 'exclusion principle violating' (EPV). The other summation restrictions can be rationalized in an analogous way. The restriction $m \neq n$ on diagram (d) need not be imposed, since the matrix element $\bar{V}_{mn}^{mn}$ vanishes for $m = n$. Similarly the restrictions $l \neq j, k \neq j$ etc. need not be imposed.

The diagrams (a) to (d) on fig. 2, corresponding to eqn. (2.46) or the 1st line of (2.23b) come from the direct part of $E^{(3)}$, there are similar terms in the renormalization part, namely in the last but one line of (2.23b). This can be reformulated to

$$-\frac{1}{2}\{\sum_{m \neq n} \bar{V}_{mn}^{mn}\}\{\sum_{i,a}\sum_{k,l} \bar{V}_{ik}^{ak}((\bar{V}_{al}^{il})_H)_H\}$$
$$= -\frac{1}{2}\sum_{i,a}\sum_{k,l}\sum_{m,n} \bar{V}_{ik}^{ak}(\bar{V}_{mn}^{mn}(\bar{V}_{al}^{il})_H)_H \tag{2.47}$$

In (2.47) the matrix elements are the same as in the last sum of (2.46), just the summation restrictions are different, in fact there are no restrictions in (2.47) (except for $m \neq n$, which does not matter in view of $\bar{V}_{mn}^{mn} = 0$). The diagrammatic representation (e) is hence as on diagram (d), but note the minus sign in front of (2.47).

The sum (2.47) factorizes into a product of two sums, and it therefore gives a contribution $\sim N^2$ to the energy for extended systems. This is typical for disconnected diagrams as (e) on fig. 2, i.e. of diagrams which consist of two separable parts with independent summation indices.

However diagram (d) of fig. 2 is disconnected as well (with not completely independent summations). If we now add (d) and (e) we get

$$-\frac{1}{2}\sum_{i,a}\sum_{k,l}\sum_{\substack{(m,n \\ m=i \text{ or } n=i)}} \bar{V}_{ik}^{ak}(\bar{V}_{mn}^{mn}(\bar{V}_{al}^{il})_H)_H = -\sum_{i,a}\sum_{k,l}\sum_{n} \bar{V}_{ik}^{ak}(\bar{V}_{in}^{in}(V_{al}^{il})_H)_H \tag{2.48}$$

where the l.h.s. of (2.48) is symbolized as diagram (f) on fig. 2. This diagram (f) is still disconnected, but the two parts have a common label (such diagrams are sometimes referred to as 'joint'[38], in contrast to (d) which is disjoint). This implies that the sum of all diagrams of type (f) has the correct $\sim N$ dependence. One can finally rewrite diagram (f) in connected form as (g) corresponding to the r.h.s. of (2.48), i.e. as a diagram that we have previously referred to as EPV (exclusion principle violating). It looks as if one gets the correct final result, if one ignores both the renormalization contributions and the Pauli-principle, and draws only connected diagrams, with a sign convention that takes care of the appropriate sign of EPV diagrams. This idea will emerge in a more direct way from Goldstone's derivation of the linked-cluster expansion in sec. 3. The other contributions to (2.23b) can be analysed in a similar way, but we skip the details.

The proof of what Brueckner called the linked-cluster theorem, sketched in this section, turned out to be elementary, but rather tedious, although we have partially used a more modern and more compact formalism than Brueckner originally did. To higher orders the proof gets even more tedious and it is not transparent on which physical concept it is based. The simplicity of the final results contrasts with its complicated derivation.

After the development that followed Brueckner's pioneering work we now understand why Brueckner's approach to many-body perturbation theory was not so successful. This is mainly due to the formulation in a Hilbert space with fixed particle number, such that the contributions to perturbation theory are expressed through matrix elements between Slater determinants. If one formulates the theory in Fock space and skips all reference to Slater determinants, things become much easier, as we shall show in sec. 4.

The first formulation of many-body perturbation theory in Fock space has been given by Goldstone[20], although it has not simply consisted in recasting Brueckner's theory to Fock space language, but has rather been done in a completely different framework. In view of its extreme impact, we must now review Goldstone's access to many-body perturbation theory, although from a modern point of view it is not the best possible one.

3. GOLDSTONE'S LINKED-DIAGRAM EXPANSION

Again we do not follow Goldstone[20] literally, but rather try to understand the essence of his derivation.

Although we know that the perturbation to be considered in many-body theory is time-independent, we study the more general case of a time-dependent perturbation $\lambda V(t)$. We must then, of course, take the time-dependent Schrödinger equation. Rather than in the wave function we are interested in the time evolution operator $U(t, t_o)$, which has the property

$$\Psi(t) = U(t, t_o)\Psi(t_o) \tag{3.1}$$

This operator satisfies the Schrödinger equation (we put $\hbar = 1$)

$$HU = i\frac{\partial U}{\partial t} \tag{3.2}$$

If H is time-independent, then the solution of (3.2) is

$$U(t, t_o) = \exp\{-iH(t - t_o)\} \tag{3.3}$$

For our purpose (with H_o time-independent and V time-dependent) it is convenient to use the 'interaction representation', in which an ordinary hermitean quantum mechanical operator A_S in the Schrödinger representation is replaced by

$$A_I = e^{-iH_o t} A_S e^{iH_o t} \tag{3.4}$$

The definition (3.4) resembles that of the Heisenberg representation, except that we now have H_o instead of H in the exponent. The time evolution operator in the interaction representation

$$U_I(t, t_o) = e^{-iH_o(t-t_o)} U_S(t, t_o) \tag{3.5}$$

satisfies the Schrödinger equation

$$\lambda V_I U_I = i\frac{\partial U_I}{\partial t} \tag{3.6}$$

One easily sees that U_I is unitary. Take the adjoint of (3.6)

$$\lambda U_I^\dagger V_I = -i\frac{\partial U_I^\dagger}{\partial t} \tag{3.7}$$

and multiply (3.6) from the left by $U_I^\dagger$, (3.7) on the right by U_I and substract the two equations

$$0 = i\frac{\partial(U_I^\dagger U_I)}{\partial t} \tag{3.8}$$

The operator $U_I^\dagger U_I$ is hence time-independent. Since for $t = t_o$, $U_I^\dagger U_I = 1$, $U_I^\dagger U_I$ it is hence equal to 1 for all t. Note that the proof of the unitarity of U_I is based on the hermiticity of V_I.

If we expand U_I in powers of λ

$$U_I = \sum_{k=0}^{\infty} \lambda^k U_I^{(k)} \tag{3.9}$$

we get the hierarchy of equations

$$\frac{\partial U_I^{(o)}}{\partial t} = 0 \tag{3.10a}$$

$$\frac{\partial U_I^{(k)}}{\partial t} = -iV_I U_I^{(k-1)} \tag{3.10b}$$

We now assume that V_I vanishes for $t < t_o$, such that for $t < t_o$ in view of (3.6) U_I is constant. The unitarity of U_I implies that $||U_I|| = 1$ and we can without loss of generality assume that $U_I = 1$ for $t < t_o$. Since (3.10a) holds for all t and $U_I = U_I^{(o)}$ for $t < t_o$, we get by integration of (3.10)

$$U_I^{(o)} = 1 \tag{3.11a}$$

$$U_I^{(1)} = -i\int_{t_o}^{t} V_I(t')dt' \tag{3.11b}$$

$$\begin{aligned} U_I^{(2)} &= (-i)^2 \int_{t_o}^{t} V_I U_I^{(1)}(t')dt' = -\int_{t_o}^{t} V_I(t')dt' \int_{t_o}^{t'} V_I(t'')dt'' \\ &= -\int_{t_o}^{t} dt' \int_{t_o}^{t} dt'' \Theta(t' - t'') V_I(t') V_I(t'') \end{aligned} \tag{3.11c}$$

with $\Theta(x)$ the Heavyside step function

$$\Theta(x) = \begin{cases} 0 & \text{for } x < 0 \\ 1 & \text{for } x > 0 \end{cases} \tag{3.11d}$$

The perturbation theory of U_I is hence formally very simple.

We now consider the case of a static perturbation V. In order to apply the formalism of time-dependent perturbation theory we multiply V by an exponential switching function $e^{\eta t}$

$$V(t) = Ve^{\eta t}; \quad \eta > 0 \tag{3.12}$$

such that $V(t)$ vanishes in the limit $t \to -\infty$ and is equal to V for $t = 0$. We choose η very small such that for a large range of time in the neighbourhood of $t = 0$, $V(t)$ is very close to V. We plan to take the limit $\eta \to 0$ finally, to recover the time-independent case. The integrals in (3.11) must then be taken from $-\infty$ to 0.

We write V in 2nd quantization form (see also eqn. (2.26))

$$V = V_{rs}^{pq} a^r a^s a_q a_p \tag{3.13}$$

with the tensor notation of the matrix elements, and with an alternative notation for creation operators

$$a^r \equiv a_r^\dagger \tag{3.14}$$

We also imply the Einstein summation convention over repeated indices.

In order to apply (3.11) we must transform (3.13) to the interaction representation. For this we need the creation and annihilation operators in the interaction representation. Since the a^r and a_r are chosen such that they create (or annihilate) an eigenorbital of H_o with orbital energy ε_r, one sees that

$$a_I^r(t) = e^{-iH_o t} a^r e^{iH_o t} = e^{-i\varepsilon_r t} a^r \tag{3.15a}$$

$$a_{rI}(t) = e^{-iH_o t} a_r e^{iH_o t} = e^{i\varepsilon_r t} a_r \tag{3.15b}$$

Insertion of (3.13) into (3.12) and transformation to the interaction representation, using (3.15) leads to

$$V_I(t) = \frac{1}{2} V_{rs}^{pq} e^{-i(\varepsilon_r + \varepsilon_s - \varepsilon_p - \varepsilon_q + i\eta)t} a^r a^s a_q a_p \tag{3.16}$$

With this V_I the integration (3.11) in the limits $(-\infty, 0)$ leads to

$$U_I^{(1)} = \frac{1}{2} V_{rs}^{pq} (\varepsilon_r + \varepsilon_s - \varepsilon_p - \varepsilon_q + i\eta)^{-1} a^r a^s a_q a_p \tag{3.17}$$

$$U_I^{(2)} = \frac{1}{4} V_{vw}^{tu} V_{rs}^{pq} (\varepsilon_r + \varepsilon_s + \varepsilon_v + \varepsilon_w - \varepsilon_p - \varepsilon_q - \varepsilon_t - \varepsilon_u + 2i\eta)^{-1} (\varepsilon_r + \varepsilon_s - \varepsilon_p - \varepsilon_q + i\eta)^{-1}$$
$$\times\, a^v a^w a_u a_t a^r a^s a_q a_p \tag{3.18}$$

Note that U_I as well as V_I are Fock space operators, i.e. no particle number is specified. For a further evaluation we need a general recipe to reformulate products of creation and annihilation operators in such a way that the final result is in normal order, i.e. that all creation operators are left from all annihilation operators. This is achieved by means of Wick's theorem [3,5], see eqn. (2.28) and the remarks ahead of it.

Before we do so we wonder what we plan to do with U_I. We are interested in the energy of our interacting n-fermion system. The wave function is

$$\Psi_I = U_I(t, -\infty)\Phi_I \tag{3.19}$$

where Φ_I is time independent and equal to the ordinary unperturbed wave function, while Ψ_I is time-dependent (but only slightly so for small η).

Let us, for a moment, go back to the Schrödinger picture. If Ψ satisfies the Schrödinger equation (and Φ the unperturbed Schrödinger equation) we can write the time-dependent energy, which in the time-independent case equals an eigenvalue E, as

$$E = \frac{< \Phi|H|\Psi >}{< \Phi|\Psi >} = \frac{< \Phi|H_o + \lambda V|\Psi >}{< \Phi|\Psi >} = E_o + \Delta E \tag{3.20}$$

$$\Delta E = \lambda \frac{< \Phi|V|\Psi >}{< \Phi|\Psi >} \tag{3.21}$$

The expression (3.21) for the interaction energy remains valid if we express all quantities in the interaction representation. We insert (3.19) into (3.21) in the interaction representation and get

$$\Delta E = \lambda \frac{< \Phi_I|V_I U_I(t,-\infty)|\Phi_I >}{< \Phi_I|U_I(t,-\infty)|\Phi_I >} \tag{3.22}$$

Now we remember that U_I satisfies (3.6), and write

$$\Delta E = \frac{i < \Phi_I|\frac{\partial U_I}{\partial t}|\Phi_I >}{< \Phi_I|U_I|\Phi_I >} = i\frac{\partial \ln < \Phi_I|U_I(t,-\infty)|\Phi_I >}{\partial t} \tag{3.23}$$

This is the 'Gell-Mann-Low' formula[24] for the interaction energy. The ΔE of (3.23) is still time-dependent. In order to get the desired interaction energy for our exponentially switched static perturbation we perform the limit $\eta \rightarrow 0$ in our switching function.

$$\Delta E = i \lim_{\eta \rightarrow 0} \frac{\partial \ln < \Phi_I|U_I(t,-\infty)|\Phi_I >}{\partial t} \tag{3.24}$$

The difficulty with this formalism is that in the limit $\eta \rightarrow 0$ neither the wave function Ψ_I nor the evolution operator exist. This limit is hence highly singular, a point to which we come back later. The limit of ΔE does exist, however, and this is the only thing that we are finally interested in.

In order to evaluate ΔE we do not need the full evolution operator U_I, but only its expectation value with respect to the unperturbed wave function. This expectation value is easily evaluated if we define Φ_I as a 'physical vacuum' and introduce the so-called particle-hole picture.

We regard the (unperturbed) state described by the (time-independent) Slater determinant Φ_I as our reference state (physical vacuum). We now omit the subscript I, because Φ_I is equal to the time-independent unperturbed wave function in the Schrödinger representation. Rather than to refer to creation of a particle (not occupied in Φ) and to the annihilation of a particle (occupied in Φ), we refer in either case to the creation of a quasiparticle, which in the second case is called a hole. We hence replace the 2nd quantization operators a_p and $a_p^\dagger$ by

$$b_p = \begin{cases} a_p & \text{if } \varphi_p \quad \text{is not occupied in } \Phi \\ a_p^\dagger & \text{if } \varphi_p \quad \text{is occupied in } \Phi \end{cases}$$

$$b_p^\dagger = \begin{cases} a_p^\dagger & \text{if } \varphi_p \quad \text{is not occupied in } \Phi \\ a_p & \text{if } \varphi_p \quad \text{is occupied in } \Phi \end{cases}$$

We say that a product of the b-operators is in normal order, if all quasi-particle creation operators are left to all annihilation operators.

If an operator in normal order (with at least one annihilation operator) acts on the physical vacuum Φ, the result is zero, because no particle and no hole is occupied in Φ, and annihilation of a particle or hole yields zero. Application of an operator in normal order, with at least one creation operator, to Φ on the left yields zero as well. The vacuum expectation value of an operator in normal order hence always vanishes, e.g. $<\Phi|b_p^\dagger b_q|\Phi>=0$.

If we apply Wick's theorem to the operator products in the particle-hole sense only full contractions (which are simple numbers rather than operators) contribute to the vacuum expectation value.

Rewriting (3.17, 18) etc. in the particle-hole sense, and using the labels $i,j,k..$ for holes, $a,b,c...$ for particles we get in terms of the antisymmetrized matrix elements (2.32) - see also (2.24) -

$$<\Phi|U_I^{(1)}|\Phi>=\bar{V}_{ij}^{ij}\frac{1}{2}(i\eta)^{-1}e^{-\eta t} \tag{3.25a}$$

$$\begin{aligned}<\Phi|U_I^{(2)}|\Phi>=\frac{1}{4}\Big\{&\bar{V}_{ij}^{ij}\bar{V}_{kl}^{kl}(2i\eta)^{-1}(i\eta)^{-1}\\&+4\bar{V}_{ia}^{ij}\bar{V}_{jk}^{ak}(2i\eta)^{-1}(\varepsilon_a-\varepsilon_j+i\eta)^{-1}\\&+2\bar{V}_{ab}^{ij}\bar{V}_{ij}^{ab}(2i\eta)^{-1}(\varepsilon_a+\varepsilon_b-\varepsilon_i-\varepsilon_j+i\eta)^{-1}\Big\}e^{-2\eta t}\end{aligned} \tag{3.25b}$$

All these expressions obviously diverge in the limit $\eta\to 0$. However we do not need these expressions directly, but rather the time derivative of $\ln<\Phi|U_I|\Phi>$.

The construction of the logarithm becomes easy if one first realizes that $<\Phi|U_I|\Phi>$ can be written as the exponential of something. The argument used by Goldstone has been based on diagrams. One can, in fact, express all contributions to (3.25a, b) etc. by diagrams. What one sees, even by inspection of the expressions (3.25) is e.g. that to $\frac{1}{2}\bar{V}_{ij}^{ij}(-i\eta)^{-1}e^{-\eta t}$ in (3.25a) there is a corresponding term $\frac{1}{8}(\bar{V}_{ij}^{ij})^2(-i\eta)^{-2}e^{-2\eta t}$ in (3.25b) and a term $\frac{1}{48}(\bar{V}_{ij}^{ij})^3(-i\eta)^{-3}e^{-3\eta t}$ in the expression for $<\Phi|U_I^{(3)}|\Phi>$ etc. All these diagrams can be summed to yield

$$\exp\left\{\frac{1}{2}\bar{V}_{ij}^{ij}(-i\eta)^{-1}e^{-\eta t}\right\} \tag{3.26}$$

The essential point is that $<\Phi|U_I|\Phi>$ contains both connected and disconnected diagrams, while in the argument of the exponential finally only connected diagrams survive. This is the key part of Goldstone's derivation. Details are not so important for a general understanding.

The final result is hence

$$<\Phi|U_I|\Phi>=\exp\{\text{sum of all connected diagrams that contribute to} <\Phi|U_I|\Phi>\} \tag{3.27}$$

Now we can easily take the logarithm. This still diverges in the limit $\eta \to 0$ as is seen from the argument of the exponential in (3.26). However differentiation with respect to t introduces a factor $-\eta$ that together with the i in (3.23) cancels the denominator $-i\eta$ such that the expression (3.23) is not singular in the limit $\eta \to 0$, and this limit (3.24) can be taken.

For the first order energy we so get from (3.26) simply

$$E^{(1)} = \frac{1}{2}\bar{V}_{ij}^{ij} = \frac{1}{2}V_{ij}^{ij} - \frac{1}{2}V_{ji}^{ij} \tag{3.28a}$$

represented by the diagrams (a) and (b) on fig. 4. One sees also from (3.25b) and from the corresponding disconnected terms to higher orders of (3.25) that the 2nd order energy becomes

$$\frac{1}{2}\bar{V}_{ba}^{ij}(\bar{V}_{ij}^{ba})_H + \bar{V}_{ia}^{ij}(\bar{V}_{jk}^{ak})_H \tag{3.28b}$$

in agreement with (2.36).

We have not demonstrated in detail that (3.27) holds generally. In fact, to show this *after* the time integration, as we sketched it here, is (except for the lowest orders) not easy, because the disconnected diagrams don't appear with the same energy denominators as the connected diagrams, such that their factorization is not obvious.

The factorization is, however, seen rather directly if one tries the factorization (3.27) before one has performed the time-integrations explicitly. One then namely sees that various diagrams with the same vertices appear with different 'time orderings' for the successive time integrations. If one adds the disconnected diagrams for all occuring time orderings the result are diagrams, in which all time integrations are independent of each other and these factorize. Consider e.g. that in one diagram the time integration over t' and t'' has to respect $-\infty < t' < t'' < t$, in another diagram with the same vertices $-\infty < t'' < t' < t$. The two diagrams added together correspond to two independent time integrations $-\infty < t' < t$ and $-\infty < t'' < t$.

If one wants to demonstrate the factorization after the time integration, one has to use the factorization theorem of Frantz and Mills [30] which amounts essentially to using identities like

$$\frac{1}{a(a+b)} + \frac{1}{b(a+b)} = \frac{1}{ab} \tag{3.29}$$

It finally turns out that the only significant advantage of the time-dependent formalism is that the derivation of eqn. (3.27) becomes very easy. Factorization after time-integration is almost indistinguishable from a time-independent formalism taken at the outset.

The construction of the correlated wave function Ψ is more tricky than that of the energy. The original approach has again been in terms of diagrams. The wave function Ψ_I given by (3.19) is obviously divergent in the limit $\eta \to 0$, as is seen e.g. from the contributions with $r = p = i$, $s = q = j$ in (3.17). One can now argue that $U_I\Phi_I$ can be written in the form

$$U_I\Phi_I = W_I e^{iK_I(t)}\Phi_I \tag{3.30}$$

where W_I is in 'intermediate normalization',

$$< \Phi_i | W_I | \phi_I >= 1 \tag{3.31}$$

while $K_I(t)$ is a simple number, not an operator. All the contributions to U_I that diverge, are contained in K_I. From (3.30) we see that

$$\lim_{\eta \to 0} \frac{\partial K_I(t)}{\partial t} = -\Delta E \tag{3.32}$$

which is consistent with

$$iK_I(t) = -it\Delta E + ia \tag{3.33}$$

with an integration a constant that happens to diverge in the limit $\eta \to 0$. So we identify $e^{iK_I(t)}$ with a phase factor, which takes care of the fact if the energy changes (from E_0 to E) also the phase factor changes. The divergent phase consistent ia is, in a way physically irrelevant.

We don't want to go into details with the derivation of (3.30). Nevertheless we note one puzzling aspect that has hardly been commented in the literature (see however ref. 47,48). The operator U_I is necessarily unitary, so is $e^{iK_I(t)}$ for real K_I. This would be consistent only with W_I unitary, not with W_I in intermediate normalization as the diagrammatic discussion suggests. We shall clarify this point in sec. 5.

What one may reproach to Goldstone's derivation in addition to the inconsistency just mentioned is that it tries to explain something simple (time independent theory) in terms of something complicated (time-dependent theory) and moreover that the time-dependent limit is singular. The big merit of this derivation is that it has introduced diagrams to many-body theory. One can nowadays hardly imagine how to do many-body theory without diagrams.

Let us now present the Goldstone diagrams for the energy.

In principle we have already shown in sec. 2 how products of matrix elements V_{rs}^{pq} with energy denominators can be represented by diagrams; see e.g. fig. 1. To be strictly conform with Goldstone's notation we make two changes from the diagrams of fig. 1. Firstly we let one line vertex now mean a primitive (not antisymmetrized) vertex, i.e. it represents e.g. V_{ij}^{ab} rather than $\bar{V}_{ij}^{ab} = V_{ij}^{ab} - V_{ij}^{ba}$.

This means the diagram (a) of fig. 1 (with antisymmetrized vertices) become now the sum of the diagrams (e) to (h) with primitive vertices on fig. 4. Similarly (b) in fig. 1 corresponds to (c) and (d) on fig. 4.

Secondly we omit the rectangular boxes that represent energy denominators on fig. 1. This is possible if there is an automatic convention where to put the energy denominators (which in Goldstone's series result from time integrations). This is indicated on fig. 3. One puts the vertices successively from bottom to top (corresponding to right to left in the formulae) and one has a new time integration (from $-\infty$ to the actual time) and hence an energy denominator from the origin up to including the new vertex.

For a convenient use of Goldstone diagrams it is noteworthy that one can read the sign of the algebraic expression corresponding to a diagram directly from the diagram. The sign rule is that every hole line (downgoing arrow) and any closed loop introduce a factor -1 each. Diagram (d) on fig. 4 has e.g. two hole lines an one closed loop, so overall a factor -1.

An advantage of the Goldstone diagrams (as compared to the Hugenholtz-Brandow diagrams with antisymmetrized vertices) is that one can easily sum over spin, since at either end of a vertex spin is conserved. As disadvantage is their large number (see below). A kind of diagrams that combine the advantages of Goldstone and Hugenholtz diagrams are those with symmetrized or antisymmetrized spinfree vertices [37].

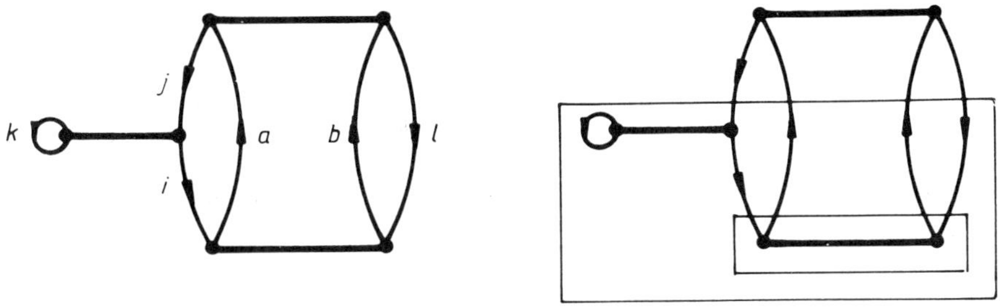

Figure 3. A conventional Goldstone diagram and the same diagram with explicitly indicated energy denominators.

The construction of the Goldstone diagrams is rather straightforward. To 1st order one has diagrams with 1 vertex and only self-contractions as (a) and (b) on fig. 4, to 2nd order one has two vertices and various possibilities as (c) to (h) on fig. 4. To k^{th} order there are k vertices. To construct all diagrams of a given order one takes a time axis from bottom to top, writes the k vertices at different times and puts particle lines (upgoing) and hole lines (downgoing) in all possible ways such that every line starts at a vertex and ends at another vertex, that at either end of a vertex there is exactly one ingoing and one outgoing line, and that the diagram is 'connected' i.e. that it does not consist of two parts each of which is a 'diagram' in itself.

To high orders book keeping becomes difficult. A good recipe is to proceed in a hierarchy of steps. The author has written a computer program for the generation of diagrams (or rather their algebraic counterparts) based on this hierarchy.

The hierarchy is conviently explained in the inverse direction. The Goldstone diagrams as on fig. 4 are fully specified. Each vertex has two distinct ends, there is one ingoing line (arrow points towards the vertex) and one outgoing line (arrow leaves the vertex) at each end of the vertex. The time-order of the vertices goes from bottom to top. Particle lines go upward, hole lines downward.

Somewhat less specified are Hugenholtz diagrams. Here one does not distinguish the two ends of a vertex. There are two ingoing and two outgoing lines per vertex. To one Hugenholtz diagram there is usual a set of Goldstone diagrams, to the Hugenholtz diagram (k) in fig. 5 correspond e.g. the 8 Goldsstone diagrams on fig. 6.

On the next step higher in the hierarchy one does no longer discriminate between particle and hole lines. One then refers to a skeleton[40] rather than a diagram. The skeleton (e) on fig. 5 corresponds e.g. to the three Hugenholtz diagrams $(i),(j),(k)$ on fig. 5.

On the top level one even ignores the time order of the vertices. Permutations of the order of the vertices with the same connection scheme characterize a skeleton class. The skeletons $(f),(g),(h)$ on fig. 5 belong e.g. to the same skeleton class (b).

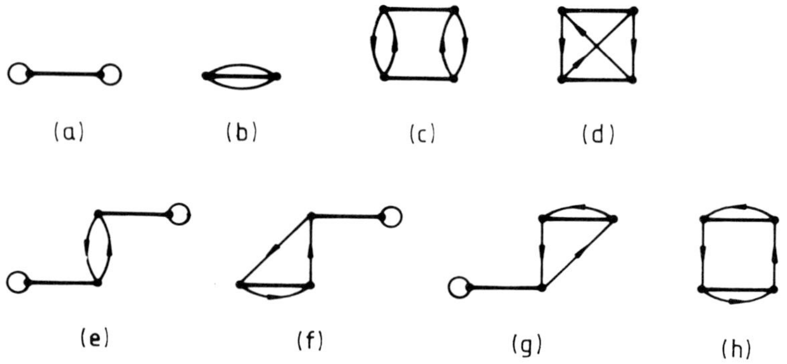

Figure 4. Goldstone diagrams (with primitive vertices) that contribute the first order and second order energies $E^{(1)}$ and $E^{(2)}$

To 3rd order there are 4 'skeleton classes', (a) to (d) on fig. 5 corresponding to 0,1,2, or 3 self-contractions respectively. We now go down in the hierarchy.

On the next level one discriminates between the time-ordering. So there are three skeletons $(f),(g),(h)$ for skeleton class (b) but only one skeleton (e) for skeleton class (a) (see fig. 5).

In the third step hole and particle lines are marked explicitly. This leads from the skeleton (e) to three 'Hugenholtz diagrams' $(i),(j),(k)$ (see fig. 5).

Finally in the forth step one distinguishes between the two ends of a vertex, i.e. one switches from antisymmetrized to primitive vertices. There are eight Goldstone diagrams shown on fig. 6 corresponding to the Hugenholtz diagram (k) on fig. 5.

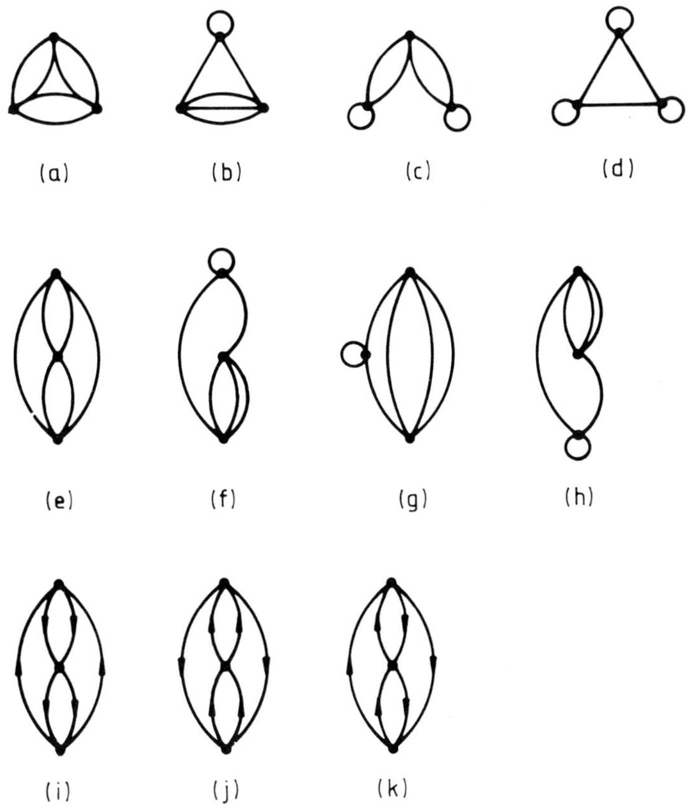

Figure 5. A hierarchy for the construction of diagrams. Diagrams (a) to (d) represent the possible skeleton classes for 3rd order diagrams (i.e. for $E^{(3)}$). No time order of the vertices, nor particle or holes are specified, antisymmetric point vertices are used. Diagram (c) is the only possible skeleton for skeleton class (a), diagrams (f) to (h) are the different possible skeleton class (b). In a skeleton the time order of the vertices is fixed. Skeletons for (c) and (d) are not indicated. Diagrams (i) to (k) are the three possible Hugenholtz diagrams with specification of particle and hole lines that correspond to the skeleton (e).

Note that in the Goldstone diagrams there is (at variance with diagrams introduced in sec. 2) no restriction on the labels. Diagrams that one might classify as 'exclusion principle violating' (EPV) have to be included.

Goldstone gives also diagrams for the wave function. We are not going to discuss them, but we have to mention a curious aspect of the nomenclature. The diagrams that contribute to the energy are all connected, those for the wave function are not. However for a wave function in intermediate normalization the wave function diagrams have the property that disconnected parts are always open, never closed. Goldstone defined diagrams as 'linked' if they are either 'connected' or 'disconnected' without a closed part. So the diagrams for the energy are both 'connected' and 'linked', while those for the wave function (in intermediate normalization) are disconnected, but nevertheless (in a counterintuitive sense) 'linked'.

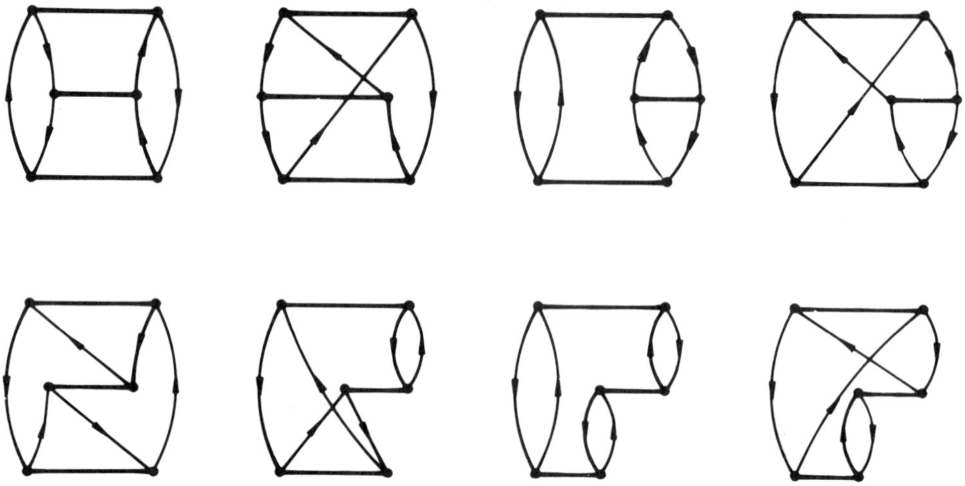

Figure 6. The eight possible Goldstone diagrams (with primitive vertices) that correspond to the Hugenholtz diagram (*k*) on Figure 5.

4. A MODERN DERIVATION OF THE GOLDSTONE EXPANSION

In Goldstone's derivation of the linked-diagram expansion the central quantity is not the wave function (as in traditional perturbation theory) but rather an operator $U(t,-\infty)$ that transforms the unperturbed to the exact wave function. This transformation is done by means of a time evolution. However, one can as well consider a time-independent operator W (called wave operator) which transforms the unperturbed wave function to the exact wave function

$$\Psi = W\Phi \tag{4.1}$$

Since (4.1) does not determine W uniquely we use a more general definition, that implies (4.1). Any non-singular operator W transforms the Hamiltonian H to an operator L

$$L = W^{-1}HW \tag{4.2}$$

(4.2) is a similarity transformation, hence L has the same eigenvalues as H, but different eigenfunctions. We now require that Φ is an eigenfunction of L. Then

$$L\Phi = E\Phi \tag{4.3}$$

$$HW\Phi = EW\Phi \tag{4.4}$$

and (4.1) holds in fact. We can rewrite (4.2) in the form (Bloch equation)

$$HW = WL \tag{4.5}$$

The requirement (4.3) still does not determine W uniquely, but we skip this problem for the moment.

It has been essential in sec. 3 that the operator U is formulated in 2nd quantization form, i.e. as a Fock space operator (although we have only considered a single n-electron state, to which U is finally applied). The connectedness of diagrams is essentially a Fock space concept and hard to formulate in an n-particle-Hilbert-space theory. We regard hence L, H and W as Fock space operators.

Now consider that the one-particle Hilbert space, from which our Fock space is constructed, is divided into two subspaces, between which there is no interaction. One may e.g. think of a system consisting of two atoms at very large distance. Assume that we know operators H_1, W_1, L_1 and H_2, W_2, L_2 defined in the various subspaces, which satisfy Bloch equations in the respective subspaces

$$H_1 W_1 = W_1 L_1 \tag{4.6a}$$

$$H_2 W_2 = W_2 L_2 \tag{4.6b}$$

Multiplication of the first equation by W_2 and the second by W_1, noting that Fock space operators defined in different subspaces commute, and finally addition of the two resulting equations leads to

$$(H_1 + H_2) W_1 W_2 = W_1 W_2 (L_1 + L_2) \tag{4.7}$$

This means the operators $H = H_1 + H_2$, $W = W_1 \cdot W_2$, $L = L_1 + L_2$ satisfy a Bloch equation in the entire Hilbert space. We conclude that H and L are 'additively separable', while W is multiplicatively separable. We can formulate the theory in terms of only additively separable quantities, if we replace W by its logarithm, i.e. if we write[11,18,49]

$$W = \exp S \tag{4.8}$$

and formulate the theory in terms of L, H and S. Then (4.2) becomes

$$L = e^{-S} H e^{S} = H + [H, S] + \frac{1}{2}[[H, S], S] + \ldots \tag{4.9}$$

Since H trivially consists of only connected diagrams, it follows from (4.9) that L consists of only connected diagrams if this is true for S. It is also straightforward to show by induction that S consists of connected diagrams only.

A pictorial explanation of this connected diagram theorem for additively separable operator is as follows. Let us artificially divide the Fock space into two subspaces as just discussed, with the only difference that there are interactions between these subspaces. Assume that disconnected diagrams are present, such that either part of a diagram consists of only labels of one of the two subspaces. We now switch off the interaction between the two subspaces, but this does not affect these disconnected diagrams. On the other hand the presence of such disconnected diagrams violates the additive separability. They must hence be absent even for interacting subspaces.

There is a certain problem with arguments of this kind related to the question whether one considers labelled or unlabelled diagrams. Disconnected diagrams with a common label behave in some sense like connected diagrams. One can classify these disconnected diagrams and connected ones as 'joint'[39] and distinguish them from disconnected diagrams without common labels to be regarded as 'disjoint'. In an n-particle Hilbert space theory where summation restrictions on the orbital labels have to be observed, the concepts joint and disjoint are more appropriate. An advantage of the Fock space formulation is that one has not to observe such restrictions - they are automatically taken care of by the anticommutation properties of the creation and annihilation operators. Hence a formulation strictly in terms of unlabelled diagrams is possible and a classification of diagrams entirely in terms of their topology is straightforward.

We can now apply perturbation theory to the Bloch equation (4.5)

$$(H_o + \lambda V) \sum_{k=0}^{\infty} \lambda^k W^{(k)} = \sum_{k=0}^{\infty} \lambda^k W^{(k)} \sum_{l=0}^{\infty} \lambda^l L^{(l)} \tag{4.10}$$

and get[37,38]

$$H_o W^{(o)} = W^{(o)} L^{(o)} \tag{4.11a}$$

$$H_o W^{(1)} + V W^{(o)} = W^{(o)} L^{(1)} + W^{(1)} L^{(o)} \tag{4.11b}$$

$$H_o W^{(2)} + V W^{(1)} = W^{(o)} L^{(2)} + W^{(1)} L^{(1)} + W^{(2)} L^{(o)} \tag{4.11c}$$

In the limit $\lambda \to 0$ one should have $W = 1$, which means that

$$W^{(o)} = 1; \quad L^{(o)} = H_o \tag{4.12a}$$

$$[H_o, W^{(1)}] + V = L^{(1)} \tag{4.12b}$$

$$[H_o, W^{(2)}] + V W^{(1)} = L^{(2)} + W^{(1)} L^{(1)} \tag{4.12c}$$

We now choose - like in sec. 3 - the particle-hole picture with Φ the physical vacuum. This physical vacuum must hence be an eigenstate of L. This means L can have no contributions that act on the vacuum i.e. that create quasiparticles (particles or holes). The constant part of L is automatically the desired eigenvalue E.

It is convenient to classify operators as

C for closed: fully contracted without external lines
B for closed from below: external lines correspond to particle or hole creation, e.g. $b_i^\dagger b_a^\dagger$
A for closed from above: external lines correspond to particle or hole annihilation, e.g. $b_i b_a$
O for open: contains both particle or hole creation and annihilation, e.g. $b_a^\dagger b_i$ (4.13)

It turns out that we only need to consider the categories C and B and that (we use the intermediate normalization $W_C = 1$ which is consistent with (4.12), which only determines W_B)

$$E = L_C; \quad W = W_B \tag{4.14a}$$
$$L^{(1)} = V_C; \quad [H_o, W^{(1)}] = W_B \tag{4.14b}$$
$$L^{(2)} = (V W^{(1)})_C; \quad [H_o, W^{(2)}] = W^{(1)} L^{(1)} - (V W^{(1)})_B \tag{4.14c}$$
$$L^{(3)} = (V W^{(2)})_C \tag{4.14d}$$

We only have to know how to invert a commutator equation. This is possible in a straightforward way and leads to energy denominators.

$$W^{(1)} = \frac{1}{2}(\bar{V}_{ab}^{ij})_H + (\bar{V}_{ib}^{ij})_H = \frac{1}{2}(\varepsilon_i + \varepsilon_j - \varepsilon_a - \varepsilon_b)^{-1}\bar{V}_{ab}^{ij} + (\varepsilon_j - \varepsilon_b)^{-1}\bar{V}_{ib}^{ij} \tag{4.15}$$

$$L^{(2)} = \frac{1}{2}V_{ij}^{ab}(V_{ab}^{ij})_H + \bar{V}_{ij}^{ab}(\bar{V}_{kj}^{ka})_H \tag{4.16}$$

$$E^{(2)} = L_C^{(2)} = (VV_{BH})_C \tag{4.17a}$$

$$E^{(3)} = L_C^{(3)} = (V(VV_{BH})_{BH})_C - (V(V_{BH}V_C)_{BH})_C \tag{4.17b}$$

One sees that the $E^{(k)}$ consist of 'direct terms' where the operations 'multiply with V' and 'invert a commutator' alternate, and renormalization terms like the 2nd contribution to $E^{(3)}$. This is formally analogous to the expression in conventional perturbation theory, but much simpler.

We now apply the connected diagram-theorem. The $E^{(k)}$ must only contain connected diagrams. However, the renormalization diagrams are necessarily disconnected. Hence they cannot contribute and the final result is

$$E^{(2)} = (VV_{BH})_{CL} \tag{4.18a}$$

$$E^{(3)} = (V(VV_{BH})_{BH})_{CL} \tag{4.18b}$$

$$E^{(4)} = (V(V(VV_{BH})_{BH})_{BH})_{CL} \tag{4.18c}$$

where the subscript L stands for linked (which has, in this context the same meaning as connected).

The diagrammatic representation is the same as that introduced at the end of sec. 3.

The use of the particle-hole picture is essential for the derivation of the linked-cluster expansion, because then the renormalization diagrams to L_c are automatically disconnected. The overall L operator contains connected renormalization diagrams that must not be ignored[37] if one wants to get more information than just L_C.

Alternatively to the derivation just given one can expand the operator S in the sense of (4.8) in powers of λ. Perturbation theory can then be formulated entirely in terms of a Lie algebra. Since a commutator of two connected operators must be connected, only connected diagrams appear and one need not appeal to the separation theorem for the cancellation of disconnected terms as in the above derivation. A slight disadvantage is that in order to establish the equivalence with Goldstone expansion the Frantz-Mill-theorem[30] must be evoked. One gets

$$S^{(1)} = V_{BH} = W^{(1)} \tag{4.19a}$$

$$S^{(2)} = [V, S^{(1)}]_{BH} \tag{4.19b}$$

$$S^{(3)} = [V, S^{(2)}]_{BH} + \frac{1}{2}[[V, S^{(1)}]S^{(1)}]_{BH} \tag{4.19b}$$

$$E^{(1)} = V_C \tag{4.20a}$$

$$E^{(3)} = [V, V_{BH}]_C \tag{4.20b}$$

$$E^{(3)} = [V, S^{(2)}]_C + \frac{1}{2}[[V, S^{(1)}], S^{(1)}]_C \tag{4.20c}$$

5. REGULARIZED TIME DEPENDENT THEORY

We want to reconsider Goldstone's derivation of many-body perturbation theory (outlined in sec. 3), in a way in which the singularities of the time-independent limit ($\eta \to 0$) are avoided. To this end we write the time-evolution operator U_I (in the interaction representation) in the form

$$U_I = W_I \exp(iK_I) \tag{5.1}$$

We insert (5.1) into (3.7) and get

$$i\left\{\frac{\partial W_I}{\partial t} + W_I \frac{\partial e^{iK_I}}{\partial t} e^{-iK_I}\right\} = \lambda V_I W_I \tag{5.2}$$

we define

$$\Delta L_I = i\frac{\partial e^{iK_I}}{\partial t} e^{-iK_I} = \frac{\partial K_I}{\partial t} - \frac{i}{2}[K_I, \frac{\partial K_I}{\partial t}] + \ldots \tag{5.3a}$$

$$L_I = H_0 + \Delta L_I \tag{5.3b}$$

such that (5.2) becomes

$$i\frac{\partial W_I}{\partial t} = \lambda V_I W_I - W_I \Delta L_I \tag{5.4}$$

Transformation from the interaction representation to the Schrödinger representation changes (5.4) to

$$[H_0, W] = W\Delta L - \lambda V W + i\frac{\partial W}{\partial t} \tag{5.5a}$$

$$HW = WL + i\frac{\partial W}{\partial t} \tag{5.5b}$$

$$W^{-1}HW = L + iW^{-1}\frac{\partial W}{\partial t} \tag{5.5c}$$

One notes that (5.5b) differs from the Bloch equation (4.5) of time-independent theory by only the extra term: $i\frac{\partial W}{\partial t}$.

We formulate the theory in terms of W and L rather than U and take care that W and L are regular in the limit $\eta \to 0$ (which U is not). Expansion of (5.4) in powers of λ leads to

$$i\frac{\partial W_I^{(1)}}{\partial t} = V_I - L_I^{(1)} \tag{5.6a}$$

$$i\frac{\partial W_I^{(k)}}{\partial t} = V_I W_I^{(k-1)} - \sum_{l=1}^{k-1} W_I^{(l)} L_I^{(k-l)} - L_I^{(k)} \tag{5.6b}$$

We choose W_I such that (unlike for U_I) no singularities arise as a result of time integration for the choice (3.12,16) in the limit of $\eta \to 0$. This is e.g. achieved if the diagonal part W_{ID} of W_I (which is not determined by (5.5a) and hence to some extent free) vanishes., i.e. if $W_I(t)$ does not contain terms W_{rs}^{pq} for $\varepsilon_p + \varepsilon_q - \varepsilon_r - \varepsilon_s = 0$. Vanishing of the diagonal part of W_I implies that

$$L_I^{(1)} = V_{ID} \tag{5.7a}$$

$$L_I^{(k)} = (V_I W_I^{(k-1)})_D = \sum_{l=1}^{k-1} (W_I^{(2)} L_I^{(k-l)})_D \tag{5.7b}$$

It is now consistent to choose W so that

$$W_{ID} = 0, \quad \lim_{\eta \to 0} \frac{\partial W}{\partial t} = 0 \tag{5.8b}$$

$$\lim_{\eta \to 0} \{[H_0, W] - W\Delta L + \lambda V W\} = 0 \tag{5.8c}$$

i.e. that for $\eta \to 0$ W is solution of the time-independent Bloch equation (4.5). We can hence proceed as in sec. 4 to get the wave operator W and the energy operator L. Actually we are only interested in the vacuum expectation value L_C and the part W_B closed from below of W and we can immediately switch to sec. 4. In fact we have not gained anything with respect to sec. 4, except for having established a connection between sections 3 and 4.

It is obvious from (5.5b) that in the limit $\eta \to 0$ and hence $\frac{\partial W}{\partial t} = 0$, L is in fact the transformed Hamiltonian (4.2) and L_C the desired ground state energy.

Like in the Goldstone derivation we have started from the time evolution operator which is necessarily unitary and we have arrived at a wave operator in intermediate normalization. This raises three questions (a) how did unitarity get lost?, (b) can one alternatively arrive at a unitary W? (c) Of which kind is the singularity of U_I and hence Ψ_I in the Goldstone derivation?

The answer is as follows. Unitarity of U_I implies in view of (5.1) that

$$W_I^\dagger W_I = \exp(-iK_I)\exp(iK_I^\dagger) \tag{5.9}$$

i.e. that W_I is unitary if K_I is hermitean and vice versa, and this further implies in view of (5.3) that L_I is hermitean and vice versa.

We can write the time dependent wave function Ψ_I for a single non-degenerate state as

$$\Psi_I = U_I \Phi_I = W_I e^{iK_I} \phi_I = W_I e^{iK_{IC}} \Phi_i \tag{5.10a}$$

$$K_{IC} = < \phi | K_I | \phi > \tag{5.10b}$$

All operators in (5.10) are time-dependent.

In the general case K_{IC} is a complex number. We can decompose it into its real and imaginary parts

$$\Psi_I = e^{i\mathrm{Re}K_{IC}} e^{-\mathrm{Im}K_{IC}} W_I \Phi_I \tag{5.11}$$

The first factor on the r.h.s. of (5.11) is a phase factor and one gets

$$1 = < \Psi_I | \Psi_I > = < \Phi_I | W_I^\dagger W_I | \Phi_I > e^{-2ImK_{IC}} \tag{5.12a}$$

$$e^{-ImK_{IC}} = <\Phi_I|W_I^\dagger W_I|\Phi_I>^{-1/2} \tag{5.12b}$$

The second factor on the r.h.s. of (5.11) is hence shown to be a normalization factor. If W_I is unitary, this normalization factor reduces to unity, which means that $\text{Im}K_{IC} = 0$, hence K_{IC} is real for all t. An alternative is to require that W_I is in intermediate normalization, i.e.

$$1 = <\Phi|W_I|\Phi_I> = e^{iK_{IC}} <\Phi_I|\Psi_I> \tag{5.13a}$$

$$iK_{IC} = \ln <\Phi_I|\Psi_I> \tag{5.13b}$$

$$E = L_{IC} = E_0 + \Delta L_{IC} = E_0 - \frac{\partial}{\partial t}K_{IC} = E_0 + i\frac{\partial}{\partial t} \ln <\Phi_I|\Psi_I> \tag{5.13c}$$

In this case the Gell-Man-Low formula (3.24) holds. The energy L_{IC} is now complex for $\eta > 0$, but it becomes real in the limit $\eta = 0$.

How can one choose between the options of a W in unitary or intermediate normalization?

The point is that eqn. (5.4) or (5.5) only determine the non-diagonal part W_N of W and that one is rather free in the choice of the diagonal part W_D.

The simplest possibility is to take $W_D = 0$, this guarantees regularity and it implies the intermediate normalization. To take W unitary and at the same to guarantee regularity is somewhat more difficult. Let us try to demand that

$$W_D = W_D^\dagger; \quad L_D = L_D^\dagger \tag{5.14}$$

The second condition is compulsory for W unitary, and the first one is at least consistent with unitarity.

We see from (5.6a) that $W_{ID}^{(1)} = 0$, hence $W_I^{(1)} = W_{IN} = V_{IH}$.

Further

$$i\frac{\partial[W_I^{(2)} + W_I^{(2)\dagger}]}{\partial t} = V_I W_I^{(1)} + W_I^{(1)\dagger} V_I - W_I^{(1)} L_I - L_I W_I^{(1)\dagger} \tag{5.15}$$

From $W^{(1)}$ one has an energy denominator $(\Delta\varepsilon + i\eta)$ from $W^{(1)\dagger}$ instead $(\Delta\varepsilon - i\eta)$. Taking the r.h.s. of (5.15) together one gets

$$(\Delta\varepsilon + i\eta)^{-1} - (\Delta\varepsilon - i\eta)^{-1} = -2i\eta[(\Delta\varepsilon)^2 + \eta^2]^{-1} \tag{5.16}$$

The time integration to get $W_I^{(2)}$ introduces a factor $(2i\eta)^{-1}$, which cancels the $2i\eta$ on the r.h.s. The hermitean part of W_I is so obtained regular in the limit $\eta \to 0$. However the antihermitean part of W_I obtained from (5.6b) will diverge unless one imposes that it equal to 0 for all t, i.e. e that (5.14) holds.

Explicitly we get in the intermediate normalization

$$W^{(2)} = (VV_{BH})_{BH} - (V_{BH}V_C)_{BH} = W_B^{(2)} \tag{5.17}$$

but in the unitary normalization

$$W^{(2)} = W_B^{(2)} - \frac{1}{2}(V_{BH}V_{BH})_C \tag{5.18}$$

One can mention that the extra terms which arise in the unitary normalization of the wave operator cannot be represented by diagrams with Goldstone's rules for energy denominators.

We can now see how in Goldstone's derivation the option for a unitary W_I has been missed. The point is that the constant a in (3.33) is not necessarily real - it is only its real part which diverges, and gives rise to a divergent phase, but it also has an imaginary part, which is regular in the limit $\eta \to 0$ and which hence implies a change in the normalization. This imaginary part of a can be ignored if one uses K_I only for the evaluation of the energy via the Gell-Mann-Low formula, because the time derivative of a vanishes.

In practice the unitary choice of W_I does not present advantages except for open-shell states, but the present discussion helps to understand how tricky the limit $\eta \to 0$ is, and that it is worthwhile wondering why in the Goldstone derivation starting from a unitary U one arrives at a W in intermediate normalization. This is so because after getting rid of the $1/\eta$ singularity in the energy expression (3.23) one has made the limit $\eta \to 0$ in all diagrams individually. This amounts to ignoring the imaginary part of L_c - which vanishes in the limit $\eta \to 0$, but which must not be ignored without checking the normalization at the same time.

Regularizations of the time-dependent theory similar to the one outlined here were discussed by Langhoff et al.[47,48] as well as by Bhattacharyya and Mukherjee[50]. Let us finally mention that our critical remarks on the treatment of a time-independent perturbation as the limit of an exponentially switched perturbation for $\eta \to 0$, only applies to stationary states. For scattering states the situation is very different, as is nicely explained in Taylor's textbook[50]. There the energy is conserved, so there is no change in the phase factor. Moreover there are two linear independent solutions of the Schrödinger equation characterized by the limits $\eta \to +0$ and $\eta \to -0$ and corresponding to different boundary conditions. The singularity for $\eta \to 0$ is a branch cut rather than a pole as it is for bound states.

CONCLUDING REMARKS

In principle many-body perturbation theory (MBPT) is nothing but application of Rayleigh-Schrödinger perturbation theory to the many-electron Hamiltonian with the electron interaction as perturbation. However the straightforward pedestrian formulation of MBPT in a Hilbert space for fixed particle number becomes extremely complicated and tedious, mainly since matrix element between Slater determinants have to be evaluated.

A very compact and surprisingly simple derivation of MBPT is possible if one does not use Slater determinants, but tries to construct a wave operator in Fock space in a particle-hole formalism. The separation theorem plays a central role according to which the energy is an additively separable quantity while the wave operator is multiplicatively separable. In a diagrammatic representation which is very convenient, additively separable quantities only consist of connected diagrams.

Goldstone's derivation of the 'linked-diagram expansion' from time-dependent theory deserves mainly a historical interest, his expansion can be derived in a much simpler way.

In this review we have only considered the case that the unperturbed Hamiltonian is that without electron interaction. The generalization to the case where the averaged Hartree-Fock type interaction is contained in H_0 as suggested by Møller and Plesset[15], is perfectly straightforward. The theory becomes even simpler, because all diagrams with self-contractions disappear.

The case of a closed-shell state, to which we have restricted our interest, is particular simple, since such a state can be regarded as a 'physical vacuum. Open-shell states are much more complicated but also more interesting in practice. There are still a few unresolved problems with open-shell states[44].

Perturbation theory is only meaningful if the expansion in powers of λ converges. This is by no means guaranteed. Therefore non-perturbative methods like e.g. the coupled-cluster approach[11,49] are often preferable. Such non-perturbative methods can often be related to perturbation theory by stating that in these methods certain important classes of diagrams are summed to infinite order.

ACKNOWLEGEMENT

The author is grateful to D. Mukherjee for the kind invitation to present this lecture at the Calcutta workshop 1988 and he thanks S. Koch for helpful comments.

REFERENCES

1. P. Nozières, 'Le problème à N corps', Dunod, Paris 1963, 'The theory of Interacting Fermi Systems', Benjamin, New York 1964
2. E.A. Hylleraas, Z. Phys. 54:347 (1929), 65:209 (1930)
3. D.R. Hartree, Proc. Cambridge Phil. Soc. 24:111, 426 (1928); J.C. Slater, Phys. Rev. 25:210 (1930)
4. V. Fock, Z. Phys. 61:126 (1930)
5. C.C.J. Roothaan, Rev. Mod. Phys. 23:69 (1961)
6. A.P. Jucys, J. Exp. Theor. Phys. (USSR) 23:129 (1952)
   J. Vizbaraité, A. Kancerevičius, A. Jucys, Opt. Spectr. (USSR) 1:9 (1956)
7. H.A. Kramers, 'Quantenmechanik (in Hand- und Jahrbuch der chemischen Physik Vol 1 (1937), english translation 'Quantum Mechanics' North-Holland, Amsterdam 1957 (Dover, New York 1964)
8. J.I. Frenkel, 'Wave mechanics, advanced general theory' Oxford, Clarendon 1934 (Dover, New York 1950)
9. H.D. Ursell, Proc. Cambridge Philos. Soc. 23:685 (1927)
10. J.E. Mayer, J. Chem. Phys. 5:67 (1937)
11. F. Coester, Nucl. Phys. 7:421 (1958)
    F. Coester, H. Kümmel, Nucl. Phys. 17:477 (1960)
    H. Kümmel, Nucl. Phys. 22:177 (1961)
12. P.O. Löwdin, Phys. Rev. 97:1474 (1955)

13. R.K. Nesbet, Proc. Roy. Soc. London A230:312 (1955)
14. I. Shavitt, in: 'Modern Theoretical Chemistry III, Methods of Electronic Structure Theory', H.F. Schaefer ed., Plenum New York (1977)
15. C. Møller, M.S. Plesset, Phys. Rev. 46:618 (1934)
16. R.J. Bartlett, D.M. Silver, J. Chem. Phys. 62:3258 (1975)
J.A. Pople, R. Krishnan, H.B. Schlegel, J.S. Binkley, Int. J. Quant. Chem. Symp. 13:225 (1979)
17. K.A. Brueckner, Phys. Rev. 97:1353 (1955), 100:36 (1955)
18. J. Hubbard, Proc. Roy. Soc. London A 240:539 (1957)
19. N.M. Hugenholtz, Physica 23:481 (1957)
20. J. Goldstone, Proc. Roy. Soc. London A239:267 (1957)
21. S. Tomonaga, Progr. Theor. Phys. 1:27 (1946), Phys Rev 74:224 (1948)
22. J. Schwinger, Phys. Rev. 74:1439 (1948), 75:651 (1949)
23. R.P. Feynman, Phys. Rev. 76:749 (1949),
24. M. Gell-Mann, F. Low, Phys. Rev. 84:350 (1951)
25. G.C. Wick, Phys. Rev. 80:268 (1951)
26. J. Čižek, J. Chem. Phys. 45:4256 (1966)
27. H.P. Kelly, Phys. Rev. 131:684 (1963), 134:1450 (1964), 144:39 (1966), Adv. Chem. Phys. 14:129 (1969)
28. A.C. Hurley, E. Lennard-Jones, J.A. Pople, Proc. Roy. Soc. London A220:446 (1953)
29. F.B. Malik, R.H. Richardson, J.Y. Shipiro, paper presented at the symposium on 'Recent Progress in Many-Body Theories', Oulu, Finland, August 1987
30. L.M. Frantz, R.L Mills, Nucl. Phys. 15:16 (1960)
31. B.H. Brandow, Rev. Mod. Phys. 39:771 (1976)
32. H. Primas, Helvet. Phys. Acta 34:331 (1961), Rev. Mod. Phys. 35:710 (1963)
33. R.J. Yaris, J. Chem. Phys. 41:2419 (1964), 42:3019 (1965)
34. E.R. Caianiello, in: Combinatorics and Renormalization in Quantum Field Theory, Benjamin, Reading, Mass. (1973)
35. C. Bloch, Nucl. Phys. 6:329 (1958)
36. D.J. Klein, J. Chem. Phys. 61:786, (1974)
F. Jørgensen, Mol. Phys. 29:1137 (1975)
37. W. Kutzelnigg, Chem. Phys. Letters 83:156 (1981), J. Chem. Phys. 77:3081 (1982), J. Chem. Phys. 80:822 (1981)
W. Kutzelnigg and S. Koch J. Chem. Phys. 77:3081 (1982),
38. W. Kutzelnigg in 'Aspects of Many-Body Effects in Molecules and Extended Systems' D. Mukherjee ed., Lecture notes in chemistry, Vol. 50, p.35 Springer Berlin 1989
39. W. Kutzelnigg, in: 'Recent Progress in Many-Body Theory', H. Kümmel and M.L. Ristig ed., Lecture notes in physics, Vol. 198 p. 361 (1984)
40. J. Paldus and J. Cizek, Adv. Quantum Chem. 9:105 (1975)
41. I. Lindgren, J. Morrison, 'Atomic Many-Body Theory', Berlin, Springer 1982
42. S. Raimes, 'Many-Electron Theory', Amsterdam, North-Holland, 1972
43. I. Lindgren, J. Quantum Chem. Symp. 12:33 (1978)
44. D. Mukherjee and S. Pal, Adv. Quantum Chem. 20:292 (1989)
45. T. Kato, 'Perturbation Theory of Linear Operators', Berlin, Springer 1966
46. J.C. Slater, Phys. Rev. 34:1293 (1929)
47. P.W. Langhoff, S.T. Epstein and M. Karplus, Rev. Mod. Phys. 44:602 (1972)
48. P.W. Langhoff and A.J. Hernandez, Int. J. Quant. Chem. Symp. 10:337 (1975)
49. J. Cižek, J. Chem. Phys. 45:4256 (1966)
R.J. Bartlett, J. Phys. Chem. 93:1697 (1989) and references therein
50. K. Bhattacharyya and D. Mukherjee, J. Phys. A 19:67 (1986)
51. J.R. Taylor, 'Scattering Theory' New York, Wiley 1972

DILEMMAS IN THE CHOICE OF MODEL SPACES SUPPORTING MAGNETIC HAMILTONIANS

Jean Paul Malrieu

Laboratoire de Physique Quantique
(U.R.A. 505 du C.N.R.S.)
Université Paul Sabatier, 118, route de Narbonne
31062 Toulouse Cedex, France

INTRODUCTION

Heisenberg Hamiltonians[1] were first introduced as phenomenological Hamiltonians to reproduce the energy splittings between the electronic states of polyradical systems, i.e. in problems involving several unpaired electrons. The foundation of such Hamiltonians has been clearly established as effective Hamiltonians [2-8], in the sense of Quasi Degenerate Perturbation Theory[9-10], concerning special types of problems and a special choice of the model space [11]. The concerned systems are half-filled band problems namely, pn e^- in pn atomic (or local) degenerate orbitals (AO), on n centers. The band must be unique and half-filled since one might put up to 2np e^- in np orbitals. The model space is uniquely defined as spanned by all the determinants in which each AO bears one and only one electron, which means that all the determinants have the same space part; then the unique degree of freedom in the spin distribution and the effective hamiltonian must be a spin-only Hamiltonian. The first purpose of this paper is to discuss the choice of that model space, showing that even for half-

Applied Many-Body Methods in Spectroscopy and Electronic Structure
Edited by D. Mukherjee, Plenum Press, New York, 1992

filled bands it lacks generality and faces numerous fundamental problems.

Other definitions of the model space are possible; one may either extend it to <u>all</u> neutral Valence Bond determinants including situations with null or double occupancies of the AOs, or reduce it to the situations where the electrons of the same atom have parallel spins. The consequences of these different choices are discussed along several criterions.

i) formal structure (simplicity, spin-only or spin-space nature of the Hamiltonian, invariance under the rotation of the axis in R^3),

ii) intruder state problems and the ways to solve them,

iii) transferability of such effective Hamiltonians from the simplest (diatomic) problem to extended systems (up to the bulk)

I. THREE DIFFERENT CHOICES OF THE MODEL SPACE : STRUCTURAL ASPECTS

The three previously mentionned model spaces will be described and their properties (as regards the invariance to the choice of the axis for instance) are discussed. The formal structure of the effective hamiltonian will be depicted from second-order developments of the Quasi-Degenerate Perturbation Theory[9-10], which is valid in the weak delocalization domain at least, according to the strategy opened by Anderson[11].

In some cases we shall develop arguments which take benefit of logical implications : if one has established the effective hamiltonian H_S^{eff} for a large model space, say S, one may reach the effective hamiltonian $H_{S'}^{eff}$ for a smaller model space S' included in S.

$$S' \cap S = S'$$

by applying the Quasi Degenerate Perturbation Theory on the "large-size" effective Hamiltonian H_S^{eff}. This relation is also valid for the recently proposed intermediate effective Hamiltonians[12-16] provided that H_S^{int} gives the exact roots associated to the small model space.

A. Traditional formulation in terms of spin-only and $\langle S_z \rangle = \pm \frac{1}{2}$ particles

The magnetic Hamiltonians or spin-effective Hamiltonians are frequently written as

$$\overline{H} = \sum_{ij} J_{ij} \; \vec{S}_i \cdot \vec{S}_j$$

where i and j represent an orthogonal set of localized atomic or nearly atomic orbitals (the case of composite particles will be treated later), J_{ij} is an effective spin coupling (or effective exchange integral) and $\vec{S}_i$ and $\vec{S}_j$ represent the spin momentum of the electrons in the orbitals i and j. In general J_{ij} is supposed to vanish when i and j are not on the same atom or located on nearest neighbours, but this is not compulsory. One sometimes writes

$$\overline{H} = \overline{H}_1 + \overline{H}_2$$

with

$$\overline{H}_1 = J_{ij} \; S_{zi} \; S_{zj}$$

$$\overline{H}_2 = J_{ij} (S_i^+ \; S_j^- + S_i^- \; S_j^+)$$

In that partition $\overline{H}_1$ simply counts the spin-alternations or spin frustrations between interacting orbitals, while $\overline{H}_2$ permutes the spins (when they are opposite) of interacting orbitals. The two preceding formulations take as zero of energy the mean energy of the configuration. Other formulations have been proposed and used, which essentially differ by the choice of the energy origin. One may for instance write[17,18]

$$\overline{H}' = \sum_{ij} \left(R_{ij} - g_{ij} \, | i\bar{j} - \bar{i}j \rangle \langle i\bar{j} - \bar{i}j | \right)$$

where $\sum_{ij} R_{ij}$ in the energy of the upper multiplet (taken from the energy of separated atoms) and $|i\bar{j}><i\bar{j}|$ stands for $a_i^+ \ a_j^+ \ a_j \ a_i$ for sake of readibility.

In the traditional formulation these Hamiltonians seem to act on spins only. Of course they implicitly act on a space of n-electronic functions, namely those in which, for n electrons and a set of n atomic orbitals, all atomic orbitals bear one and only one electron. Although they are formulated in terms of spin, these Hamiltonians act on a very precise space of n- electronic determinants, which all have the same space part and differ by the spin distribution. Considering n electrons in n atomic orbitals (moreover supposed to be orthogonal) define a very specific and narrow domain : the half-filled band problem. One should add that they usually concern a single-band ; it is not easy to consider for instance a set of two bands, for instance the s and p bands on carbon, with a mean half-filling of the two bands, since the energies of the s orbitals and p orbitals are different, which can introduce some difficulties in the pure spin Hamiltonian.

Heisenberg Hamiltonians therefore concern a very restrictive domain of the matter. One may consider them as effective Hamiltonians, in the sense of Quasi Degenerate Perturbation Theory and Wave Operator formulations defined on model spaces where each valence atomic orbital bears one electron and only one. If

$$P_0 = \sum_{I=i,N} |\phi_I><\phi_I|$$

$$\phi_I = |a \ldots n| \prod_{i=1,n} \gamma_i$$

where a, i, n are valence Orthogonal AOs (OAO) and γ an α or β spin function.

Then $\bar{H}$ is the effective Hamiltonian associated with the model space P_0. In the Bloch's formulation[6] this means that

$$\bar{H}P_0\psi_m = \epsilon_m P_0\psi_m \qquad m=1,N$$

with $$H\psi_m = \epsilon_m\psi_m$$

H being the exact Hamiltonian, N is the number of determinants ϕ_I belonging to the model space. Since $\bar{H}$ and H do not couple States of different $<S_z>$ values, considering the lowest value of $<S_z>$ (0 or ½), N is limited to $C_n^{n/2}$ (or $C_n^{(n-1)/2}$).

However the fact that $\bar{H}$ must give N exact eigenvalues of H and the projections of the corresponding eigenvectors onto the model space is not sufficient to define $\bar{H}$. Actually $\bar{H}$ is defined by the choice of two isodimensional subspaces, the model space of projector P_0, which is known, and a yet undefined target space, a stable subspace of H. The one-to-one correspondance between the model space and the target space is not straightforward[8]. Let us call P the projector on the target space,

$$P = \sum_{m=i,n} |\psi_m><\psi_m|$$

$$P = \Omega\ P_0$$

where Ω is the wave-operator.

Then traditional procedures, such as Quasi Degenerate Perturbation Theory, are supposed to lead unambiguously to a meaningfull target space when they converge. One may expect that P consists of the N eigenvectors of H which have the largest components onto the model space, i.e. that

$$\|P_0\psi_m\| > \|P_0\psi_n\| \text{ when } \begin{cases} P\psi_m = \psi_m \\ P\psi_n = 0 \end{cases}$$

but there is no proof for that conjecture and some of the difficulties which we shall discuss are linked to the uncertainties of that correspondance.

There is in principle no difficulty to define P_0 when one considers a system in which each atom contributes for one orbital only. The definition of P_0 is more difficult beside this extremely peculiar case. Let us consider for instance [19] a two-dimensional (2-D) system in which each atom introduces two orbitals of p character in which one puts two electrons (this example is taken for simplicity, but one might consider a three-dimensional problem, i.e. the p band of s^2p^3 atoms such as N, P, As, which would be more realistic). The $<S_z>=0$ determinants of the isolated atoms belong to two categories. In the first one

$$\phi_1 = x\bar{y}\ ,\ \phi_2 = \bar{x}y$$

each AO bears one and only one electron, as required in Heisenberg Hamiltonians, (and as occurs in the $<S_z> = \pm 1$ determinants xy and $\bar{x}\ \bar{y}$). In the second one, built of

$$\phi_3 = x\bar{x},\ \phi_4 = y\bar{y}$$

there is one empty and one doubly occupied AO. The eigenstates of the problem are one triplet ϕ_T and three singlet states

$$\phi_T = \left(x\bar{y} + \bar{x}y\right)/\sqrt{2} \text{ with energy } E = E_0$$

$$\phi_{S_1} = \left(x\bar{y} - \bar{x}y\right)/\sqrt{2}\quad E = E_0 + 2K$$

$$\phi'_{S_1} = \left(x\bar{x} - y\bar{y}\right)/\sqrt{2}\quad E = E_0 + 2K$$

$$\phi_{S_2} = \left(x\bar{x} + y\bar{y}\right)/\sqrt{2}\quad E_0 = E_0 + 4K$$

since the Hamiltonian matrices are

$$
\begin{array}{c}
\begin{array}{cc} \phi_1 & \phi_2 \end{array} \\
\begin{vmatrix} E_0 + K & K \\ K & E_0 + K \end{vmatrix}
\end{array}
\quad \text{and} \quad
\begin{array}{c}
\begin{array}{cc} \phi_3 & \phi_4 \end{array} \\
\begin{vmatrix} E_0 + 3K & K \\ K & E_0 + 3K \end{vmatrix}
\end{array}
$$

In these equations E_0 is the energy of the Triplet state, i.e. of $|xy|$, K is the monocentric exchange integral and one takes benefit of the relation between coulomb and exchange integrals

$$
\begin{aligned}
J_{xx} = \langle xx|r_{12}^{-1}|xx\rangle &= \langle xy|r_{12}^{-1}|xy\rangle + 2\ K_{xy} \\
&= J_{xy} + 2\ K_{xy}
\end{aligned}
$$

The important point is the degeneracy between a state ϕ_{s_1} belonging to the traditional model space and a state ϕ'_{s_1} which does not belong to it. Now one may notice first that if one changes the axis of the coordinates, i.e. rotate the orthogonal references axis,

$$
\begin{aligned}
x' &= x \cos\varphi + y \sin\varphi \\
y' &= -x \sin\varphi + y \cos\varphi
\end{aligned}
$$

the $\langle S_z\rangle = \pm 1$ components are unvariant

$$|x'y'| = |xy|$$

while the set $\{\phi_1, \phi_2\}$ is not changed into $\left\{\phi'_1, \phi'_2\right\}$

$$
\begin{aligned}
\phi'_1 = x'\ \bar{y}\ &= \cos^2\varphi\ x\bar{y} - \sin^2\varphi\ y\bar{x} + \sin\varphi\cos\varphi\ \left(y\bar{y} - x\bar{x}\right) \\
&= \cos^2\varphi\ \phi_1 + \sin^2\varphi\ \phi_2 + \sin\varphi\cos\varphi\ (\phi_4 - \phi_3)
\end{aligned}
$$

For a half-filled band problem in which each atom brings p (p>1) electrons in p degenerate orbitals, the model space on which the Heisenberg Hamiltonians for $\langle S_z\rangle = \pm 1/2$ particles is defined is not invariant in a change of axis, and hence the Heisenberg Hamiltonian defined on that model space is axis-dependant.

This remark questions the interest of these Heisenberg Hamiltonians except when a specific choice of axis is directly imposed. One may consider for instance the case of the polyines C_nH_2 which are linear, with two orthogonal π systems[19]

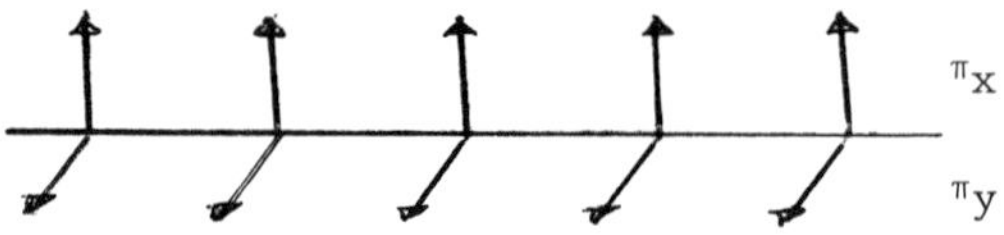

for which the choice of the z axis is natural and that of x, y is unimportant, the Heisenberg Hamiltonian remaning unchanged under a rotation around the z axis. This problem has been studied in our laboratory with satisfactory results regarding the lowest states.

As another example one may consider the half-filled p band (s^2p^3 atoms) in a cubic structure, where again, the choice of the axis seems natural and imposed by the lattice [20]. However in such a problem one cannot take into account the spin interactions between non-nearest neighbours since from

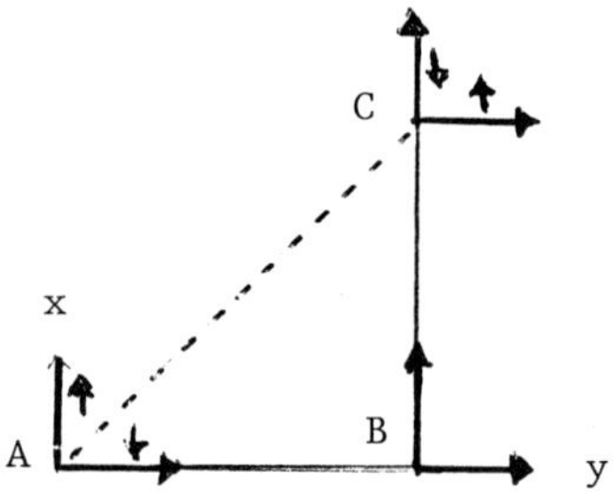

the determinant $|\ldots x_a \bar{y}_a \ldots \bar{x}_c y_c \ldots|$ it is impossible to make a local change of axis $\{x, y\} \rightarrow \{x', y'\}$ without going out of the model space. The study of non cubic clusters is also forbidden.

The Heisenberg Hamiltonian is <u>hermitian</u> (by hypothesis) ; in the Bloch formulation, this is only true at second order if the zeroth order Hamiltonian H_0 is degenerate in the model space, i.e. for the acetylene six-dimensional example :

$<\phi_1|H_0|\phi_1> = <\phi_3|H_0|\phi_3> = <\phi_5|H_0|\phi_5>$ while

$$<\phi_1|H|\phi_1> - <\phi_3|H|\phi_3> = <\phi_1|H|\phi_1> - <\phi_5|H|\phi_5> = 2K$$

In that case the Heisenberg Hamiltonian matrix will take the form

$$\begin{array}{cccccc}
\phi_1 & \phi_2 & \phi_3 & \phi_4 & \phi_5 & \phi_6 \\
\\
-g_x-g_y & & g_y & g_x & & \\
 & -g_x-g_y & g_x & g_y & & \\
 & & 2K-g_x-g_y & & -K & -K \\
 & & & 2K-g_x-g_y & -K & -K \\
 & & & & 2K & \\
 & & & & & 2K
\end{array}$$

where K is the K_{xy} intraatomic direct (ferromagnetic) exchange, and $-g$, the interatomic effective (antiferromagnetic) exchange, reflects the effect of the coupling between neutral and ionic VB configurations, as well known from Anderson's derivation[(11)]. To the second order, the QDPT gives the correct form of the Hamiltonian and provides an estimate of g

$$g_x = + \frac{2t_x^2}{U} > 0$$

where t is the interatomic hopping integral and U is the neutral/ionic energy difference. Of course g may involve higher order contributions. For large interatomic distances $|g|$ decreases and may become small in comparison to K, while the reverse may happen at small interatomic distances. When g tends to zero, the ${}^1\Sigma_g^+$ ground state tends to the atomic limit i.e.

$$\psi \rightarrow \frac{1}{\sqrt{3}} (\phi_1 + \phi_2 - (\phi_3 + \phi_4 + \phi_5 + \phi_6)/2)$$

while when $K \ll g$

$$\psi \rightarrow (\phi_1 + \phi_2 - \phi_3 - \phi_4)/2$$

Higher-order contributions would introduce four-body terms such as $<\phi_1|\bar{H}|\phi_2>$ and modify the value of g, as will be discussed later.

B. Enlarging the model space to all neutral VB structures

Of course it is possible to enlarge the model space and to include in it all the neutral VB determinants of the half-filled band problem, i.e. those in which each atom bears p and only p electrons in its p AOs, whatever their spatial distribution. Now some AOs may have zero electrons and others may bear two electrons. This model space will now be *unvariant under changes of the axis*.

For the acetylene molecule for instance, restricted to its Π systems, this extended model space represents ten $<S_z> = 0$ determinants with 4 electrons in the form π AOs ; six of them are those which support the classical Heisenberg Hamiltonians

$$\phi_1 = |x_A y_A \bar{x}_B \bar{y}_B|, \quad \phi_2 = |\bar{x}_A \bar{y}_A x_B y_B|$$

$$\phi_3 = |x_A \bar{y}_A \bar{x}_B y_B|, \quad \phi_4 = |\bar{x}_A y_A x_B \bar{y}_B|$$

$$\phi_5 = |x_A \bar{y}_A x_B \bar{y}_B|, \quad \phi_6 = |\bar{x}_A y_A \bar{x}_B y_B|$$

and four imply double occupancies of orbitals

$$\phi_7 = |x_A \bar{x}_A y_B \bar{y}_B|, \quad \phi_8 = |y_A \bar{y}_A x_B \bar{x}_B|$$

$$\phi_9 = |x_A \bar{x}_A x_B \bar{x}_B|, \quad \phi_{10} = |y_A \bar{y}_A y_B \bar{y}_B|$$

These states may be called "pseudo ionic" since they are globally neutral but ionic in each subsystem.

If one goes to this 10-dimensional model space, including all the neutral VB structures, H^{eff} is expected to take the form.

Φ_1	Φ_2	Φ_3	Φ_4	Φ_5	Φ_6	Φ_7	Φ_8	Φ_9	Φ_{10}
$-g'_x-g'_y$		g'_x	g'_y			$+g''$	$+g''$		
	$-g'_x-g'_y$	g'_y	g'_x			$+g''$	$+g''$		
		$2K-g'_x-g'_y$		$-K$	$-K$	$-g''$	$-g''$		
			$2K-g'_y-g'_x$	$-K$	$-K$	$-g''$	$-g''$		
				$2K$					
					$2K$				
						$4K-g'_x-g'_y$		K	K
							$4K-g'_x-g'_y$	K	K
								$4K$	
									$4K$

This structure is evident from the second-order development. Again the hermiticity is only guaranteed at order 2 in the Bloch's expansion if H_0 is degenerate in the full model space. In this matrix g" again reflects the effect of the coupling with ionic structures and it is easy to show that at the 2nd order

$$g'' = 2\,\frac{t_x\ t_y}{U} = \sqrt{g'\ g'_y}$$

In the above matrix, we have introduced g' instead of g, since g and g' are only indentical at 2nd order. Actually for the strict Heisenberg Hamiltonian (6-dimensional model space) g incorporates 4th-order corrections of the type

$$\Phi_1 \to \Phi_{ionic} \to \Phi_7 \text{ or } \Phi_8 \to \Phi'_{ionic} \to \Phi_1,\ \Phi_3 \text{ or } \Phi_4$$

which are proportional to $\dfrac{t^4}{U^2\,.\,4K}$, while in g' the corresponding fourth-order contributions are Exclusion Principle Violating contributions and, since Φ_7 (or Φ_8) now belong to the model space, the corresponding terms are proportional to $-\ t^4/U^3$.

It is interesting to give an algebraic form to this generalized effective Hamiltonian ; it may be written as

$$H^{eff}_{Neutral} = \sum_A \left[K_A \sum_{i_A<j_A} |i_A \bar{j}_A - \bar{i}_A j_A><i_A \bar{j}_A - \bar{i}_A j_A| \right.$$

$$\left. + K_A \left(\sum_{i_A} |i_A \bar{i}_A><i_A \bar{i}_A| + \sum_{j_A \neq i_A} |i_A \bar{i}_A><j_A \bar{j}_A| \right) \right]$$

$$+ \sum_A \sum_{<B} \left[R_{AB} - \sum_{i_A} \sum_{i_B} g'_{i_A i_B} |i_A \bar{i}_B - \bar{i}_A i_B><i_A \bar{i}_B - \bar{i}_A i_B| \right.$$

$$-\sum_{i_A} \sum_{j_A} \sum_{i_B} \sum_{j_B} g''_{i_A j_A i_B j_B} \left(|\bar{j}_A i_B><i_A \bar{j}_B| \right.$$

$$\left. \left. + |j_B \bar{i}_B><\bar{i}_A j_B| - |\bar{j}_A \bar{i}_B><\bar{i}_A \bar{j}_B| - |j_A i_B><i_A j_B| \right) \right]$$

$$\text{with} \qquad g''_{i_A j_A i_B j_B} = \sqrt{g'_{i_A i_B} \; g'_{j_A j_B}}$$

Notice that such an Hamiltonian should :

- be invariant under a unitary change of the basis AO (change of axis) ;

- deliver new roots. In particular for the acetylene problem the Heisenberg Hamiltonian does not provide roots of Δ symmetry, while the extended effective Hamiltonian gives them, spanned by model vectors of the type $|x_A \bar{x}_A \; x_B \bar{y}_B|$, which are VB-neutral, for the xy component, and by model vectors of the type $|x_A \bar{x}_A \, . \, y_B \bar{y}_B|$ for the $x^2 - y^2$ component.

C. Restricting the model space to atomic high-spin situations use of composite particles

Heisenberg Hamiltonians are frequently presented as acting on spin-particles with $S_z \neq ½$; they are supposed to act on sets of electrons on the same site with parallel spins, due to a strong ferromagnetic local coupling, i.e. two electrons for $S_z = 1$, three electrons for $S_z = 3/2$ and so on. Moving back to the Atomic Orbitals, this means that for $S_z = k/2$ each atom I brings either the product $|i_1 \ldots i_k|$ or the product $|\bar{i}_1 \ldots \bar{i}_k|$ to the model functions with a supposedly well-defined space part. Then for a 2n-centre problem, the model space is only C^n_{2n}-dimensional, as for the elementary problem with one electron and one AO per atom.

One may notice that in this picture, the model space is invariant under the rotation of the axis, as noticed for the atom, and this is a strong advantage. However it cannot provide a detailed spectroscopy ; for the two-center problem, the effective hamiltonian provides only *two* states, namely an in-phase and an out-of-phase combination of two determinants

$$|i_1 \ldots i_k \; \bar{j}_1 \ldots \bar{j}_k| \pm |\bar{i}_1 \ldots \bar{i}_k \; j_1 \ldots j_k| = \Phi_1 \pm \Phi_2$$

There is strictly no reasons to impose now that one of them has the same effective energy as the upper multiplet, the $\langle S_z \rangle = k$ component of which is

$$|i_1 \ldots i_k \; j_1 \ldots j_k|$$

as one might be tempted to assume, since the $\langle S_z \rangle = 0$ component of this upper multiplet has all its coefficients equal to $C\binom{k}{2k}^{-1/2}$. This is true for Φ_1 and Φ_2 ; these values are very low and it is sure that there are other states of larger projections on the model subspace $\{\Phi_1, \Phi_2\}$. So that the target space cannot involve the upper multiplet.

Actually if one applies the Degenerate Perturbation Theory to that model space one may see immediately that

i) there are 2nd order diagonal contributions, as for the simplest case. If all orbitals are of different symmetries, the resulting energy lowering will be

$$-g = -\sum_{d=1,k} 2\frac{t_d^2}{U}$$

where $t_d = \langle i_d | F | j_d \rangle$ is the hopping integral for the orbitals of symmetry d.

ii) there are important diagonal fourth-order corrections since at the second order on the wave-function one will reach *neutral* VB structures which do not belong to the model space but are only at distance 2(k-1)K, where K is the monocentric exchange integral. These configurations result from Φ_1 (or Φ_2) by one spin exchange between the two atoms

$$d \quad \Phi_{i_d \to j_d} \qquad t_d$$

$$\Phi_1 \qquad\qquad \Phi = a^+_{i_d}\, a^+_{j_d}\, a_{i_d}\, a_{j_d}\, \Phi_1$$

$$t_d \quad \Phi_{i_d \to j_d} \qquad t_d$$

and have a coefficient proportional to $\dfrac{t_d^2}{U\ (k-1)\ K}$

Notice that the 2nd-order wave operator also leads to other neutral determinants involving empty and doubly occupied AO.

$$\Phi_1 \xrightarrow{t_d} \Phi_{i\,d \to j\,d} \xrightarrow{t_c} \Phi' = a^+_{i_c}\, a_{j_c}\, a^+_{j_d}\, a_{i_d}\, \Phi_1$$

which lie at an energy 2(k - 1) K + 4K = 2(k + 1) K above Φ_1. The perturbation has no chance to converge if

$$\max \left(\frac{t_d^2}{U(k-1)\ K} \right) > \tfrac{1}{2}$$

which means that (k - 1) K must be larger them $2\dfrac{t^2}{U}$, which is the elementary effective antiferromagnetic interatomic exchange concerning the orbitals of largest overlap (changed sign).

iii) the effective coupling between Φ_1 and Φ_2 is only obtained at order 2k since to change all spins the 2k electrons must travel through the bond ; the lowest order contribution is then proportional to $\left(\prod_{d=1,k} t_d^2 \right) / U^{k-1}$ but there are so many possibilities of travels, through states of various ionicities that its evaluation is a problem per se. Moreover it is not sure that these processes, in which all the electrons remain in their initial symmetry are the lowest ones. Other processes combining electron transfers in the channels of largest amplitude with two electron displacements on the same atom (through the K integral) may be of larger amplitude.

Anyway the non-diagonal terms are expected to be of much smaller amplitude and the effective hamiltonian may be written

$$\bar{H} = \sum_{ij} \Big[R_{ij} - g_{ij} \Big(|i_1 \ldots i_k \ \bar{j}_1 \ldots \bar{j}_k\rangle\langle i_1 \ldots i_k \ \bar{j}_1 \ldots \bar{j}_k \Big) \Big]$$

$$+ |\bar{i}_1 \ldots \bar{i}_k \ j_1 \ldots \bar{j}_k\rangle\langle \bar{i}_1 \ldots \bar{i}_k \ j_1 \ldots j_k |)$$

$$+ g'_{ij} \Big(|i_1 \ldots i_k \ \bar{j}_1 \ldots \bar{j}_k\rangle\langle \bar{i}_1 \ldots \bar{i}_k j_1 \ldots j_k | \Big)$$

$$+ |\bar{i}_1 \ldots \bar{i}_k \ j_1 \ldots j_k\rangle\langle i_1 \ldots i_k \ \bar{j}_1 \ldots \bar{j}_k |]$$

This is the general form of Heisenberg Hamiltonian acting on spins larger them ½

$$\bar{H} = \sum_{ij} J_{ij} \ S_{zi} \ S_{zj} + J'_{ij} \left(S^+_i \ S^-_j + S^-_i \ S^+_j \right)$$

with $|J'| \leq |J|$; the preceding discussion makes clear the origin of this *anisotropy* of the Heisenberg Hamiltonian.

II. EFFECTIVE OR INTERMEDIATE HAMILTONIANS?

So far one had accepted the idea that both the Heisenberg Hamiltonian and the extended Neutral VB effective Hamiltonian could be obtained by the classical Quasi Degenerate Perturbation expansion. Except for symmetry reasons (when all eigenvectors of H^{eff} are of different symmetries) the exact (fully converged) Bloch effective Hamiltonian has no reason to be hermitian. It is always possible to insure hermiticity by considering the des Cloizeaux' expansion[7], the eigenvectors of which are $S^{-1/2}$ orthogonalized projections of the eigenvectors onto the model space. Instead of using such a symmetrical orthogonalization, hierarchized orthogonalization (of Schmidt type) of these projections might be considered as well[(19)].

However numerical studies [(19)] on the acetylene Π problem have shown that both the Bloch and the des Cloizeaux exact (from their spectral definitions) effective Hamiltonians defined on the 6-dimensional model space completely deviate from what is expected for an Heisenberg Hamiltonian (four-body operators having as large amplitudes as the two body ones). On the contrary the hierarchized symmetrization, orthogonalizing the 2nd vector $P_0\left(\psi\left({}^1\Sigma_g^+\right)\right)$ to the first one of the same symmetry by a Schmidt orthogonalization led to an effective Hamiltonian which was quite similar to an Heisenberg Hamiltonian (the four-body terms being smaller than 10 % of the two-body terms).

The failure of the classical QDPT approaches is not necessarily universal ; these approaches remain certainly valid at large interatomic distances when $K >> \frac{2t^2}{U}$, i.e. when the various asymptotic eigenstates of the atom are not strongly mixed by the interatomic delocalization. However one may reasonably desire to build effective magnetic Hamiltonians for systems where intraatomic exchange is relatively weak. The theory of intermediate Hamiltonians[(12-16)], recently developped in our group, happens to provide a convenient tool to build effective hamiltonians keeping the main logics of the Heisenberg Hamiltonian (hermiticity, weakness of four-body terms). The failure of the classical QDPT approaches is due to the action of *intruder states*. If one wants to build an Heisenberg Hamiltonian (with a model space in which each orbital bears one and only one electron) the neutral VB structures with zero and two electrons in some AOs act as intruder states when 2K is not significantly larger than $2t^2/U$; for instance in the above example (acetylene π systems), ϕ_7 and ϕ_8 generate eigenstates which are below (or close to) the states generated by ϕ_5 and ϕ_6 which, having parallel spins in both π subsystems, do not permit any electron delocalization and will necessarily have vanishing components in the lowest eigenstates at short interatomic distances.

Similarly if one wants to build the generalized effective Hamiltonian, including all neutral VB structures, some of them, such as Φ_9 and Φ_{10} in the same example, generate eigenstates which become degenerate with some ionic or Rydberg excited states. Φ_9 and Φ_{10} actually introduce four electrons in the same subsystem, preventing any delocalization to take place and introducing strong repulsive effects. They are therefore subject to crossing with outer states. Actually they will rapidly enter into the continuum of unbound states and the choice of corresponding target vectors seems hopeless. This is a typical case where one cannot maintain the basic claims of effective Hamiltonian theory i.e. a one-to-one correspondance between isodimensional model and target spaces. The size and content of the model space is imposed by internal consistence criterions, and except for the asymptotic region, it will become impossible to find a relevant corresponding target space and to avoid discontinuities (due to avoided crossings with intruder states[8]).

Ounce again the necessity of leaving the effective Hamiltonian Theory doctrine is not essentially a problem of non-convergence of the Quasi Degenerate Perturbation Theory (a real practical problem which may be avoided by finding the exact solutions of large CI matrices), but a more fundamental and conceptual problem.

III. TRANSFERABILITY PROBLEMS

The effective Hamiltonians have two possible uses :

i) they may simply be employed to concentrate and rationalize information through a convenient model.

ii) one may define effective interactions from an accurate study of a small problem (let say a diatom) and use them to study large (eventually infinite) systems. The transfer of effective operators from small to large systems is not always legitimate, as we would like to show here.

From a pragmatic point of view one may quote the

surprising efficiency of Heisenberg Hamiltonians in terms of $<S_z> = \pm 1/2$ particles. Extracting the information from the ethylene molecule lowest eigenstates, our group proposed an r-dependant Heisenberg Hamiltonian for π electrons of conjugated molecules which proved to be incredibly successfull for both ground and lowest excited states properties[19,21]. This is a case where each atom brings only one electron in one orbital, i.e. the optimal situation, and the surprise essentially comes from the fact that one does not work in the $|4t|<U$ regime where perturbation converges. The possible relevance of the magnetic models in the domain of weak correlation and oppositely of the Gutzwiller type[22-25] transformations of the single determinantal wave function in the strongly correlated domain is a fascinating problem[26,27] but it seems to work for lattices which are free of the Hubbard transition, i.e. for alternant lattices (without odd-membered rings).

As a dramatic illustration one may quote a recent work treating the dimerisation of the polyacetylene 1-D chain from the above mentionned Heisenberg Hamiltonian, the metal-insulator transition is treated as a spin Peierls distortion [28]. Going one step further one may mention a work on the p-half filled band of Phosphorus. An r-dependant Heisenberg Hamiltonians in terms of $<S_z> = \pm 1/2$ particles has been extracted from the spectroscopy of the P_2 diatom. It proves to give reliable results on Cubic P_8 cluster, and one may even treat from it the metal-insulator phase transition of Phosphorus (cubic $\rightarrow$ orthorhombic), which is nicely reproduced as a first-order phase transition at the correct volume[20].

The same phenomenon could not be treated with an Heisenberg Hamiltonian in terms of $<S_z> = \pm 3/2$ particles as extracted from P_2, the cohesive energy then being far too large. This is easy to understand, by moving back to the simplest polyine problem ($<S_z> = \pm 1$ particles). The effective operator responsible for the spin exchange between adjacent atoms is much larger on the diatom than on the infinite system, as may be easily understood from Fig. 1. Refering again to the Heisenberg Hamiltonian in terms of $<S_z> = \pm 1/2$ particles to establish the Heisenberg Hamiltonian in terms of $<S_z> = \pm 1$ particles, it is easy to see that the neutral VB determinants which belong to the $<S_z> = \pm 1/2$ model space have much higher energies in the condensed system than on the diatom. The effect is much more dramatic when the number of neighbour increases. So that *one cannot hope to transfer an Heisenberg Hamiltonian in terms of composite particles from the diatom to the condensed systems*, which prevents most of its interest.

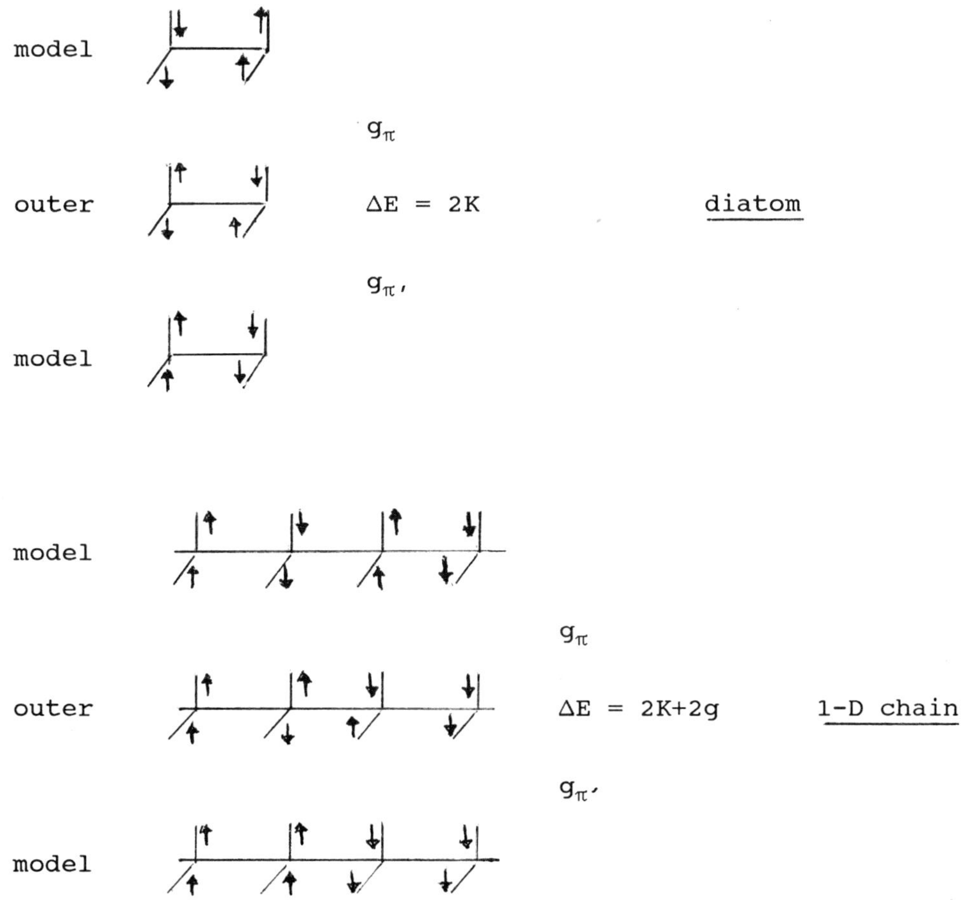

Figure 1. Non transferability of off-diagonal operators for Heisenberg Hamiltonians acting on $<S_z> \neq \pm 1/2$ spins as illustrated on the acetylene/polyine π systems problem.

CONCLUSION

The present discussion of the half-filled band problem leads to the following conclusions

- the most grounded model space would include all neutral VB determinants (including "pseudo ionic") ; this space is large and the corresponding effective Hamiltonian is not a pure-spin Hamiltonian.
- an Heisenberg Hamiltonian in terms of $<S_z> = \pm 1/2$ particles is axis-dependant, which limits its applicability for bonds with several electrons and several AOs per atom. It nevertheless may be very useful.
- the Heisenberg Hamiltonians acting on $<S_z> = \pm 1/2$ composite particles, although simple and axis-independant are of poor interest, due to a non-transferability of the operators from the diatom to condensed systems.
- in all cases (except for $<S_z> = \pm 1/2$ particles with one electron per atom) the effective Hamiltonians cannot be considered as obtained from Bloch or des Cloizeaux definitions. Dramatic intruder state problems compell one to use the intermediate Hamiltonians approaches.

BIBLIOGRAPHY

1) C. Herring, Magnetism 2B, 1 (1962)
2) B.H. Brandow, in Effective Interactians and Operators in Nuclei (Ed. B.R. Barret) Lecture Notes in Physics Vol. 40, p. 1, Springer Berlin (1975)
3) I. Lindgren and J. Morrison, Atomic Many-Body Theory, p. 200, Springer, Berlin (1982)
4) H. Feschbach, Ann. Rev. Nucl. Sci. B, 49 (1958) ; Ann. Phys., N.Y., J, 357 (1958) ; ibid., 19, 287 (1962)
5) P.O. Lowdin, J. Math. Phys. 3, 969 (1962)
6) C. Bloch, Nucl. Phys. 6, 329 (1962)
7) J. des Cloizeaux, Nucl. Phys. 20, 321 (1960)
8) Ph. Durand and J.P. Malrieu, in Ab-initio Methods in Quantum Chemistry (Ed. K.P. Lawley) Vol. 1, p. 352, J. Wiley, New-York (1987)
9) J.H. Van Vleck, Phys. Rev. 33, 467 (1929)

10) K. Suzuki and R. Okamoto, Progr. Theor. Phys., 72, 534 (1984)

11) P.W. Anderson, Solid State Phys. 14, 99 (1963)

12) J.P. Malrieu, Ph. Durand and J.P. Daudey, J. Phys. A, Math. Gen. 18, 809 (1985)

13) S. Evangelisti, Ph. Durand and J.L. Heully, Phys. Rev. A. 43, 1258 (1991)

14) A.V. Zaitsevskii and A.I. Dement'ev, J. Phys. B 23, L517 (1990)

15) J.L. Heully, S. Evangelisti and Ph. Durand, Int. J. Quant. Chem., in press

16) S. Koch, Theor. Chim. Acta, in press

17) J.P. Malrieu and D. Maynau, J. Amer. Chem. Soc. 104, 3021, 3029 (1982)

18) M. Said, D. Maynau, J.P. Malrieu and M.A. Garcia-Bach, J. Amer. Chem. Soc. 106, 571 (1984)

19) D. Maynau, M.A. Garcia-Bach and J.P. Malrieu, J. Physique 47, 207 (1986)

20) A. Pelegatti and J.P. Malrieu, J. Chim. Phys. 87, 941 (1990)

21) M. Said, D. Maynau and J.P. Malrieu, J. Amer. Chem. Soc. 106, 580 (1984)

22) M.C. Gutzwiller, Phys. Rev. Lett. 10, 159 (1963)

23) P. Horsch and P. Fulde, Z. Physik B 36, 23 (1979)

24) G. Stolhoff and P. Fulde, J. Chem. Phys. 73, 4548 (1980)

25) B. Oujia, M.B. Lepetit, D. Maynau and J.P. Malrieu, Phys. Rev. A 39, 3289 (1989)

26) T.A. Kaplan, P. Horsh and P. Fulde, Phys. Rev. Lett. 49, 889 (1982)

27) B. Oujia, M.B. Lepetit and J.P. Malrieu, Chem. Phys. Lett. 158, 559 (1989)

28) M.A. Garcia-Bach, P. Blaise and J.P. Malrieu, to be published

RECENT DEVELOPMENTS IN THE CALCULATION OF MOLECULAR AUGER SPECTRA

F. Tarantelli, A. Sgamellotti and **L.S. Cederbaum**(*)

Dipartimento di Chimica, Università di Perugia, I–06100 Perugia (Italy)
(*) Theoretische Chemie, Universität Heidelberg, D–6900 Heidelberg (Germany)

I. INTRODUCTION

The spectroscopy and dynamics of doubly charged molecular cations in the gas phase are nowadays intensively studied by means of several complementary experimental techniques. In the oldest exploited class of experiments dications are produced by decay of core ionized molecules via electron emission. This is the basis of Auger electron spectroscopy (AES) [1], where the number of emitted electrons is measured as a function of their kinetic energy. In recent developments, vibrationally resolved Auger spectra have been obtained [2–4] and additional information on the nature of the dicationic states populated can be gained via the coincidence detection of Auger electrons and the sufficiently long–lived dications or their fragmentation products [5,6]. In a second important class of techniques the doubly charged cations are produced directly from their neutral parent species. In double charge transfer spectroscopy [7] this is achieved by collision of protons impinging on the target molecules. The one–step two–electron removal channel can be identified by its specific pressure dependence and the energy loss of the detected H^- ions is measured. Dicationic species can also be obtained by electron impact, and their lowest lying states observed by translational energy loss spectroscopy [8], or by collisional charge stripping from singly charged precursors [9]. Finally, in recent years, new powerful techniques have been successfully established, [10–12] based on direct double photoionization from synchrotron radiation, followed by the coincidence detection of photoelectrons and/or photoions (either long–lived dications or charged fragments).

All these and other experimental developments represent an engaging challenge to the theoretical study of dications because the number of directly or indirectly observable dicationic states is very large (compared, say, to neutral or singly ionized species) and increases more rapidly than the square of the number of electrons with the size of the molecule. This is particularly true for AES measurements, where no strict selection rules exist. Auger spectral bands are usually broad features which result from the convolution of many electronic transition, and the Auger decay probability for each of them depends on a final state wavefunction which includes the

Applied Many-Body Methods in Spectroscopy and Electronic Structure
Edited by D. Mukherjee, Plenum Press, New York, 1992

escaping electron in the continuum of the dicationic field [13–15]. The high density of state produces strong correlation effects on the transition energies and the number of states which acquire significant intensity in the spectrum largely exceeds the number of two–hole (2h) states of an independent–particle model. Hole localization effects, vibrational broadening and vibronic coupling further complicate the picture. The aid of reliable theoretical calculations is therefore very important for the interpretation of Auger spectra,and substantial efforts have been devoted through the years, paralleling the developments in *ab initio* theoretical and computational chemistry, to increase the accuracy and the scope of the calculations. For an overview of early calculations see ref.[16]. Recent developments have concerned, for example, the *ab initio* study of the vibrational fine structure of well resolved Auger bands [2,4,15] and the evaluation of transition rates from scattering calculations [14]. However, only for small systems, rarely larger than triatomics, accurate calculations of a sufficient number of double ionization transitions can effectively be carried out by conventional *ab initio* methods like Multiconfiguration Self Consistent Field (MCSCF) and Configuration Interaction (CI).

A generally more effective balance between accuracy and efficient computation of many electronic transitions is provided by Green's function methods. These formalisms have been very successful in the investigation of molecular single ionization [17] in connection with photoelectron spectroscopy and, in recent years, we have used a Green's function approach [18–21] to study molecular double ionization and Auger spectroscopy [22–26]. The most advantageous feature of Green's function based techniques is that they provide a framework for the ***direct*** calculation of ionization (or excitation) spectra, without resorting to the calculation of the wavefunction and energy of each individual final state. This implies that the perturbation expansion of the Green's function implicitly accounts for the cancellation of analytically equal perturbation terms in the initial and final states [18]. Therefore, to obtain results of a given accuracy (in perturbation theory terms) smaller configuration spaces are needed, and with a weaker dependence on molecular size, than with the usual wavefunction approaches. Other Green's function based methods for the calculation of double ionization spectra include the first order method of Liegener [27] and the non–perturbative approach of Graham and Yeager [28]. We have instead implemented a second order scheme [18] drawn from the Algebraic Diagrammatic Construction (ADC) formalism [18–21,29]. The ADC is a general theory providing size–consistent order–by–order approximations to the Green's function. In the language of diagrammatic perturbation theory, the n–th order of ADC, ADC(n), comprises all the diagrams up to n–th order of the propagator expansion and, in addition, series of selected diagrams up to infinite order. The perturbation series is intended around the Hartree–Fock ground state of the neutral molecule. The ADC formulation of the Green's function leads to a symmetric eigenvalue problem in the space of the dicationic configurations, whose solution yields the double ionization energies and pole strengths. The size–consistency of the ADC scheme is a further distinguishing feature of the method, which ensures its applicability, in principle, to large molecular system.

In the following sections we shall briefly review the basic theory of the two–particle Green's function and the development of the ADC formalism. We shall then discuss the theoretical treatment of a few key aspects of molecular Auger spectroscopy (hole localization effects, vibrational structure, transition rates) in the context of the Green's function method and with the aim to develop an efficient theoretical tool for the routine reproduction and interpretation of the spectra of polyatomic systems. The main guideline behind our approach is, therefore, to minimize the limitations imposed by the calculation of the large number of observable states and, in fact, to introduce approximations which exploit the inherently limited resolution of the spectra with respect to this high density of states. In a final chapter we shall review some representative applications, documenting the range of applicability and potentiality of the method.

II. THEORY

II.1 – The Green's function ADC method

The ADC theory has originally been developed [18,29] as a general constructive approach to the derivation of order–by–order terms in the perturbation expansion of the Green's function. It consists in formulating a formal non–diagonal spectral representation of the propagator and evaluating the perturbation terms of this ansatz by comparison with the diagrammatic expansion [30] of the propagator itself. A general algebraic approach to the perturbation theory of Green's functions, independent of the diagrammatic expansion and based instead on effective hamiltonian theory, has subsequently been developed [20] and the ADC equations have been shown to coincide with this theory at least up to third order. Finally, a closed form expression for the transformation between the diagonal and ADC representations of the propagators has recently been derived [21]. We would like here to give an overview of the fundamental aspects of the theory, referring the reader to the literature for more technical details.

For the study of double ionization (or electron attachment) processes the quantity of interest is the two–body Green's function [30]

$$G_{ij,kl}(t_1,t_2;t_3,t_4) = -\langle\Psi_0^N|T\{a_i(t_1)a_j(t_2)a_l^\dagger(t_3)a_k^\dagger(t_4)\}|\Psi_0^N\rangle \tag{1}$$

where $|\Psi_0^N\rangle$ is the molecular N–particle ground state of energy E_0^N (which we shall assume nondegenerate), $a_i(t)$ and $a_i^\dagger(t)$ are annihilation and creation operators in Heisenberg representation, defined in some chosen one–particle basis, and T is Wick's time ordering operator. In the limit of simultaneous creation of two particles at time t' and their annihilation at time t we obtain the particle–particle propagator, defined by

$$\Pi(t,t') = iG(t,t;t',t') \tag{2}$$

The Fourier transform of Eq.(2) to energy space leads to

$$\begin{aligned}\Pi_{ij,kl}(\omega) = &\langle\Psi_0^N|a_ia_j(\omega - H + E_0^N)^{-1}a_l^\dagger a_k^\dagger|\Psi_0^N\rangle \\ &-\langle\Psi_0^N|a_l^\dagger a_k^\dagger(\omega + H - E_0^N)^{-1}a_ia_j|\Psi_0^N\rangle\end{aligned} \tag{3}$$

where H is the hamiltonian of the system. Inserting complete sets of {N+2}– and {N–2}–particle states in the first and second terms of Eq.(3), respectively, the double ionization and electron attachment spectra appear explicitly in the expression for $\Pi(\omega)$:

$$\Pi_{ij,kl}(\omega) = \sum_{f\in\{N+2\}} \frac{x_{f,ij}^{*}\, x_{f,kl}}{\omega - E_f^{N+2} + E_0^N} - \sum_{f\in\{N-2\}} \frac{x_{f,ij}^{*}\, x_{f,kl}}{\omega + E_f^{N-2} - E_0^N} \tag{4a}$$

$$\begin{aligned}x_{f,ij} &= \langle\Psi_f^{N+2}|a_j^\dagger a_i^\dagger|\Psi_0^N\rangle \quad \text{for } f\in\{N+2\} \\ x_{f,ij} &= \langle\Psi_0^N|a_j^\dagger a_i^\dagger|\Psi_f^{N-2}\rangle \quad \text{for } f\in\{N-2\}\end{aligned} \tag{4b}$$

Here $|\Psi_f^{N\pm2}\rangle$ and $E_f^{N\pm2}$ are the (N±2)–particle states and energies, respectively. As Eq.(4a) shows, the matrix function $\Pi(\omega)$ has poles at energy values equal (within sign) to each double electron attachment and ionization energy of the system. The two terms on the rhs of Eq.(4a) can be treated independently and thus we can limit the analysis to the second term, describing the double ionization spectrum. We rewrite this term in matrix notation as

$$\mathbf{P}(\omega) = \mathbf{X}^{\dagger}(\omega\mathbf{1} - \boldsymbol{\Omega})^{-1}\mathbf{X} \tag{5}$$

where the matrix **X** has elements $X_{f,ij} = x_{f,ij}$ and **Ω** is the diagonal matrix of (negative) double ionization potentials (DIPs). It is immediately seen that the amplitudes $x_{f,ij}$ appearing in the residues at the poles of **P** are related to the double ionization transition moments. The latter can be written

$$T_f = \langle \Psi_0^N | D | \Psi_f^{N-2} \rangle \tag{6}$$

where D is any operator inducing the transition:

$$D = \sum_{i>j} D_{ij} a_j^{\dagger} a_i^{\dagger} \tag{7}$$

with matrix elements $D_{ij}=-D_{ji}$. The exact nature of D depends of course on the physical mechanism of the specific double ionization process considered. Using Eq.(4b) the transition moment can be written

$$T_m = \sum_{i>j} x_{m,ij} D_{ij} \tag{8}$$

To evaluate perturbatively the propagator, one defines a partitioning of the hamiltonian, $H=H_0+V$, where the unperturbed one–body operator H_0 is diagonal in the chosen one–particle basis and is conveniently assumed to be the neutral ground state Fock operator. The perturbation analysis may then be carried out equivalently from the representation of $\mathbf{P}(\omega)$ given by the second term of Eq.(3), or from Eq.(5). In Eq.(3) the perturbation expansion of the neutral ground state $|\Psi_0^N\rangle$ induces a transformation of the operator E_0^N-H such that the transformed operator is block diagonal in the basis of the unperturbed {N–2}-particle configurations. The structure and characteristics of this transformation have been studied in detail in ref.[20]. In the ADC formulation [18,19], we analyze a nondiagonal representation of Eq.(5) in the space of the unperturbed dicationic configurations:

$$\mathbf{P}(\omega) = \mathbf{f}^{\dagger}(\omega\mathbf{1} - \boldsymbol{\Gamma})^{-1}\mathbf{f} \tag{9}$$

Of course, Eqs.(5) and (9) are related by the unitary tranformation **Y** which diagonalizes **Γ**:

$$\boldsymbol{\Gamma}\mathbf{Y} = \mathbf{Y}\boldsymbol{\Omega}\ ; \qquad \mathbf{Y}^{\dagger}\mathbf{Y}=\mathbf{Y}\mathbf{Y}^{\dagger}=\mathbf{1} \qquad \mathbf{X} = \mathbf{Y}^{\dagger}\mathbf{f} \tag{10}$$

A straightforward configuration interaction approach would immediately lead to identify the matrix **Γ** in Eq.(9) with the representation of the operator E_0^N-H in the space of the {N–2}-particle unperturbed configurations, obtained by inserting complete sets of dicationic configurations at left and right of the resolvent in Eq.(3). The perturbation expansion of the ground state would then determine, at each order n, the relevant configuration space in **Γ**. As is well known and we shall illustrate later, this CI configuration space is large and fast–growing with n. As mentioned above, the fundamental step in the ADC is to require instead that **Γ** be the representation of a transformed hermitian operator E_0^N-H', with $H'=T^{\dagger}HT$. The unitary operator T is dictated by the Møller–Plesset expansion of the ground state in such a way that in **Γ** the *smallest* configuration space effectively required at any given order of perturbation theory is decoupled from the rest. Alternatively [21], one may regard Γ as the representation of E_0^N–H over a set of *intermediate* states $|\Phi_j^{N-2}\rangle$ which are related to the true states via

$$|\Psi_f^{N-2}\rangle = \sum_j Y_{jf}^* |\Phi_j^{N-2}\rangle \tag{11}$$

The minimality, or "compactness", property of the ADC configuration space is a distinguishing feature of the method which is at the origin of its computational efficiency. In terms of required excitation classes, it can be shown [18,21] that the ADC configuration space grows only half as fast as the CI one with the order of perturbation (at the comparatively little price, of course, that some matrix elements of Γ are not linear in the perturbation). It is clear that this, in turn, translates into a much weaker dependence of the space size on the size of the molecular system under study. We shall further illustrate this point below.

It was originally shown [18] that the required properties of the ADC scheme are automatically fulfilled if one determines the perturbation terms of the arrays Γ and **f** by comparison with the diagrammatic perturbation expansion of the propagator, with the additional provision that Γ be rendered "maximally non–diagonal". This procedure shows also that the ADC expansion fully conserves the well–behaved structure of the diagrammatic series, without occurrences of potentially diverging terms. The diagrammatic construction is carried out by first expressing Γ as the sum of two terms **K**+**C**, where **K** is the diagonal matrix of zero order ionization energies and **C** is the *effective interaction* matrix, whose expansion begins in first order. One then expands the inverse matrix $(\omega\mathbf{1} - \Gamma)^{-1}$ in powers of $(\omega\mathbf{1} - \mathbf{K})^{-1}$ and inserts in the resulting series for $\mathbf{P}(\omega)$ the expansions

$$\begin{aligned} \mathbf{C} &= \mathbf{C}^{(1)} + \mathbf{C}^{(2)} + \mathbf{C}^{(3)} + \cdots \\ \mathbf{f} &= \mathbf{f}^{(0)} + \mathbf{f}^{(1)} + \mathbf{f}^{(2)} + \cdots \end{aligned} \tag{12}$$

for **C** and for the matrix of *effective amplitudes* **f**. This leads to

$$\begin{aligned} \mathbf{P}(\omega) = {} & \mathbf{f}^{(0)\dagger}(\omega\mathbf{1} - \mathbf{K})^{-1}\mathbf{f}^{(0)} + \mathbf{f}^{(0)\dagger}(\omega\mathbf{1} - \mathbf{K})^{-1}\mathbf{C}^{(1)}(\omega\mathbf{1} - \mathbf{K})^{-1}\mathbf{f}^{(0)} \\ & + \mathbf{f}^{(1)\dagger}(\omega\mathbf{1} - \mathbf{K})^{-1}\mathbf{f}^{(0)} + \mathbf{f}^{(0)\dagger}(\omega\mathbf{1} - \mathbf{K})^{-1}\mathbf{f}^{(1)} + \cdots \end{aligned} \tag{13}$$

which can then be compared with the diagrammatic perturbation series for $\mathbf{P}(\omega)$ through any finite order to determine explicitly the terms in Eq.(12). It should be noted that, since the ADC ansatz Eq.(9) involves the inversion of an energy dependent matrix, the eigenvalues of Γ at the n–th order are not exact n–th order ionization energies and include implicitly additional terms up to infinite order. In fact, it can be shown that these terms represent the summation of partial infinite diagram series of $\mathbf{P}(\omega)$. The constructive procedure, which involves the lengthy evaluation of the rapidly growing number of diagrams contributing to $\mathbf{P}(\omega)$, has been carried out explicitly for the second [18] and third order [19] ADC approximations to the particle-particle propagator. A much more efficient approach to the determination of the perturbation series for **f** and **C** has evolved from the direct study of the properties and structure of the operator T [20] or, equivalently, of the transformation matrix **Y** [21]. This analysis can be rendered totally independent of the diagrammatic expansion of $\mathbf{P}(\omega)$. It turns out [21] that the perturbation terms in Eq.(12) can in fact be written systematically in closed form as a function of the ground state density matrices.

As described above, the configuration space of the matrix Γ is the space of the unperturbed {N–2}–particle configurations. In particle–hole notation, these configurations are classified as two–hole (2h), three–hole–one–particle (3h1p), etc., with respect to the Hartree–Fock ground state of the neutral system. The law of growth of the ADC configuration space [18] states that at perturbation orders 2n and 2n+1 (n=0,1,...) the highest excitation class involved is (n+2)h–np. Thus, the configuration space grows by one class only at each even order of perturbation theory. The construction principle of the ADC ensures that at order n the DIPs and spectroscopic amplitudes of dicationic states perturbatively derived from 2h configurations are reproduced consistently through (and beyond) n–th order. When correlation and relaxation effects are neglected, these are the only states with nonzero spectroscopic amplitudes and are referred to, therefore, as *main* states. Higher excited, *satellite*, states, which appear in the spectrum in consequence of many–body effects, are computed at correspondingly decreasing orders of accuracy, according to their perturbation theory genealogy. At even orders, first order accuracy for the highest excited states can easily be ensured by including in Γ the matrix elements of E_0^N–H over the highest excitation class of configurations. In strict ADC these matrix elements would appear at the next odd order. It is interesting to note that, as a result of this design strategy, the ADC theory treats the dicationic states at a varying degree of accuracy according to their "importance" in the double ionization spectrum. From the computational standpoint this feature simplifies as much as possible the evaluation of the matrix elements of Γ.

According to the above, in the second order ADC(2) scheme that we have implemented so far, the active configuration space comprises all the 2h and 3h1p configurations. The Γ matrix elements in the (small) 2h space are of second order in V, while those between the 2h and 3h1p classes and those within the 3h1p space are first order, CI–like, matrix elements. The eigenvalue problem for Γ yields double ionization energies correct to second order of perturbation theory for main states and to first order for satellite states. In the *same* configuration space ADC(3), which is currently being implemented, delivers main state ionization energies correct up to third order. It is instructive to compare this result with what is obtained in the familiar CI or wavefunction perturbation theory approaches. Expanding the wavefunction of a main state in the same configuration space as above, one can easily show that the resulting energy does not even reach second order accuracy. To achieve such accuracy, the CI expansion must employ a configuration space including 4h2p configurations, as well as 2h2p configurations for the ground state. A simple analysis [18], which we do not repeat here, shows that the effect of the 4h2p dicationic configurations and of the 2h2p neutral ones is implicitly accounted for by the second order term of the ADC effective interaction matrix **C**. Thus, up to third order, the ADC eigenvalue problem has a third power dependence on the size of the molecule, and only linear on the size of the basis set, while the CI dependence, already in second order, goes as the fourth power of the number of electrons and the square of the basis set size. It is clear that this permits the applicability of ADC to comparatively much larger molecular systems and the efficient computation of wide regions of the double ionization spectrum, containing very many states, in a single diagonalization step. Of course, the CI approach allows, in principle, to obtain results of high accuracy for any individual state by using relaxed or MCSCF orbital bases and configuration spaces appropriately selected for that state, but this procedure very rapidly becomes impracticable for computing the whole double ionization spectrum of all but the smallest molecules. The main characteristics of ADC appear thus particularly well suited for the theoretical investigation of molecular Auger spectra. In this respect, a further advantage of the method, especially in view of its application to relatively large systems, is its size–consistency.

II.2 – Hole localization effects in doubly ionized states

The spatial distribution of the two–hole density in doubly ionized molecular states is an important factor which affects in significant ways their dynamics and spectroscopy. In the Auger process, it influences directly the matrix elements between the initial, highly localized, core–hole state and the final states, and thus the intensities of the Auger peaks. Also, the perturbation expansion of the energy of a dicationic state contains a leading hole–hole repulsion term, in first order, which varies enormously with the two–hole overlap. Similarly, the hole density distribution is the most important factor controlling the dicationic fragmentation patterns observed in various coincidence spectroscopies [5,6,10–12]. In extreme cases, e.g., fluorides [6,26,31–34], and through a mechanism analogous to that of double *core* vacancies in molecules with equivalent atoms [35], a very strong atomic localization of the two valence holes takes place [26]. In such situations *partner* localized states are observed with the two holes either located each at another of two equivalent atomic sites or both at the same site. From the theoretical viewpoint, it is useful to remark that, contrary to the case of single vacancy localization which can be described by a symmetry–broken single-determinant wavefunction, two–hole localization effects are in general inaccessible in a molecular orbital picture [26]. We shall later return with more detail on the strong effects of such hole localization in Auger spectroscopy.

The occurrence of valence hole localization in molecular dicationic states has generally been discussed only qualitatively [6,31–33,36–38], based on the spectroscopy of fragmentation products or estimated hole–hole repulsion energies deduced from Auger spectra. To provide a more quantitative tool for the theoretical analysis of double–hole density distribution and localization effects, we have proposed [26] a two–hole atomic population analysis in analogy with the familiar Mulliken approach. Starting from the transformation from the molecular orbitals (MO) to the atomic orbital (AO) basis functions:

$$a_i^\dagger = \sum_p C_{pi} a_p^\dagger \tag{14}$$

where **C** is the Hartree–Fock eigenvector matrix, we rewrite the propagator amplitudes as

$$X_{f,ij} = \sum_{p>q} U_{pq,ij} \langle \Psi_0^N | a_q^\dagger a_p^\dagger | \Psi_f^{N-2} \rangle \tag{15}$$

Here $U_{pq,ij} = C_{pi}C_{qj} - C_{qi}C_{pj}$ and the matrix **U** satisfies

$$\sum_{\substack{p>q \\ r>s}} U^*_{pq,ij}\, O_{pq,rs}\, U_{rs,kl} = \delta_{ik}\delta_{jl} - \delta_{il}\delta_{jk} \tag{16}$$

where

$$O_{pq,rs} = S_{pr}S_{qs} - S_{ps}S_{qr} \tag{17}$$

and **S** is the AO overlap matrix. We now consider the total *pole strength* of a dicationic state $|\Psi_f^{N-2}\rangle$

$$P_f = \mathbf{X}_f \mathbf{X}_f^\dagger = \sum_{i>j} |X_{f,ij}|^2 \tag{18}$$

with the aid of Eq.(16) we obtain immediately

$$P_f = \mathbf{D}_f^\dagger\, \mathbf{O}\, \mathbf{D}_f = \sum_{p>q} Q_{pq,f} \tag{19}$$

where

$$\mathbf{D}_f = \mathbf{U}\mathbf{X}_f^\dagger \tag{20}$$

and

$$Q_{pq,f} = D^{*}_{pq,f} \sum_{r>s} O_{pq,rs}\, D_{rs,f} \tag{21}$$

The term $Q_{pq,f}$ defined by Eq.(21) may be interpreted as the contribution of the AO hole pair p,q to the pole strength. This provides a well-defined way to analyze the pole-strengths in terms of *localized* atomic contributions. The sum of terms $Q_{pq,f}$ where both p and q refer to basis functions centered on a given atom A is the "one-site pole-strength" of that atom, and measures the extent to which the dicationic state can be described as having both holes localized on atom A. Similarly, the "two-site" character of a state for each pair of atoms A and B, describing the localized component with one hole localized on A and the other on B, is measured by the sum of terms $Q_{pq,f}$ where p and q refer to basis functions centered on A and B, respectively. Thus, the predominance of one of these contributions for a given state indicates that the two vacancies are strongly localized in space (either at the same or each at another atomic center, according to the dominating component). States for which more than one component is significantly present are characterized instead as having correspondingly delocalized holes. To simplify the analysis, since the 2h configurations give by far the largest contributions to the spectroscopic amplitudes (the sole contributions for an uncorrelated ground state), it is sufficient to limit the analysis to the 2h components of the pole strength, by restricting to $j < i \leq N$ the sums in Eq.(18) and in the definition of $\mathbf{D}_f$, Eq.(20). We may further approximate P_f at the lowest order as the sum of square 2h coefficients of the ADC eigenvectors, by using, in Eq.(21), $\mathbf{D}_f$ defined by:

$$D_{pq,f} = \sum_{i>j} U_{pq,ij} Y_{ij,f} \tag{22}$$

II.3 – Vibrational effects on Auger bandshapes

Due to the removal of two valence electrons in the final states, a pronounced vibrational structure can generally be expected in Auger spectra. Owing to the small spacing among the electronic states, the vibrational progressions are most often densely overlapped and not well resolvable but the different vibrational broadening of the bands may still strongly affect the resulting Auger profiles. In addition, the intensity maxima of the envelopes of the vibrational cross section distributions may occur at energy positions which can be significantly shifted with respect to the vertical transition energies. The theoretical study of the Auger vibrational fine structure is much complicated by the fact that it reflects the combined effect of nuclear motion in both the short-lived decaying core ionized state and the final states. Such detailed calculations [2,4,15] are required to accurately reproduce the vibrational progressions of a usually small number of isolated Auger transitions, which are evidenced in the high resolution spectra of small molecules. For larger polyatomic systems, or whenever a considerable number of dicationic states contribute to the spectrum, individual vibrational progressions cannot usually be resolved and their calculation is very impractical. In most such cases it suffices and is much simpler [39] to compute reliable estimates of the vibrational widths and energy shifts (vibrational centers of gravity) of the Auger transitions. In the assumption that the bandshapes are approximately symmetric, and if non-adiabatic couplings do not complicate the picture exceedingly, these parameters provide sufficient information, together with the electronic transition rates and the core-hole lifetime, to reconstruct band vibrational envelopes theoretically. The center of gravity and width of a band are related to the two lowest order moments of the vibrational cross section distribution. To determine them, we review in the following the necessary moment analysis.

Let the final dicationic state $|\Psi_f^{N-2}\rangle$ have vibrational levels $|n_f\rangle$ of energies E_{nf}. The corresponding band in the Auger spectrum, as a function of the Auger electron energy E, is given by [2,40]

$$\sigma_f(E) \sim \frac{\Gamma}{2\pi} \sum_{n_f} \left| \sum_{n_c} \frac{\langle n_f | n_c \rangle \langle n_c | 0 \rangle}{E - (E_{nc} - E_{nf}) + i\Gamma/2} \right|^2 \tag{23}$$

where $|0\rangle$ is the ground vibrational state of the target molecule, $|n_c\rangle$ denote the vibrational states of the intermediate, core ionized, molecule, and E_{nc} the corresponding energies. $2\pi/\Gamma$ is the lifetime of the core hole. The dependence of Γ on the internuclear dynamics is neglected [40]. Expressions similar to Eq. (23) hold also for other decay processes [2,4,15,40,41] involving core ionized or excited molecules. Therefore, the following considerations apply to a whole class of processes, which includes, e.g., X–ray emission, resonant Auger and other autoionization processes. The term "Auger" is used symbolically for all of them.

Introducing the hamiltonians H_f and H_c for the nuclear dynamics in the final dicationic and core hole states, respectively, we may rewrite Eq. (23) to obtain:

$$\sigma_f(E) \sim \frac{\Gamma}{4\pi} \int_0^{\infty} e^{-\Gamma T/2}\, dT \int_{-T}^{T} e^{iEt} \langle 0 | e^{iH_c(T-t)/2}\, e^{iH_f t}\, e^{-iH_c(T+t)/2} | 0 \rangle\, dt \tag{24}$$

In deriving Eq. (24) we have made use of the fact that $|n_c\rangle$ and $|n_f\rangle$ are eigenstates of H_c and H_f, respectively. Eq. (24) is useful for discussing and evaluating the Auger spectrum in a time–dependent picture. It is amenable to techniques for wavepacket propagation and, hence, is particularly suited for computations in cases where dissociation processes in the intermediate or final electronic states are of relevance.

In the limit of infinite lifetime of the core–hole, i.e., $\Gamma \to 0$, the lorentzians in Eq.(23) become δ–functions and Eq. (24) simplifies somewhat into

$$\sigma_f(E) \sim \frac{\Gamma}{4\pi} \int_0^{\infty} e^{-\Gamma T/2}\, dT \int_{-\infty}^{\infty} e^{iEt} \langle 0 | e^{iH_c(T-t)/2}\, e^{iH_f t}\, e^{-iH_c(T+t)/2} | 0 \rangle\, dt \tag{25}$$

where $\Gamma \to 0$ must be performed after the integration. It should be noted that a finite lifetime leads to interesting interference effects [2,4,15,40,41] if the vibrational levels in the intermediate electronic state overlap owing to their lifetime broadening. The gross features of the bands in the spectrum are, however, hardly affected at all by these interference effects, which influence only the fine structure.

The energy moments of Eq.(25) provide relevant information on the spectrum. The center of gravity of the band in the spectrum is given by the first moment

$$\langle E \rangle = \int E\, \sigma_f(E)\, dE \tag{26a}$$

and the width of the band is related to $\langle E^2 \rangle - \langle E \rangle^2$, where $\langle E^2 \rangle$ is the second moment of $\sigma_f(E)$:

$$\langle E^2 \rangle = \int E^2\, \sigma_f(E)\, dE \tag{26b}$$

After some algebra, these moments can be expressed by

$$\langle E\rangle = \Gamma \int_0^{\infty} e^{-\Gamma T} \langle 0|e^{iH_cT} (\Delta H_{cf})\, e^{-iH_cT}|0\rangle \, dT \tag{27a}$$

$$\langle E^2\rangle = \Gamma \int_0^{\infty} e^{-\Gamma T} \langle 0|e^{iH_cT} (\Delta H_{cf})^2\, e^{-iH_cT}|0\rangle \, dT \tag{27b}$$

where ΔH_{cf} is the difference between the core–hole and final state hamiltonians:

$$\Delta H_{cf} = H_c - H_f \tag{27c}$$

It can be seen from the above equations that the moments of the spectrum depend explicitly on the core–hole state dynamics, but their dependence on final state dynamics is only via ΔH_{cf}.

Eqs.(27) are exact in the limit of infinite lifetime of the core vacancy, and their evaluation requires the knowledge of H_c and ΔH_{cf}. We are now interested in deriving simple and useful formulas for the positions and widths of the bands in the Auger spectrum due to vibrational motion. To proceed, we initially concentrate on a diatomic system and assume that the potential curves involved in the process are harmonic oscillators of frequency ω, shifted with respect to each other in energy and coordinates. After some lengthy but straightforward algebra, we obtain (for $\Gamma \to 0$) the final expression of the first moment:

$$\langle E\rangle = \Delta - 2\beta\kappa/\omega \tag{28}$$

where κ is the slope of the core hole state, again evaluated at the equilibrium geometry of the target system. Notice that we use everywhere atomic units ($\hbar$=1). For a diatomic molecule, the quantities appearing in Eq.(28) read explicitly:

$$\Delta = E_c(R_0) - E_f(R_0)$$

$$\kappa = \sqrt{\frac{1}{2\mu\omega}} \left[\frac{\delta E_c(R)}{\delta R}\right]_{R_0} \tag{29}$$

$$\beta = \sqrt{\frac{1}{2\mu\omega}} \left[\frac{\delta[E_c(R) - E_f(R)]}{\delta R}\right]_{R_0}$$

where μ is the reduced mass of the molecule, $E_c(R)$ and $E_f(R)$ are the electronic potential energy curves for the core–hole and final state, respectively, and R_0 is the equilibrium geometry of the target system. For the second moment we similarly obtain

$$\langle E^2\rangle - \langle E\rangle^2 = \beta^2(1 + 2\kappa^2/\omega^2) \tag{30}$$

Let us now briefly discuss the relevant equations (28)–(30). The first important consideration is that these equations *are also useful when the potential curves are non–harmonic* [17a,42], and even when the final dicationic states are repulsive and do not support bound states: ω is then the frequency of the target molecule. The gross features of the spectrum like the center of gravity and the width are determined by the *short–time* behavior of the wavepacket of the vibrational ground state transferred to the excited states. This behavior is, in turn, determined by the local changes of the potential energy surfaces around R_0, since there is only little time for the wavepacket to move. The details of the band, of course, depend on the details of the surfaces. Only when a potential energy curve is strongly varying with R in the Franck–Condon zone, as, e.g., in the presence of a narrow avoided crossing, we might be forced to discard the present explicit equations.

The expression for the width of a band in the Auger spectrum requires some clarification. Eqs.(28) and (30) have been derived in the limit of infinite lifetime for the core hole. While the position $\langle E\rangle$ of a band does not change if we assume a finite lifetime, i.e., if the δ-functions are replaced by lorentzians or gaussians, the width does change slightly. Choosing for simplicity a gaussian instead of a lorentzian with the same full width at half maximum (fwhm) Γ, the variance, Eq.(30), becomes

$$\langle E^2\rangle - \langle E\rangle^2 = \beta^2(1 + 2\kappa^2/\omega^2) + \frac{\Gamma^2}{8\ln 2} \tag{31}$$

Assuming now that the whole electronic band in the spectrum takes on the appearance of a gaussian, then the position of this gaussian is given by $\langle E\rangle$ in Eq.(28) and its fwhm W reads

$$W = \sqrt{8\ln 2(\langle E^2\rangle - \langle E\rangle^2)} \tag{32}$$

Γ in Eq.(31) may be chosen to include also the experimental resolution of the apparatus. If the spectral band has a highly asymmetric shape, Eq.(32) can still be used to estimate its width. It should be remembered, however, that in this case $\langle E\rangle$ does not correspond to the maximum of the band, but rather to its center of gravity. The asymmetry of the band can of course also be determined by computing the third moment of the spectrum.

In case the potential energy curve of the core ionized state is parallel to that of the target system (i.e., $\kappa=0$), then the position of the band is just given by the vertical energy difference Δ. Hence, it can be computed at a fixed geometry, namely that of the ground state of the target molecule. The width of the band is just given by β^2. In another limiting case, when the energy curves of the final dicationic and the core ionized states are parallel to each other, one finds $\beta=0$. In this situation the position of the band is again given by Δ, but its width reduces to $W=\Gamma$ (see Eqs.(31) and (32)). In other words, the band consists of a single vibrational component, whose width is solely determined by the lifetime broadening of the core–hole. This very interesting result is independent from the form of the potential energy curve of the target system and follows from the basic Eq.(23) also for any non–harmonic curve.

To each Auger spectrum of a given system there corresponds a specific value of κ. Therefore, the width in Eqs.(30) and (31) is seen to be composed of a factor, β^2, which is "state specific", i.e., changes from band to band of a given spectrum and an amplification factor $(1+2\kappa^2/\omega^2)$ which is "spectrum specific". Similar considerations hold for the shift of the band positions, $\langle E\rangle-\Delta$, due to the nuclear dynamics.

Finally, we would like to analyze briefly the case of a polyatomic system, for which Eqs.(23) to (27) are, of course, also valid. For molecules with more than just a few atoms, there is very little hope to evaluate the basic Eq.(23) and, therefore, even rather crude equations for the position and width of the Auger bands would be very helpful. Assuming again the picture of shifted harmonic oscillators, now for M possible totally symmetric vibrational modes of normal frequencies ω_j (j=1,...,M), a derivation analogous to the one previously described leads to

$$\langle E\rangle = \Delta - 2\sum_{j=1}^{M} \beta_j\kappa_j/\omega_j \tag{33}$$

for the center of gravity of the band under investigation. Here, Δ is the same as in Eq.(29) and the β_j and κ_j are energy derivatives with respect to the dimensionless

$$\kappa_j = \frac{1}{\sqrt{2}} \left[\frac{\delta E_c(Q)}{\delta Q_j}\right]_{Q=Q_0}$$

$$\beta_j = \frac{1}{\sqrt{2}} \left[\frac{\delta[E_c(Q)-E_f(Q)]}{\delta Q_j}\right]_{Q=Q_0} \tag{34}$$

The equation for the bandwidth is slightly more involved. We obtain

$$\langle E^2\rangle - \langle E\rangle^2 = \sum_{j=1}^{M} \beta_j{}^2 + 2\sum_{j=1}^{M} \left(\beta_j \kappa_j / \omega_j\right)^2 \tag{35}$$

This relation holds if the frequencies ω_i are all different and the lifetime of the core hole is long. For degenerate modes, the summation sign in the last term of Eq.(35) must be shifted *into* the brackets, giving rise to cross terms which couple the various degenerate degrees of freedom. Interestingly, such mode–mixing effects are always present for intermediate and short lifetimes as long as quasidegeneracy (in the sense that $|\omega_i - \omega_j| \lesssim \Gamma$) holds.

In polyatomic molecules one may encounter conical intersections of potential energy surfaces, and the corresponding strong non–adiabatic effects may greatly influence the appearance of the spectra [42]. In such situations, one can also easily derive a basic equation similar to Eq.(23) and then use simple, but powerful, hamiltonians to evaluate the moments, along similar lines as done here. Such hamiltonians are well known for the Jahn–Teller case [43] and general conical intersection situations [42].

II.4 – A statistical approach to Auger spectra

A central question for the theoretical intepretation of Auger spectroscopy concerns the evaluation of the Auger electronic intensities of the dicationic states. The full treatment of the scattering process which leads from the initial (ground) state through the intermediate (core–hole) state to a final state in which the outgoing electron interacts with a dicationic remainder is very complex. One usually considers the Auger decay as decoupled from the core ionization (two–step process) and assumes the Auger electron $|\varphi_u\rangle$ to be fast and strongly orthogonal to the dicationic state. By approximating the core–hole state as $a_c|\Psi_0^N\rangle$ one then obtains from Wentzel formula [44,45] the basic working expression

$$T_f = \sum_{i>j} x_{f,ij} \langle ij||uc\rangle \tag{36}$$

for the transition amplitude of the dicationic state $|\Psi_f^{N-2}\rangle$, where $\langle ij||uc\rangle$ is an antisymmetrized coulomb two–electron integral among the one–particle spinorbitals of specified indeces. Eq.(36) is immediately recognized to be of the form generically given by Eq.(8). The main difficulty presented by Eq.(36) is the evaluation of the continuum orbital $|\varphi_u\rangle$. For small molecules, having an atomiclike spectrum, the calculation of $|\varphi_u\rangle$ has often been avoided by approximating the matrix elements in Eq.(36) with atomic integrals [46]. Other more rigorous theoretical investigations [13,14] have devoted attention to this problem, using different methods, for first row hydrides and diatomics, but it is clear that such scattering calculations cannot be routinely carried out for larger polyatomic molecules. On the other hand, Auger spectra exhibit rather different qualitative features depending on the size of the system, and this fact should be carefully considered before selecting the most appropriate theoretical approach. When a very small number of well isolated dicationic states contribute to the spectrum there is an essentially one–to–one correspondence between the observed peaks and the individual electronic states. To compute the spectra one needs in this case an accurate knowledge of the transition energies and rates. The removal of two valence electrons has the consequence that the potential energy surfaces of the dicationic states are generally very different from that of the initial state, and often dissociative. Thus, a considerable vibrational broadening of the Auger bands can usually be expected. High resolution experiments may show a well resolved vibrational fine structure of some bands and the theoretical and computational techniques have developed to reproduce these features quite accurately [2,4,15]. This state of things changes drastically for many–electron polyatomic molecules. As we have pointed out, here the situation is typically characterized by a strong redistribution of intensity over a large number of states, due to final state configuration interaction effects. Each of these components is broadened by nuclear motion effects and even the core hole lifetime broadening is of the same magnitude as the electronic state spacings. Furthermore, the closely spaced electronic states can be vibronically coupled, causing substantial non–adiabatic effects to produce additional nontrivial intensity redistributions. As a result the overall Auger spectral profiles arise from typically very broad and overlapped bands containing very many contributions which cannot be resolved further. It seems clear that the limit is soon reached where the interpretation of Auger spectra based on the computation of individual electronic/vibrational transition rates not only is unfeasible but loses meaning altogether. It can in fact be argued that to reproduce satisfactory Auger bandshapes from theory in such conditions it is sufficient to compute the envelope of the energy distribution of the dicationic states, weighting only the principal factors which control their relative transition probabilities and widths. These considerations are at the basis of what we refer to as a *statistical approach* to Auger spectra [25].

Since for the uncorrelated ground state only the 2h configurations contribute to the spectroscopic amplitudes $x_{f,ij}$, it seems sufficient to consider only these components at the lowest order, coinciding with the 2h coefficients of the ADC eigenvectors. Owing to the strong space localization of the core hole, if the electron vacancies in the final states are fully delocalized we may simply approximate the Auger intensity ratios as proportional to the total 2h pole strength of the states, thus to the sum of square 2h ADC components. If significant localization effects occur in the final states, then the coulomb matrix elements in Eq.(36) may vary by orders of magnitude depending on the localization of the holes with respect to the initial core vacancy [47]. In fact, the main qualitative effect of the coulomb matrix elements in Eq.(36) may be seen as that of selecting the localized components of the pole strength describing two holes at the same atomic site where the core ionization takes place. In general, it is then reasonable to approximate the Auger intensities as proportional to these "one–site" components, which can easily be extracted by using the two–hole population analysis, as described above. The case of delocalized states is easily seen to be automatically accounted for in this way, since ideal delocalization is characterized by one–site components in fixed ratio to the total pole strengths. Spin adaption of Eq.(14) shows that the differences of the integrals $\langle ij|uc\rangle$ and $\langle ij|cu\rangle$ enters the expression for triplet states, while their sum appears for singlets. On the grounds that the two integrals should not differ much in magnitude (and sign) it can be expected that triplet states will generally have a small (relative) Auger intensity compared to singlets. The simplest way to account for this fact is to reduce the triplet pole strength by some fixed factor, which in our applications we have chosen equal to 1/3.

With the above described estimates for the relative Auger rates, and using the computed vibrational widths and energy shifts, we can easily reconstruct approximate band profiles by, e.g., gaussian convolutions. As mentioned, the whole procedure is designed to be particularly suitable for density of states sufficiently high to make a statistical approach meaningful, i.e., for large molecular systems, where the Green's function method is most efficient and advantageous. The following illustrative applications show that these limiting conditions are approached sooner than expected and that the method may be satisfactorily useful also for molecules smaller than anticipated.

III. ILLUSTRATIVE APPLICATIONS

III.1 – Auger spectra of hydrocarbons

The Auger spectra of hydrocarbon series provide interesting examples to study double ionization spectra for different chemical bonding situations and molecular sizes. We investigated by ADC(2) [23–25], using double zeta plus polarization basis sets [48], the spectra of acetylene, ethylene, ethane and benzene, as typical representatives of systems with single, double, triple and aromatic carbon–carbon bonds. Few theoretical investigations beyond the one–particle level had previously been reported on the acetylene and ethylene spectra [27c,49]. Already the double ionization spectrum of a molecule the size of ethane, due to low symmetry, represents a serious computational problem for conventional *ab initio* methods, and no calculations beyond the Hartree–Fock level were previously available. In the case of benzene, for which no other theoretical studies are known to us, we were able to compute over 220 double ionization transitions, a result which is probably beyond the reach of any other approach at a comparable level of accuracy. In the hydrocarbons investigated hole localization effects do not occur or are weak. In addition, the limited resolution of the available Auger spectra does not require explicit consideration of vibrational effects on the bandshapes and peak positions. Therefore these systems are among the simplest examples which allow a systematic assessment of the quality of the statistical approach to Auger intensities.

Acetylene

We have computed [25] the double ionization spectrum of acetylene, C_2H_2, up to 70 eV, resulting in 286 double ionization transitions. The complexity of the spectrum already for such a small molecule appears clearly by considering that, although only 14 computed states are dominated by 2h character ("main" states), as many as 85 states have nonnegligible 2h contributions and are thus used in our scheme to reproduce the Auger spectrum. The vertical DIP's and composition of the main and first few satellite states are reported in Table 1 and 2, respectively. The experimental [50] and theoretical Auger spectra of C_2H_2 are shown in Fig. 1 on the double ionization energy scale. The accuracy in the absolute energy calibration of the experimental spectrum has been questioned and appears to have an error bar of several electron volts [51]. The reproduction in Fig.1 has simply been shifted to align the first theoretical and experimental peaks at low DIP. The theoretical spectrum was generated in this case by simple convolution of the 2h pole strengths with gaussian functions of fixed full width at half maximum (chosen at 4.7 eV to fit the observed bandshapes) and centered at the vertical DIPs. Although this constitutes the simplest possible level of approximation within our statistical approach, ignoring hole localization and vibrational effects, it is already seen to give a satisfactory reproduction of the spectral features. At the limited resolution of the experimental measurement, three visible bands appear in the spectrum in the energy region up to 55 eV, where the main states are computed. The first two bands at lower DIP contain relatively few and intense contributions, but the density of states increases substantially at higher energy as satellite states start to appear. The third band in the spectrum comprises many unresolved satellite contributions which give it a very broad appearance and the fourth theoretical band, which is entirely due to satellite states, is barely visible in the experimental spectrum as a broad peakless structure. This shows that at higher double ionization energies increasingly stronger intensity redistribution effects over very large numbers of states take place which can only be accounted for at higher levels of theory.

The most evident discrepancy between the theoretical and observed spectra of C_2H_2 is due to the low computed relative intensity of the first peak. Subsequent calculations [52], including the two–hole population analysis, have demonstrated that this error is essentially *entirely* due to neglect of hole localization effects. The theoretical spectrum obtained by convolution of the C^{-2} components of the pole

Table 1. Vertical DIP and composition (square leading coefficients) of the dicationic main states of acetylene, computed by ADC(2). The 2h components are underlined.

No.	STATE	DIP [eV]	COMPOSITION
1	$^3\Sigma_g^-$	31.35	$\underline{0.85}\ 1\pi_u^{-2}$, $0.05\ 1\pi_u^{-3}2\pi_u^{+1}$
2	$^1\Delta_g$	32.47	$\underline{0.86}\ 1\pi_u^{-2}$, $0.05\ 1\pi_u^{-3}2\pi_u^{+1}$
3	$^1\Sigma_g^+$	33.24	$\underline{0.86}\ 1\pi_u^{-2}$
4	$^3\Pi_u$	36.75	$\underline{0.81}\ 3\sigma_g^{-1}1\pi_u^{-1}$, $\underline{0.01}\ 2\sigma_g^{-1}1\pi_u^{-1}$
5	$^1\Pi_u$	37.64	$\underline{0.81}\ 3\sigma_g^{-1}1\pi_u^{-1}$, $\underline{0.01}\ 2\sigma_g^{-1}1\pi_u^{-1}$
6	$^3\Pi_g$	38.15	$\underline{0.78}\ 2\sigma_u^{-1}1\pi_u^{-1}$
7	$^1\Pi_g$	39.62	$\underline{0.79}\ 2\sigma_u^{-1}1\pi_u^{-1}$, $0.05\ 3\sigma_g^{-1}1\pi_u^{-2}1\pi_g^{+1}$
8	$^1\Sigma_g^+$	43.30	$\underline{0.63}\ 3\sigma_g^{-2}$, $\underline{0.21}\ 2\sigma_u^{-2}$, $\underline{0.01}\ 2\sigma_g^{-1}3\sigma_g^{-1}$, $0.05\ 3\sigma_g^{-2}1\pi_u^{-1}2\pi_u^{+1}$
9	$^3\Sigma_u^+$	43.81	$\underline{0.85}\ 2\sigma_u^{-1}3\sigma_g^{-1}$, $0.06\ 2\sigma_u^{-1}3\sigma_g^{-1}1\pi_u^{-1}2\pi_u^{+1}$
10	$^3\Pi_u$	44.58	$\underline{0.61}\ 2\sigma_g^{-1}1\pi_u^{-1}$, $\underline{0.02}\ 3\sigma_g^{-1}1\pi_u^{-1}$, $0.16\ 2\sigma_u^{-1}1\pi_u^{-2}1\pi_g^{+1}$, $0.06\ 2\sigma_g^{-1}1\pi_u^{-2}2\pi_u^{+1}$
11	$^1\Sigma_u^+$	48.39	$\underline{0.60}\ 2\sigma_u^{-1}3\sigma_g^{-1}$, $\underline{0.12}\ 2\sigma_g^{-1}2\sigma_u^{-1}$, $0.07\ 3\sigma_g^{-2}1\pi_u^{-1}1\pi_g^{+1}$
12	$^1\Sigma_g^+$	48.62	$\underline{0.29}\ 2\sigma_u^{-2}$, $\underline{0.17}\ 2\sigma_g^{-1}3\sigma_g^{-1}$, $\underline{0.14}\ 3\sigma_g^{-2}$, $\underline{0.05}\ 2\sigma_g^{-2}$, $0.16\ 2\sigma_u^{-1}3\sigma_g^{-1}1\pi_u^{-1}1\pi_g^{+1}$, $0.06\ 2\sigma_u^{-1}3\sigma_g^{-1}1\pi_u^{-1}2\pi_g^{+1}$
13	$^3\Sigma_u^+$	52.82	$\underline{0.52}\ 2\sigma_g^{-1}2\sigma_u^{-1}$, $0.08\ 3\sigma_g^{-1}1\pi_u^{-2}3\sigma_u^{+1}$, $0.07\ 3\sigma_g^{-1}1\pi_u^{-2}4\sigma_u^{+1}$, $0.06\ 3\sigma_g^{-2}1\pi_u^{-1}1\pi_g^{+1}$, $0.05\ 2\sigma_g^{-1}2\sigma_u^{-1}1\pi_u^{-1}2\pi_u^{+1}$
14	$^1\Sigma_u^+$	54.24	$\underline{0.47}\ 2\sigma_g^{-1}2\sigma_u^{-1}$, $\underline{0.05}\ 2\sigma_u^{-1}3\sigma_g^{-1}$, $0.08\ 3\sigma_g^{-1}1\pi_u^{-2}3\sigma_u^{+1}$, $0.07\ 3\sigma_g^{-1}1\pi_u^{-2}4\sigma_u^{+1}$, $0.06\ 2\sigma_u^{-1}1\pi_u^{-2}4\sigma_g^{+1}$

Table 2. Vertical DIP and composition (square leading coefficients larger than 0.1) of the lowest few dicationic satellite state of acetylene, computed by ADC(2).

No.	STATE	DIP [eV]	COMPOSITION
1	$^1\Sigma_u^-$	39.91	$0.70\ 1\pi_u^{-3}1\pi_g^{+1}$, $0.28\ 1\pi_u^{-3}2\pi_g^{+1}$
2	$^3\Delta_u$	40.23	$0.71\ 1\pi_u^{-3}1\pi_g^{+1}$, $0.28\ 1\pi_u^{-3}2\pi_g^{+1}$
3	$^3\Sigma_u^+$	40.47	$0.71\ 1\pi_u^{-3}1\pi_g^{+1}$, $0.27\ 1\pi_u^{-3}2\pi_g^{+1}$
4	$^3\Sigma_u^-$	43.88	$0.70\ 1\pi_u^{-3}1\pi_g^{+1}$, $0.23\ 1\pi_u^{-3}2\pi_g^{+1}$
5	$^3\Pi_g$	44.81	$0.69\ 3\sigma_g^{-1}1\pi_u^{-2}1\pi_g^{+1}$, $0.24\ 3\sigma_g^{-1}1\pi_u^{-2}2\pi_g^{+1}$
6	$^3\Pi_g$	45.56	$0.69\ 3\sigma_g^{-1}1\pi_u^{-2}1\pi_g^{+1}$, $0.25\ 3\sigma_g^{-1}1\pi_u^{-2}2\pi_g^{+1}$
7	$^3\Pi_g$	45.79	$0.71\ 3\sigma_g^{-1}1\pi_u^{-2}1\pi^{+1}$, $0.24\ 3\sigma_g^{-1}1\pi_u^{-2}2\pi_g^{+1}$
8	$^1\Delta_u$	45.80	$0.74\ 1\pi_u^{-3}1\pi_g^{+1}$, $0.21\ 1\pi_u^{-3}2\pi_g^{+1}$
9	$^1\Pi_g$	45.93	$0.69\ 3\sigma_g^{-1}1\pi_u^{-2}1\pi^{+1}$, $0.25\ 3\sigma_g^{-1}1\pi_u^{-2}2\pi_g^{+1}$
10	$^3\Pi_u$	46.43	$0.71\ 2\sigma_u^{-1}1\pi_g^{-2}1\pi_u^{+1}$, $0.24\ 2\sigma_u^{-1}1\pi_u^{-2}2\pi_g^{+1}$
11	$^1\Pi_u$	46.82	$0.51\ 2\sigma_u^{-1}1\pi_u^{-2}1\pi_g^{+1}$, $0.22\ 2\sigma_g^{-1}1\pi_u^{-1}$, $0.19\ 2\sigma_u^{-1}1\pi_u^{-2}2\pi_g^{+1}$
12	$^3\Pi_g$	47.12	$0.68\ 3\sigma_g^{-1}1\pi_u^{-2}1\pi_g^{+1}$, $0.23\ 3\sigma_g^{-1}1\pi_u^{-2}2\pi_g^{+1}$
13	$^1\Pi_g$	47.17	$0.72\ 3\sigma_g^{-1}1\pi_u^{-2}1\pi_g^{+1}$, $0.26\ 3\sigma_g^{-1}1\pi_u^{-2}2\pi_g^{+1}$
14	$^3\Pi_u$	47.28	$0.73\ 2\sigma_u^{-1}1\pi_u^{-2}1\pi_g^{+1}$, $0.25\ 2\sigma_u^{-1}1\pi_u^{-2}2\pi_g^{+1}$
15	$^1\Sigma_u^+$	47.57	$0.73\ 1\pi_u^{-3}1\pi_g^{+1}$, $0.18\ 1\pi_u^{-3}2\pi_g^{+1}$
16	$^1\Pi_g$	48.44	$0.68\ 3\sigma_g^{-1}1\pi_u^{-2}1\pi_g^{+1}$, $0.23\ 3\sigma_g^{-1}1\pi_u^{-2}2\pi_g^{+1}$
17	$^3\Pi_u$	48.53	$0.72\ 2\sigma_u^{-1}1\pi_u^{-2}1\pi_g^{+1}$, $0.24\ 2\sigma_u^{-1}1\pi_u^{-2}2\pi_g^{+1}$
18	$^3\Pi_g$	48.81	$0.39\ 1\pi_u^{-3}4\sigma_u^{+1}$, $0.32\ 1\pi_u^{-3}3\sigma_u^{+1}$, $0.16\ 1\pi_u^{-3}5\sigma_u^{+1}$
19	$^3\Pi_u$	48.89	$0.58\ 2\sigma_u^{-1}1\pi_u^{-2}1\pi_g^{+1}$, $0.20\ 2\sigma_u^{-1}1\pi_u^{-2}2\pi_g^{+1}$, $0.11\ 2\sigma_g^{-1}1\pi_u^{-1}$

strengths produces bandshapes in almost exact intensity ratios. The two–hole population analysis shows indeed that in the outermost dicationic states of C_2H_2, involving double ionization out of the 1π orbital, the two electron holes are fully delocalized over the carbon atoms, while at higher double ionization energy, where the less bonding and CH bonding electrons are involved, hole localization takes place to some extent. In particular, we found [52] that some states with significant total pole strength have a two–site component two to three times larger than the spectrally relevant one–site projection. This obviously enhances the relative intensity of the first peak compared to the profiles in Fig.1. As a general remark, we should also re–stress that our convolution *averaging* procedure for computing the spectral profiles has a justification of statistical nature and is thus expected to work well in regions of high density of states. When the observed Auger bands are due to few isolated electronic transitions, as typically occurs only in the outer valence region of very small

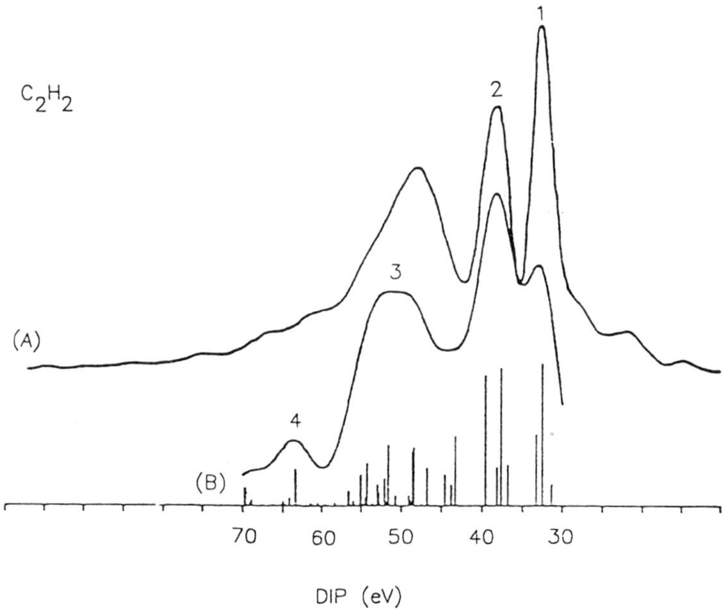

Fig. 1 Experimental (A) and theoretical (B) Auger spectra of acetylene. The bar spectrum shows the individual 2h pole strength contributions, which are convoluted by gaussian functions.

molecules, quantitatively accurate intensity ratios can only be obtained by computing the individual electronic transition probabilities.

Ethylene

For the double ionization spectrum of C_2H_4 our ADC(2) calculations [24] yielded 666 vertical DIP values and pole strengths in the energy interval from 29 to 65 eV. Of these, only 27 correspond to main states. The DIPs and 2h composition of the latter are reported in the first columns of Table 3. Satellite states start appearing in the computed spectrum already at 39.1 eV. In the region where main states appear, below 47 eV, we computed a total of 27 satellites. Their number grows very rapidly at higher energy, reaching 57 already below 50 eV. The experimental [50] and theoretical Auger spectra of ethylene are shown in Fig.2. Also in this case the experimental profile has been shifted in energy to obtain the correct alignment, since its absolute energy scale is affected by substantial uncertainty in the calibration [51]. On the other hand the vertical DIP values obtained by ADC(2) should be considered of very good accuracy: the lowest DCT peak of ethylene is measured at 29.4 ± 0.5 eV [49]. Due to spin conservation this peak shold be associated to the lowest singlet state of the dication, which we compute at 29.46 eV.

The Auger spectrum of C_2H_4, at the reported experimental resolution of 2.7 eV, shows only six clearly distinct peaks, numbered in the figure. Again here it is thus clear that each band must arise as the envelope of several underlying components. As in the case of acetylene, our theoretical spectrum was obtained by simple gaussian convolution of the computed total pole strengths, of width fixed at 2.5 eV and centered at the vertical DIPs. Fig.2 shows clearly that our procedure reproduces all the details of the spectrum with remarkable accuracy, again with the exception of the relative height of the first peak. The computed peak positions are compared with the available experimental data [50,53] in the last three columns of Table 3. The

Table 3. Vertical DIP and leading 2h components of the dicationic main states of ethylene computed by ADC(2). The last three columns report the theoretical and experimental maxima of the Auger peaks.

ADC(2)				Auger[a]	
State	DIP(eV)	2h composition	Peak(eV)	ref.[53]	ref.[50]
^1A_g	29.46	0.85 $1b_{3u}1b_{3u}$			
^3A_u	30.65	0.84 $1b_{3g}1b_{3u}$	30.60	30.6(257.9[b])	30.6(257.0[b]
^1A_u	31.19	0.85 $1b_{3g}1b_{3u}$			
$^3B_{3u}$	32.78	0.82 $3a_g1b_{3u}$			
$^3B_{1g}$	33.73	0.73 $1b_{2u}1b_{3u}$			
$^1B_{3u}$	33.81	0.83 $3a_g1b_{3u}$			
^1A_g	33.93	0.79 $1b_{3g}1b_{3g}$, 0.05 $1b_{2u}1b_{2u}$, 0.01 $3a_g3a_g$			
$^1B_{1g}$	34.87	0.78 $1b_{2u}1b_{3u}$			
$^3B_{3g}$	34.96	0.82 $3a_g1b_{3g}$	34.30	35.2	35.0
$^3B_{1u}$	35.92	0.87 $1b_{2u}1b_{3g}$			
$^1B_{3g}$	36.31	0.79 $3a_g1b_{3g}$, 0.05 $2b_{1u}1b_{2u}$			
$^3B_{2g}$	36.87	0.72 $2b_{1u}1b_{3u}$			
$^3B_{2u}$	38.31	0.76 $1b_{2u}3a_g$, 0.02 $2b_{1u}1b_{3g}$			
^1A_g	38.37	0.72 $3a_g3a_g$, 0.05 $1b_{2u}1b_{2u}$, 0.03 $2b_{1u}2b_{1u}$, 0.01 $2a_g3a_g$			
$^1B_{1u}$	38.40	0.62 $1b_{2u}1b_{3g}$, 0.06 $2b_{1u}3a_g$			
$^1B_{2u}$	38.57	0.59 $1b_{2u}3a_g$, 0.22 $2b_{1u}1b_{3g}$, 0.01 $2a_g1b_{2u}$	38.50	38.7	39.0
$^1B_{2g}$	38.88	0.78 $2b_{1u}1b_{3u}$			
$^3B_{2u}$	39.03	0.75 $2b_{1u}1b_{3g}$			
^1A_g	40.85	0.57 $1b_{2u}1b_{2u}$, 0.06 $2b_{1u}2b_{1u}$, 0.03 $3a_g3a_g$, 0.03 $1b_{3g}1b_{3g}$			
$^3B_{1u}$	41.27	0.81 $2b_{1u}3a_g$			
$^3B_{3u}$	42.30	0.64 $2a_g1b_{3u}$			
$^3B_{3g}$	42.42	0.70 $2b_{1u}1b_{2u}$			
$^1B_{3g}$	43.07	0.37 $2b_{1u}1b_{2u}$, 0.32 $2a_g1b_{3g}$, 0.01 $3a_g1b_{3g}$	43.20	42.8	43.0
$^1B_{2u}$	43.39	0.40 $2b_{1u}1b_{3g}$, 0.09 $1b_{2u}3a_g$, 0.05 $2a_g1b_{2u}$			
$^1B_{1u}$	44.41	0.60 $2b_{1u}3a_g$, 0.05 $2a_g2b_{1u}$, 0.04 $1b_{2u}1b_{3g}$			
$^1B_{3u}$	46.70	0.69 $2a_g1b_{3u}$			
^1A_g	47.19	0.35 $2b_{1u}2b_{1u}$, 0.22 $2a_g3a_g$, 0.05 $2a_g2a_g$, 0.03 $3a_g3a_g$, 0.01 $1b_{2u}1b_{2u}$	47.00	47.0	46.6
			51.30	51.1	
		pure satellite peaks	54.60		53.6
			61.10		

[a]The spectra are on the DIP energy scale, and shifted so that the first peak matches the first theoretical ADC(2) peak.

[b]Kinetic energy of the actual Auger peak.

theoretical spectrum exhibits 8 peaks instead of the experimental 6 and it is apparent that the observed broad sixth band is an unresolved composition of the sixth and seventh theoretical peaks. Notice that the latter are exclusively due to satellite states (see the table). The last theoretical peak at 61.1 eV, which is similarly a pure satellite state feature, is also not clearly visible in the experimental spectrum: in this energy region the Auger spectrum shows the typical essentially featureless structure arising when the density of states is so high to produce significant higher order intensity redistribution effects of electronic and vibronic nature which are not accounted for at the ADC(2) level of theory. The data in Table 3 evidence that the agreement between all the theoretical and observed energies of the peaks is typically within few tenths of eV. Interestingly, the two sets of reported experimental relative energies agree with each other also within similar bounds with the exception of the sixth band, where the discrepancy is as large as 2.5 eV. It seems clear that the origin of this inconsistency lies in the substantial broadening of the band which, as discussed above, our calculations show to be composed of two major unresolved components. Indeed, the value of 51.1 eV reported for the sixth band in ref.[53] is in excellent agreement with the first computed component (51.3 eV), while the value of ref.[50] (53.6 eV) reflects clearly the presence of the second component at 54.6 eV.

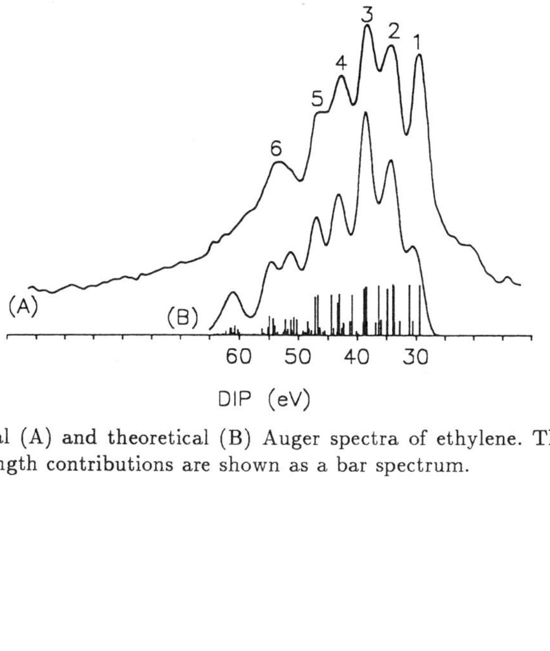

Fig. 2. Experimental (A) and theoretical (B) Auger spectra of ethylene. The individual ADC(2) 2h pole strength contributions are shown as a bar spectrum.

As noted above, and surprisingly for a molecule of the size of etyhylene, the density of dicationic states appears to be sufficient for the spectral profile, at the given resolution, to be essentially controlled only by the energy distribution of the pole strengths, averaging out the details of the individual state Auger transition rates. Indeed, except for the first band in Fig.2, the theoretical peak intensity ratios follow the observed profile with sufficient accuracy to allow an unquestionable interpretation. As found in the case of acetylene, the discrepancy in the intensity of the first peak essentially disappears [52] by accounting for the weak hole localization effects which are totally absent for the very lowest lying dicationic states and start appearing at higher energy. Also in this case, these effects produce dicationic states with a larger two-site character, i.e., with the two holes significantly localized on different carbon atoms, thereby decreasing their relative Auger intensity.

Ethane

The double ionization spectrum computed by ADC(2) for the saturated ethane molecule [25] appears to have very different characteristics from the spectra of the unsaturated analogues acetylene and ethylene. The ionization energies and composition of the 30 main states computed below 60 eV are shown in Table 4. The data evidence a strong configuration mixing among 2h configurations and, by contrast, only minor contributions from 3h1p space, involving exclusively excitations ot the relatively high lying $3e_g$ and $3e_u$ virtual orbitals. The strong interaction between 2h and 3h1p configurations which was found in the unsaturated systems already at low energy and due to the presence of low lying non–diffuse π orbitals is here lacking completely. On the other hand, a higher density of 2h configurations characterizes the ethane dication, explaining the substantial 2h–2h coupling computed in the spectrum. Strong 2h configuration mixing may also be a typical indication of significant hole localization effects, as will be further discussed below. The lowest lying satellite states of C_2H_6 are displayed in Table 5. Consistently with the picture of the electronic structure outlined above, satellite states begin to appear in the spectrum of ethane at substantially higher ionization energies (around 48 eV) than for acetylene or ethylene (39 eV) and are expected to contribute very little to the principal features observed in the Auger spectrum.

Table 4. Vertical DIP and composition (square leading coefficients) of the 30 dicationic main states of ethane, computed by ADC(2). The 2h components are underlined.

No.	STATE	DIP [eV]	COMPOSITION
1	$^3A_{2g}$	31.61	$\underline{0.83}\ 1e_g^{-2},\ \underline{0.04}\ 1e_u^{-2}$
2	^1E_g	32.01	$\underline{0.64}\ 1e_g^{-2},\ \underline{0.17}\ 3a_{1g}^{-1}1e_g^{-1},\ \underline{0.06}\ 1e_u^{-2}$
3	^3E_g	32.03	$\underline{0.84}\ 3a_{1g}^{-1}1e_g^{-1}$
4	$^1A_{1g}$	32.61	$\underline{0.67}\ 1e_g^{-2},\ \underline{0.12}\ 1e_u^{-2},\ \underline{0.07}\ 3a_{1g}^{-2}$
5	^1E_g	32.98	$\underline{0.66}\ 3a_{1g}^{-1}1e_g^{-1},\ \underline{0.14}\ 1e_g^{-2},\ \underline{0.03}\ 1e_u^{-2},$
			$\underline{0.02}\ 2a_{2u}^{-1}1e_u^{-1}$
6	$^1A_{1u}$	33.52	$\underline{0.87}\ 1e_u^{-1}1e_g^{-1}$
7	^3E_u	33.62	$\underline{0.87}\ 1e_u^{-1}1e_g^{-1}$
8	$^1A_{1g}$	33.64	$\underline{0.72}\ 3a_{1g}^{-2},\ \underline{0.05}\ 1e_g^{-2},\ \underline{0.04}\ 1e_u^{-2},\ \underline{0.01}\ 2a_{2u}^{-2},$
			$0.05\ 3a_{1g}^{-2}1e_g^{-1}3e_g^{+1},\ 0.05\ 1e_u^{-1}3a_{1g}^{-2}3e_u^{+1}$
9	$^3A_{2u}$	33.69	$\underline{0.87}\ 1e_u^{-1}1e_g^{-1}$
10	^3E_u	34.98	$\underline{0.80}\ 1e_u^{-1}3a_{1g}^{-1},\ \underline{0.03}\ 2a_{2u}^{-1}1e_g^{-1}$
11	^1E_u	35.61	$\underline{0.76}\ 1e_u^{-1}3a_{1g}^{-1},\ \underline{0.07}\ 2a_{2u}^{-1}1e_g^{-1}$
12	$^3A_{1u}$	35.99	$\underline{0.82}\ 1e_u^{-1}1e_g^{-1}$

The resulting theoretical Auger spectrum of ethane, shown in Fig.3 under the experimental one [54], was again been simply obtained by convolution of the total 2h pole strengths centered at the vertical DIPs. In this case, for reasons to be addressed below, we used lorentzian convoluting functions (of fwhm 3.0 eV) instead of gaussians. The figure visualizes with clarity the essential characteristics of the spectrum outlined above and the qualitative differences with the case of the unsaturated hydrocarbons. The bar spectrum, representing the individual component pole strengths computed, shows that the spectral density of states does not increase continuously with increasing double ionization energy, but rather tends to decrease in most regions of higher DIP. This reflects the lack of satellite state contributions together with the obvious increase in 2h state spacings in the inner valence region. This characteristics

Table 4

No.	STATE	DIP [eV]	COMPOSITION
13	^1E_u	37.44	0.75 $1e_u^{-1}1e_g^{-1}$, 0.08 $2a_{2u}^{-1}1e_g^{-1}$
14	$^3A_{2g}$	37.98	0.79 $1e_u^{-2}$, 0.03 $1e_g^{-2}$
15	^1E_g	39.16	0.72 $1e_u^{-2}$, 0.07 $1e_g^{-2}$, 0.02 $2a_{2u}^{-1}1e_u^{-1}$, 0.01 $2a_{1g}^{-1}1e_g^{-1}$
16	$^1A_{2u}$	39.21	0.74 $1e_u^{-1}1e_g^{-1}$, 0.06 $2a_{2u}^{-1}3a_{1g}^{-1}$, 0.03 $2a_{1g}^{-1}2a_{2u}^{-1}$
17	$^3A_{2u}$	39.55	0.81 $2a_{2u}^{-1}3a_{1g}^{-1}$, 0.05 $2a_{2u}^{-1}3a_{1g}^{-1}1e_g^{-1}3e_g^{+1}$
18	^3E_u	39.72	0.78 $2a_{2u}^{-1}1e_g^{-1}$, 0.02 $1e_u^{-1}3a_{1g}^{-1}$, 0.02 $2a_{1g}^{-1}1e_u^{-1}$
19	$^1A_{1g}$	40.46	0.64 $1e_u^{-2}$, 0.11 $1e_g^{-2}$, 0.06 $2a_{2u}^{-2}$, 0.01 $2a_{1g}^{-2}$
20	^1E_g	42.20	0.44 $2a_{2u}^{-1}1e_u^{-1}$, 0.36 $2a_{1g}^{-1}1e_g^{-1}$
21	^1E_u	42.31	0.53 $2a_{2u}^{-1}1e_g^{-1}$, 0.14 $2a_{1g}^{-1}1e_u^{-1}$, 0.06 $1e_u^{-1}3a_{1g}^{-1}$, 0.05 $1e_u^{-1}1e_g^{-1}$
22	^3E_g	42.37	0.64 $2a_{2u}^{-1}1e_u^{-1}$, 0.15 $2a_{1g}^{-1}1e_g^{-1}$
23	$^1A_{2u}$	42.84	0.71 $2a_{2u}^{-1}3a_{1g}^{-1}$, 0.04 $2a_{1g}^{-1}2a_{2u}^{-1}$, 0.04 $1e_u^{-1}1e_g^{-1}$, 0.05 $2a_{2u}^{-1}3a_{1g}^{-1}1e_g^{-1}3e_g^{+1}$
24	$^3A_{1g}$	44.20	0.76 $2a_{1g}^{-1}3a_{1g}^{-1}$, 0.05 $2a_{1g}^{-1}3a_{1g}^{-1}1e_g^{-1}3e_g^{+1}$
25	^3E_g	44.42	0.61 $2a_{1g}^{-1}1e_g^{-1}$, 0.15 $2a_{2u}^{-1}1e_u^{-1}$
26	$^1A_{1g}$	46.43	0.48 $2a_{1g}^{-1}3a_{1g}^{-1}$, 0.21 $2a_{2u}^{-2}$, 0.05 $2a_{1g}^{-2}$
27	^3E_u	46.94	0.73 $2a_{1g}^{-1}1e_u^{-1}$, 0.01 $2a_{2u}^{-1}1e_g^{-1}$
28	^1E_g	49.96	0.29 $2a_{1g}^{-1}1e_g^{-1}$, 0.25 $2a_{2u}^{-1}1e_u^{-1}$
29	$^1A_{1g}$	51.65	0.35 $2a_{2u}^{-2}$, 0.21 $2a_{1g}^{-1}3a_{1g}^{-1}$, 0.02 $2a_{1g}^{-2}$, 0.05 $3a_{1g}^{-1}1e_g^{-2}4a_{1g}^{+1}$
30	$^1A_{1g}$	59.87	0.50 $2a_{1g}^{-2}$, 0.03 $2a_{2u}^{-2}$

Table 5. Vertical DIP and composition (square coefficients larger than 0.6) of the dicationic satellite states of ethane lying below 50 eV, computed by ADC(2).

No.	STATE	DIP [eV]	COMPOSITION
1	^3E_u	47.47	0.26 $3a_{1g}^{-1}1e_g^{-2}5a_{2u}^{+1}$, 0.17 $3a_{1g}^{-1}1e_g^{-2}3a_{2u}^{+1}$, 0.11 $3a_{1g}^{-1}1e_g^{-2}4a_{2u}^{+1}$, 0.10 $1e_g^{-3}3a_{2u}^{+1}$, 0.10 $1e_g^{-3}5a_{2u}^{+1}$
2	^3E_g	47.74	0.57 $1e_g^{-3}4a_{1g}^{+1}$, 0.09 $1e_u^{-2}1e_g^{-1}4a_{1g}^{+1}$, 0.06 $1e_u^{-1}1e_g^{-2}4a_{2u}^{+1}$
3	$^3A_{1u}$	47.96	0.34 $3a_{1g}^{-1}1e_g^{-2}5a_{2u}^{+1}$, 0.18 $3a_{1g}^{-1}1e_g^{-2}3a_{2u}^{+1}$, 0.17 $3a_{1g}^{-1}1e_g^{-2}4a_{2u}^{+1}$, 0.07 $1e_g^{-3}3e_u^{+1}$
4	$^3A_{2u}$	47.97	0.29 $3a_{1g}^{-1}1e_g^{-2}5a_{2u}^{+1}$, 0.17 $3a_{1g}^{-1}1e_g^{-2}3a_{2u}^{+1}$, 0.12 $3a_{1g}^{-1}1e_g^{-2}4a_{2u}^{+1}$, 0.07 $1e_g^{-3}3e_u^{+1}$
5	^3E_u	48.13	0.24 $1e_g^{-3}3a_{2u}^{+1}$, 0.20 $1e_g^{-3}5a_{2u}^{+1}$, 0.11 $3a_{1g}^{-1}1e_g^{-2}5a_{2u}^{+1}$, 0.10 $1e_g^{-3}4a_{2u}^{+1}$, 0.06 $3a_{1g}^{-1}1e_g^{-2}3a_{2u}^{+1}$
6	$^1A_{1u}$	48.53	0.25 $1e_g^{-3}3e_u^{+1}$, 0.17 $3a_{1g}^{-1}1e_g^{-2}5a_{2u}^{+1}$, 0.12 $1e_g^{-3}2e_u^{+1}$, 0.09 $3a_{1g}^{-1}1e_g^{-2}4a_{2u}^{+1}$, 0.07 $3a_{1g}^{-1}1e_g^{-2}3a_{2u}^{+1}$
7	^3E_g	48.53	0.18 $1e_g^{-3}2e_g^{+1}$, 0.18 $1e_g^{-3}3e_g^{+1}$, 0.13 $3a_{1g}^{-1}1e_g^{-2}4a_{1g}^{+1}$, 0.07 $1e_u^{-1}1e_g^{-2}3e_u^{+1}$
8	$^3A_{2g}$	48.53	0.21 $1e_g^{-3}2e_g^{+1}$, 0.20 $1e_g^{-3}3e_g^{+1}$, 0.12 $3a_{1g}^{-1}1e_g^{-2}4a_{1g}^{+1}$, 0.08 $1e_u^{-1}1e_g^{-2}3e_u^{+1}$
9	^3E_u	48.54	0.24 $1e_g^{-3}3e_u^{+1}$, 0.18 $1e_g^{-3}2e_u^{+1}$, 0.06 $3a_{1g}^{-1}1e_g^{-2}3a_{2u}^{+1}$
10	$^3A_{1g}$	48.56	0.25 $1e_g^{-3}2e_g^{+1}$, 0.21 $1e_g^{-3}3e_g^{+1}$, 0.07 $1e_u^{-1}1e_g^{-2}3e_u^{+1}$, 0.07 $3a_{1g}^{-1}1e_g^{-2}4a_{1g}^{+1}$
11	$^3A_{1u}$	48.57	0.22 $1e_g^{-3}3e_u^{+1}$, 0.19 $1e_g^{-3}2e_u^{+1}$, 0.13 $3a_{1g}^{-1}1e_g^{-2}3a_{2u}^{+1}$, 0.09 $3a_{1g}^{-1}1e_g^{-2}5a_{2u}^{+1}$
12	$^3A_{2u}$	48.65	0.22 $1e_g^{-3}3e_u^{+1}$, 0.18 $1e_g^{-3}2e_u^{+1}$, 0.11 $3a_{1g}^{-1}1e_g^{-2}3a_{2u}^{+1}$, 0.09 $3a_{1g}^{-1}1e_g^{-2}5a_{2u}^{+1}$
13	^1E_g	48.67	0.52 $1e_g^{-3}4a_{1g}^{+1}$, 0.07 $1e_g^{-3}2e_g^{+1}$, 0.06 $1e_u^{-2}1e_g^{-1}4a_{1g}^{+1}$, 0.06 $1e_g^{-3}3e_g^{+1}$

No.	STATE	DIP [eV]	COMPOSITION
14	^1E_u	48.75	0.36 $1e_g^{-3}3a_{2u}^{+1}$, 0.29 $1e_g^{-3}5a_{2u}^{+1}$, 0.12 $1e_g^{-3}4a_{2u}^{+1}$, 0.06 $1e_u^{-2}1e_g^{-1}3a_{2u}^{+1}$
15	^3E_u	48.82	0.29 $1e_u^{-1}1e_g^{-2}4a_{1g}^{+1}$, 0.19 $1e_g^{-3}4a_{2u}^{+1}$, 0.11 $1e_g^{-3}3a_{2u}^{+1}$, 0.06 $1e_u^{-2}1e_g^{-1}4a_{2u}^{+1}$
16	$^3A_{2g}$	49.11	0.41 $1e_u^{-1}3a_{1g}^{-1}1e_g^{-1}5a_{2u}^{+1}$, 0.27 $1e_u^{-1}3a_{1g}^{-1}1e_g^{-1}3a_{2u}^{+1}$, 0.12 $1e_u^{-1}3a_{1g}^{-1}1e_g^{-1}4a_{2u}^{+1}$
17	$^1A_{1u}$	49.28	0.27 $3a_{1g}^{-1}1e_g^{-2}5a_{2u}^{+1}$, 0.25 $3a_{1g}^{-1}1e_g^{-2}3a_{2u}^{+1}$, 0.15 $1e_g^{-3}3e_u^{+1}$, 0.09 $1e_g^{-3}2e_u^{+1}$, 0.06 $3a_{1g}^{-1}1e_g^{-2}4a_{2u}^{+1}$
18	$^1A_{2g}$	49.30	0.35 $1e_g^{-3}3e_g^{+1}$, 0.22 $1e_g^{-3}2e_g^{+1}$, 0.07 $3a_{1g}^{-1}1e_g^{-2}4a_{1g}^{+1}$, 0.07 $1e_u^{-1}1e_g^{-2}3e_u^{+1}$
19	^3E_g	49.44	0.13 $1e_u^{-1}1e_g^{-2}5a_{2u}^{+1}$, 0.13 $3a_{1g}^{-1}1e_g^{-2}4a_{1g}^{+1}$, 0.10 $1e_u^{-1}1e_g^{-2}3a_{2u}^{+1}$, 0.10 $1e_u^{-1}1e_g^{-2}4a_{2u}^{+1}$, 0.09 $1e_u^{-1}3a_{1g}^{-1}1e_g^{-1}5a_{2u}^{+1}$, 0.06 $1e_u^{-1}3a_{1g}^{-1}1e_g^{-1}3a_{2u}^{+1}$
20	^3E_u	49.51	0.32 $3a_{1g}^{-1}1e_g^{-2}3e_u^{+1}$, 0.19 $3a_{1g}^{-1}1e_g^{-2}2e_u^{+1}$, 0.07 $1e_u^{-1}3a_{1g}^{-1}1e_g^{-1}3e_g^{+1}$
21	^3E_g	49.55	0.17 $3a_{1g}^{-1}1e_g^{-2}4a_{1g}^{+1}$, 0.13 $1e_u^{-1}1e_g^{-2}3a_{2u}^{+1}$, 0.09 $1e_u^{-1}1e_g^{-2}5a_{2u}^{+1}$, 0.08 $1e_u^{-1}3a_{1g}^{-1}1e_g^{-1}5a_{2u}^{+1}$, 0.07 $3a_{1g}^{-1}1e_g^{-2}3e_g^{+1}$
22	^1E_u	49.56	0.18 $1e_g^{-3}3e_u^{+1}$, 0.14 $1e_g^{-3}4a_{2u}^{+1}$, 0.11 $1e_u^{-1}1e_g^{-2}4a_{1g}^{+1}$, 0.11 $1e_g^{-3}2e_u^{+1}$
23	$^3A_{1g}$	49.58	0.28 $1e_u^{-1}3a_{1g}^{-1}1e_g^{-1}5a_{2u}^{+1}$, 0.15 $1e_u^{-1}3a_{1g}^{-1}1e_g^{-1}3a_{2u}^{+1}$, 0.14 $1e_u^{-1}3a_{1g}^{-1}1e_g^{-1}4a_{2u}^{+1}$
24	$^3A_{1g}$	49.76	0.34 $3a_{1g}^{-1}1e_g^{-2}4a_{1g}^{+1}$, 0.11 $3a_{1g}^{-1}1e_g^{-2}3e_g^{+1}$, 0.09 $3a_{1g}^{-1}1e_g^{-2}2e_g^{+1}$
25	$^3A_{2g}$	49.91	0.49 $3a_{1g}^{-1}1e_g^{-2}4a_{1g}^{+1}$
26	$^3A_{2g}$	49.99	0.23 $3a_{1g}^{-1}1e_g^{-2}4a_{1g}^{+1}$, 0.13 $1e_u^{-1}1e_g^{-2}3e_u^{+1}$, 0.11 $1e_u^{-1}1e_g^{-2}2e_u^{+1}$

of the distribution of states suggested the use of less rapidly decaying convoluting functions, lorentzians instead of gaussians, to reproduce the observed profile. The choice of either function was of course largely irrelevant for the unsaturated hydrocarbons. The experimental spectrum appears as a single very broad structure, made up of several largely unresolved bands, which would be of impossible interpretation without reliable theoretical aid. Aside from isolated discrepancies in the relative intensities (but notice that no background subtraction appears to have been carried out in the experimental profile [54], affecting in particular the high DIP side of the spectrum), the theoretical spectrum satisfactorily reproduces all the features and the correspondences are evidenced in the figure by vertical dotted lines.

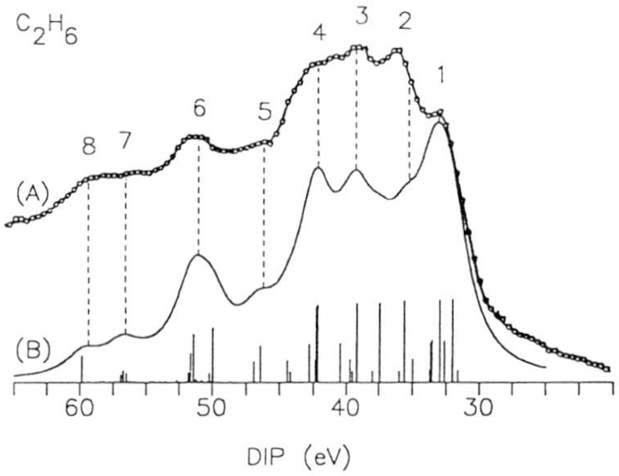

Fig. 3. Experimental (A) and theoretical (B) Auger spectra of ethane. The individual ADC(2) 2h pole strength contributions are shown as a bar spectrum.

The theoretical to experimental intensity comparison shows that also in this case the deviations affects essentially the low DIP side of the spectrum. Interestingly, however, the computations give in this case a relative intensity of the first peak which is higher than observed, contrary to what we found for acetylene and ethylene. Although not yet supported by actual calculations, consideration of hole localization effects in the ethane dication should help to clarify this behavior. In the unsaturated hydrocarbons C_2H_2 and C_2H_4, the calculations have shown that hole localization effects are totally absent in the outer valence double ionization region, involving removal of two π electrons, and start to play a role at higher energy. In ethane the π bond between the carbon atoms is replaced by CH bonds, with the further implication that the carbon–carbon distance is longer. In general, the conditions under which two–site hole localization is most favored occur upon removal of pairs of degenerate (or nearly so) electrons which can be described by spatially non–overlapping distributions, whereby hole–hole repulsion dominates and drives the mechanism. It thus seems clear that in ethane double hole localization should take place more pronouncedly in the very outer valence region, producing states with some appreciable two–site character. This effect would indeed tend to decrease the relative Auger intensity at the low DIP end of the spectrum with respect to Fig.3. It should also be observed that the effects of vibrational broadening on the outer valence Auger bandshapes are in general more likely to be important in a saturated system, where ionization is invariably accompanied by bond rupture.

Benzene

Already without consideration of correlation effects, the number of valence dicationic states of C_6H_6 is 150. This gives an idea of the computational problem presented by the calculation of the double ionization spectrum of a molecule of this size at the level of theory afforded by ADC(2). The standard single–reference CI calculation of the energy and wavefunction of a dicationic state including (part of) the 4h2p configuration space with a double zeta plus polarization basis set involves hamiltonian matrix sizes of the order of several hundred thousands. By contrast, the dimension of the ADC(2) configuration spaces (*unique* for each state symmetry) is at most of order 20000. A qualitative idea of the extent of correlation effects affecting the spectrum may be obtained by considering that of the 226 dicationic states we computed in the outer valence region (from 23 to 40 eV DIP) over 140 have significant 2h components and thus carry Auger intensity. Notice that the total 2h projection of these states amounts to only 45% of the full valence 2h space, which comprises, as noted above, 150 uncorrelated states. In Table 6 we have reported the DIP and composition of the lowest 35 computed states, evidencing the strong final state configuration mixing taking place already at low double ionization energy. Due to the typical strong increase in density of satellite states in the inner valence region, it is easy to extrapolate that many thousands of dicationic states give significant contributions to the Auger spectrum of this system.

The theoretical outer valence part of the Auger spectrum of C_6H_6 is displayed in Fig.4, below the corresponding part of the experimental [1a] profile (in the original kinetic energy scale). As usual, the computed spectrum is a gaussian convolution of the vertical transitions weighted by their 2h pole strength and with individual fwhm of 1 eV. The experimental spectrum shows only eight distinctly resolved peaks in the lower energy half (higher kinetic energy), emerging from a unique broad structure which extends for about 20 eV. At higher energy no further peaks are visible, confirming the nearly continuous intensity distribution. The convolution is remarkably accurate in reproducing the number and relative energy positions of the peaks, showing once again that, provided many dicationic states can be efficiently computed with enough accuracy, the inherent complexity of Auger spectra does afford a simple statistical description, circumventing the need of explicitly computing the transition rates. Our calculations permit the assigment of the spectrum of benzene. This is shown in Table 7, where the observed and computed relative peak energies and their 2h composition (expressed as percentage of the peak heights) are reported. It can be seen in the table that the computed band maxima agree with experiment to within 0.1–0.3 eV. The composition reported summarizes all the contributions of each 2h configuration, regardless of the spin/space symmetry of the actual electronic states underlying a given peak. This gives a useful interpretation of the spectrum and evidences its complexity. It is seen that, except the first two peaks which can be described almost purely in terms of double ionization out of the two outermost orbitals, all the peaks result from many transitions having approximately equal weight, and a description in terms of individual states is manifestly impossible. The configuration mixing tends to become more pronounced as the ionization energy increases, accompanying the occurrence of many satellite states, as expected for an unsaturated (aromatic) hydrocarbon. The apparent inversion of this trend for the last peak must be considered an artifact of the calculation due to the lack of computed states at higher energy: this peak has a too low intensity and no visible counterpart in the experiment, appearing to be just a low energy component of the peak at 16.6 eV (relative energy). In fact, states involving ionization of one or both electrons out of the three innermost valence orbitals lie outside the energy range covered by the calculations.

As seen above, the pole strength convolution procedure appears to be entirely satisfactory for predicting the occurence and energy spacing of the peaks in the Auger spectrum. It can be concluded that the emergence of peaks in energy regions of moderate to high density of states is largely controlled by the clustering of 2h contributions and can be evaluated without knowledge of individual state transition

Table 6. Vertical DIP (eV) and composition (square leading coefficients) of the lowest 35 doubly ionized states of benzene. The 2h components are underlined.

State	Energy	Composition
$^3A_{2g}$	23.34	$\underline{0.83}$ $1e_{1g}^{-2}$
$^1E_{2g}$	23.96	$\underline{0.81}$ $1e_{1g}^{-2}$
$^1A_{1g}$	24.59	$\underline{0.78}$ $1e_{1g}^{-2}$, $\underline{0.08}$ $1a_{2u}^{-2}$
$^3E_{1u}$	26.07	$\underline{0.75}$ $1a_{2u}^{-1}1e_{1g}^{-1}$, 0.11 $1e_{1g}^{-3}1e_{2u}^{1}$
$^3B_{1g}$	26.19	$\underline{0.80}$ $3e_{2g}^{-1}1e_{1g}^{-1}$
$^3E_{1g}$	26.30	$\underline{0.80}$ $3e_{2g}^{-1}1e_{1g}^{-1}$
$^3B_{2g}$	26.41	$\underline{0.80}$ $3e_{2g}^{-1}1e_{1g}^{-1}$
$^1B_{1g}$	26.44	$\underline{0.79}$ $3e_{2g}^{-1}1e_{1g}^{-1}$, $\underline{0.02}$ $1b_{2u}^{-1}1a_{2u}^{-1}$
$^1E_{1g}$	26.55	$\underline{0.79}$ $3e_{2g}^{-1}1e_{1g}^{-1}$, $\underline{0.02}$ $3e_{1u}^{-1}1a_{2u}^{-1}$
$^1B_{2g}$	26.63	$\underline{0.80}$ $3e_{2g}^{-1}1e_{1g}^{-1}$, $\underline{0.01}$ $2b_{1u}^{-1}1a_{2u}^{-1}$
$^1A_{1u}$	28.08	$\underline{0.78}$ $3e_{1u}^{-1}1e_{1g}^{-1}$
$^3A_{1u}$	28.22	$\underline{0.78}$ $3e_{1u}^{-1}1e_{1g}^{-1}$
$^3E_{2u}$	28.32	$\underline{0.67}$ $3e_{1u}^{-1}1e_{1g}^{-1}$, $\underline{0.11}$ $1a_{2u}^{-1}3e_{2g}^{-1}$, $\underline{0.01}$ $2b_{1u}^{-1}1e_{1g}^{-1}$
$^1E_{2u}$	28.39	$\underline{0.50}$ $3e_{1u}^{-1}1e_{1g}^{-1}$, $\underline{0.26}$ $1a_{2u}^{-1}3e_{2g}^{-1}$, $\underline{0.04}$ $1b_{2u}^{-1}1e_{1g}^{-1}$
$^1E_{1u}$	28.54	$\underline{0.72}$ $1a_{2u}^{-1}1e_{1g}^{-1}$, 0.09 $1e_{1g}^{-3}1e_{2u}^{1}$
$^3A_{2u}$	28.74	$\underline{0.75}$ $3e_{1u}^{-1}1e_{1g}^{-1}$, $\underline{0.02}$ $3a_{1g}^{-1}1a_{2u}^{-1}$

$^3E_{2u}$	28.86	$\underline{0.45}$ $1b_{2u}^{-1}1e_{1g}^{-1}$, $\underline{0.23}$ $1a_{2u}^{-1}3e_{2g}^{-1}$, $\underline{0.07}$ $2b_{1u}^{-1}1e_{1g}^{-1}$, $\underline{0.04}$ $3e_{1u}^{-1}1e_{1g}^{-1}$
$^1E_{2u}$	29.06	$\underline{0.28}$ $1b_{2u}^{-1}1e_{1g}^{-1}$, $\underline{0.19}$ $1a_{2u}^{-1}3e_{2g}^{-1}$, $\underline{0.18}$ $3e_{1u}^{-1}1e_{1g}^{-1}$, $\underline{0.14}$ $2b_{1u}^{-1}1e_{1g}^{-1}$
$^3B_{2u}$	29.09	0.84 $1e_{1g}^{-3}1e_{2u}^{1}$, 0.07 $1e_{1g}^{-3}2e_{2u}^{1}$
$^3A_{2g}$	29.09	$\underline{0.79}$ $3e_{2g}^{-2}$
$^3E_{2u}$	29.33	$\underline{0.29}$ $2b_{1u}^{-1}1e_{1g}^{-1}$, $\underline{0.27}$ $1b_{2u}^{-1}1e_{1g}^{-1}$, $\underline{0.18}$ $1a_{2u}^{-1}3e_{2g}^{-1}$, 0.07 $3e_{2g}^{-1}1e_{1g}^{-2}1e_{2u}^{1}$
$^1A_{2u}$	29.37	$\underline{0.75}$ $3e_{1u}^{-1}1e_{1g}^{-1}$, $\underline{0.05}$ $3a_{1g}^{-1}1a_{2u}^{-1}$
$^1E_{2g}$	29.41	$\underline{0.77}$ $3e_{2g}^{-2}$, $\underline{0.02}$ $1b_{2u}^{-1}3e_{1u}^{-1}$, $\underline{0.02}$ $2b_{1u}^{-1}3e_{1u}^{-1}$
$^1A_{1g}$	29.60	$\underline{0.74}$ $3e_{2g}^{-2}$, $\underline{0.04}$ $3e_{1u}^{-2}$, $\underline{0.02}$ $2b_{1u}^{-2}$, $\underline{0.01}$ $1b_{2u}^{-2}$
$^1E_{2u}$	29.65	$\underline{0.35}$ $2b_{1u}^{-1}1e_{1g}^{-1}$, $\underline{0.33}$ $1b_{2u}^{-1}1e_{1g}^{-1}$, $\underline{0.06}$ $1a_{2u}^{-1}3e_{2g}^{-1}$, 0.06 $3e_{2g}^{-1}1e_{1g}^{-2}1e_{2u}^{1}$, $\underline{0.02}$ $2e_{2g}^{-1}1a_{2u}^{-1}$
$^3B_{1u}$	29.70	0.81 $1e_{1g}^{-3}1e_{2u}^{1}$, 0.07 $1a_{2u}^{-2}1e_{1g}^{-1}1e_{2u}^{1}$, 0.05 $1e_{1g}^{-3}2e_{2u}^{1}$
$^3E_{2u}$	30.02	$\underline{0.38}$ $2b_{1u}^{-1}1e_{1g}^{-1}$, $\underline{0.19}$ $1a_{2u}^{-1}3e_{2g}^{-1}$, 0.12 $3e_{2g}^{-1}1e_{1g}^{-2}1e_{2u}^{1}$, $\underline{0.07}$ $3e_{1u}^{-1}1e_{1g}^{-1}$, $\underline{0.03}$ $1b_{2u}^{-1}1e_{1g}^{-1}$
$^1A_{1g}$	30.58	$\underline{0.58}$ $1a_{2u}^{-2}$, 0.20 $1a_{2u}^{-1}1e_{1g}^{-2}1e_{2u}^{1}$, $\underline{0.07}$ $1e_{1g}^{-2}$
$^1E_{1g}$	30.88	$\underline{0.43}$ $3a_{1g}^{-1}1e_{1g}^{-1}$, $\underline{0.33}$ $3e_{1u}^{-1}1a_{2u}^{-1}$
$^3E_{1g}$	30.94	$\underline{0.48}$ $3a_{1g}^{-1}1e_{1g}^{-1}$, $\underline{0.28}$ $3e_{1u}^{-1}1a_{2u}^{-1}$
$^3B_{1u}$	30.95	$\underline{0.82}$ $3e_{1u}^{-1}3e_{2g}^{-1}$
$^3B_{2u}$	30.98	$\underline{0.81}$ $3e_{1u}^{-1}3e_{2g}^{-1}$
$^1E_{2u}$	31.02	$\underline{0.26}$ $2b_{1u}^{-1}1e_{1g}^{-1}$, $\underline{0.20}$ $1a_{2u}^{-1}3e_{2g}^{-1}$, 0.13 $3e_{2g}^{-1}1e_{1g}^{-2}1e_{2u}^{1}$, $\underline{0.11}$ $1b_{2u}^{-1}1e_{1g}^{-1}$, $\underline{0.11}$ $3e_{1u}^{-1}1e_{1g}^{-1}$
$^3E_{1u}$	31.14	$\underline{0.68}$ $3e_{1u}^{-1}3e_{2g}^{-1}$, $\underline{0.12}$ $1b_{2u}^{-1}3e_{2g}^{-1}$
$^3E_{1u}$	31.27	0.72 $1e_{1g}^{-3}1e_{2u}^{1}$, $\underline{0.09}$ $1a_{2u}^{-1}1e_{1g}^{-1}$

rates. Since this procedure does not consider in any way the spatial hole density distribution of the states, it seems clear that an essentially delocalized picture of the two holes holds with sufficient approximation for the hydrocarbons studied. However, as we have discussed for acetylene, ethylene and ethane, weak hole localization effects, which can effectively be accounted for by a simple two–hole population analysis, do play a role in determining the correct intensity ratio of the peaks, especially between the lowest energy part of the spectrum and the rest. The spectrum of benzene in Fig.4 shows similar evidence, with a too low computed intensity for the first peak as we have typically found also in the other unsaturated systems. In the next example we shall examine a typical case where, unlike the hydrocarbon series, hole localization effects play a dominant role in controlling Auger intensities and bandshapes.

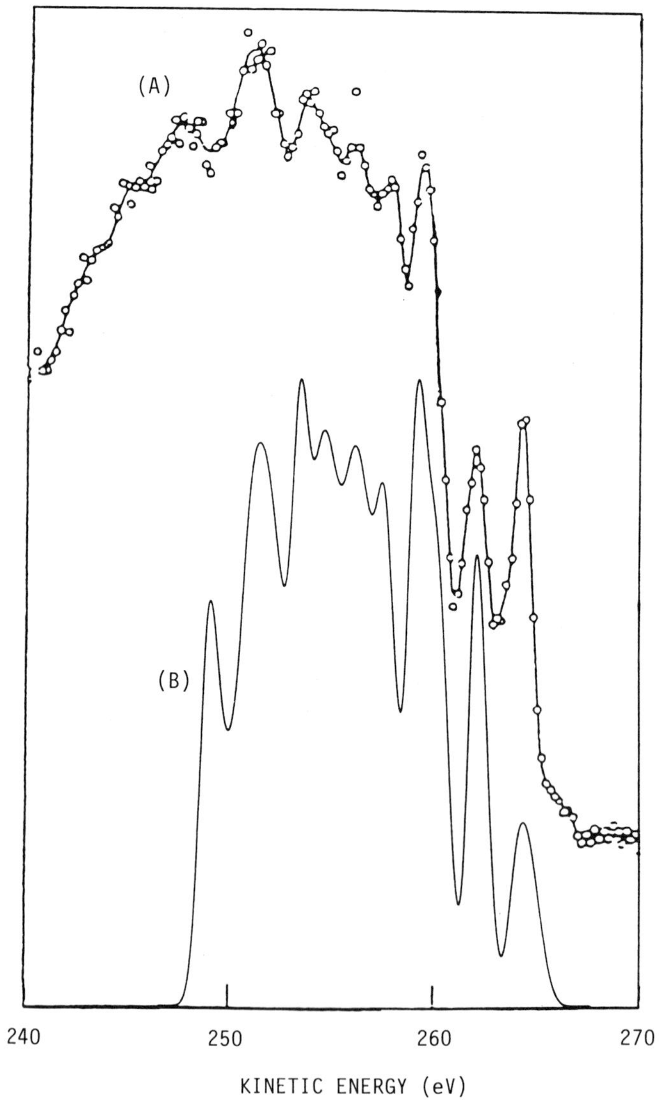

Fig. 4. Experimental (A) and theoretical (B) Auger spectra of benzene

Table 7. Experimental [54] and theoretical relative energies (eV) of the Auger peaks of benzene. The DIP of the first peak in given in parentheses. Also shown is the computed percent 2h composition of the peaks.

Exp.	ADC(2)	
Energy	Energy	Composition
0.0(26.1)	0.0(24.3)	$1e_{1g}^{-2}$(95) $1a_{2u}^{-2}$(5)
2.1	2.2	$3e_{2g}^{-1}1e_{1g}^{-1}$(94) $1a_{2u}^{-1}1e_{1g}^{-1}$(4)
4.8	5.0	$3e_{2g}^{-2}$(35) $3e_{1u}^{-1}1e_{1g}^{-1}$(24) $1b_{2u}^{-1}1e_{1g}^{-1}$(15) $2b_{1u}^{-1}1e_{1g}^{-1}$(11) $1a_{2u}^{-1}3e_{2g}^{-1}$(8) $1a_{2u}^{-1}1e_{1g}^{-1}$(3) $3a_{1g}^{-1}1a_{2u}^{-1}$(1)
6.6	6.8	$3e_{1u}^{-1}3e_{2g}^{-1}$(36) $3a_{1g}^{-1}1e_{1g}^{-1}$(14) $3e_{1u}^{-1}1a_{2u}^{-1}$(11) $1a_{2u}^{-2}$(8) $2b_{1u}^{-1}1e_{1g}^{-1}$(7) $1a_{2u}^{-1}3e_{2g}^{-1}$(5) $1b_{2u}^{-1}3e_{2g}^{-1}$(5)
		$1b_{2u}^{-1}1e_{1g}^{-1}$(3) $3e_{1u}^{-1}1e_{1g}^{-1}$(3) $2b_{1u}^{-1}3e_{2g}^{-1}$(3) $1b_{2u}^{-1}1a_{2u}^{-1}$(1)
8.2	8.1	$3e_{1u}^{-1}3e_{2g}^{-1}$(23) $2b_{1u}^{-1}1a_{2u}^{-1}$(19) $1b_{2u}^{-1}1a_{2u}^{-1}$(13) $3e_{1u}^{-1}1a_{2u}^{-1}$(9) $2e_{2g}^{-1}1e_{1g}^{-1}$(9) $1b_{2u}^{-1}3e_{2g}^{-1}$(8) $3a_{1g}^{-1}1e_{1g}^{-1}$(6)
		$2b_{1u}^{-1}3e_{2g}^{-1}$(5) $3e_{1u}^{-2}$(2) $3a_{1g}^{-1}2b_{1u}^{-1}$(1)
	9.6	$2e_{2g}^{-1}1e_{1g}^{-1}$(23) $3e_{1u}^{-2}$(20) $2b_{1u}^{-1}3e_{2g}^{-1}$(12) $1b_{2u}^{-1}3e_{1u}^{-1}$(11) $3a_{1g}^{-1}3e_{2g}^{-1}$(10) $2b_{1u}^{-1}3e_{1u}^{-1}$(9) $3a_{1g}^{-1}1a_{2u}^{-1}$(4)
		$1b_{2u}^{-1}3e_{2g}^{-1}$(2) $2b_{1u}^{-2}$(2) $3a_{1g}^{-1}1e_{1g}^{-1}$(2) $3e_{1u}^{-1}3e_{2g}^{-1}$(2) $3a_{1g}^{-1}3e_{1u}^{-1}$(2) $1b_{2u}^{-2}$(1)
10.4	10.7	$2e_{2g}^{-1}1e_{1g}^{-1}$(18) $2b_{1u}^{-1}1b_{2u}^{-1}$(16) $3a_{1g}^{-1}1a_{2u}^{-1}$(13) $1b_{2u}^{-1}3e_{1u}^{-1}$(11) $1b_{2u}^{-2}$(11) $2e_{2g}^{-1}3e_{2g}^{-1}$(8) $2b_{1u}^{-1}3e_{1u}^{-1}$(6)
		$3a_{1g}^{-1}3e_{2g}^{-1}$(4) $2b_{1u}^{-2}$(4) $1b_{2u}^{-1}1a_{2u}^{-1}$(2) $3e_{1u}^{-2}$(2)
12.9	12.7	$2e_{1u}^{-1}1e_{1g}^{-1}$(22) $3a_{1g}^{-1}1b_{2u}^{-1}$(18) $2e_{2g}^{-1}3e_{2g}^{-1}$(11) $3a_{1g}^{-1}2b_{1u}^{-1}$(7) $2e_{2g}^{-1}1a_{2u}^{-1}$(7) $2e_{2g}^{-1}3e_{1u}^{-1}$(6) $2b_{1u}^{-2}$(6)
		$3a_{1g}^{-1}3e_{1u}^{-1}$(4) $1b_{2u}^{-2}$(3) $2e_{2g}^{-1}1e_{1g}^{-1}$(3) $1a_{2u}^{-1}3e_{2g}^{-1}$(1) $2b_{1u}^{-1}1b_{2u}^{-1}$(1) $3a_{1g}^{-2}$(1) $3e_{1u}^{-2}$(1)
	15.1	$2e_{2g}^{-1}3e_{1u}^{-1}$(30) $2e_{1u}^{-1}3e_{2g}^{-1}$(24) $2e_{1u}^{-1}1e_{1g}^{-1}$(14) $2e_{2g}^{-1}2b_{1u}^{-1}$(6) $2e_{2g}^{-1}1b_{2u}^{-1}$(6) $3a_{1g}^{-1}1b_{2u}^{-1}$(4)
		$3a_{1g}^{-1}2b_{1u}^{-1}$(4) $2a_{1g}^{-1}1a_{2u}^{-1}$(3) $2e_{2g}^{-1}1a_{2u}^{-1}$(1) $2e_{2g}^{-1}3e_{2g}^{-1}$(1) $3e_{1u}^{-1}1a_{2u}^{-1}$(1)
16.6		

III.2 – Strong double hole localization effects: boron trifluoride

BF_3 is a strongly ionic molecule, in which most of the outer valence electron density is located on the three equivalent fluorine atoms. Furthermore, there is no direct bonding between these atoms, which are far apart from each other. Already in the neutral ground state, therefore, the electronic structure of the system can adequately be described in a localized picture, with three ionic σ bonds between boron and the fluorines and three nonbonding, nonoverlapping, electron distributions (lone pairs) concentrated around the fluorine atoms. In the molecular orbital picture, these localized orbitals give rise to symmetry adapted (delocalized) combinations. In such a situation it is expected that a valence ionization event creates a symmetry breaking driving force, leading to an effective localization of the positive charge. In single ionization, the localizing mechanism can be interpreted in terms of enhanced relaxation effects, similar to the case of core ionization [55]. Upon double ionization, an additional large term arises because of hole–hole repulsion, leading to low–lying states with two holes localized on different fluorine atoms and high–lying counterparts with two holes localized on the same atom. BF_3 represents therefore an ideal case to study hole localization effects on Auger spectra. It is worthwhile to comment that hole localization constitutes an additional complication in molecular double ionization spectra which inherently requires explicit consideration of electron correlation. It is easy to see that while the low–lying dicationic states with two holes localized at different atomic centers can be approximately described at the Hartree–Fock level by enforcing symmetry breaking of the wavefunction (in analogy to the case of single hole localization), a variational independent–particle approach will even qualitatively fail to describe at all the high–lying states with both holes localized on the same center [26].

The computed [26] DIP and 2h composition of the lowest main dicationic states of BF_3 lying below 60 eV are reported in Table 8. The total number of states computed by ADC(2) has having significant 2h character exceeded 100 in the outer valence energy region considered (from 37 to 70 eV). Some interesting observations can already be made from the results in the table, underlining the implications of hole localization. First of all one notices a strong 2h configuration mixing, except for few triplet states which remain largely described by a single configuration. As an example, the 2h components corresponding to double ionization of the fluorine out of plane lone pairs are bold faced in the table to evidence this feature. Straighforward symmetry considerations show immediately that singlet 2h configurations involving these orbitals are effectively combinations of localized configurations in which the two holes are either localized on only one fluorine atom or at two different ones. On the contrary, triplet configurations involving these lone pairs (one out of plane lone pair per fluorine atom) can obviously only describe a situation in which each hole is on a different fluorine. This explains that, for localization to take place, we must find a strong mixing of these configurations in the singlet states but only a much weaker mixing (except of other correlation effects) in the triplet states, as indeed observed. Still considering, for simplicity, only the dicationic states mainly arising from ionization of the out of plane lone pairs, we notice that these states tend to form two distinct groups, well separated in energy by roughly 10–13 eV. This is visualized more clearly in Fig.5, where the vertical bars represent the bold faced contributions of Table 8. This is a typical consequence of strong hole localization, which gives rise to low lying two–site states and their higher energy partners of the one–site type. The energy spread of each group is small compared to the gap between the two groups, which is dominated by the large difference in hole–hole repulsion. By taking the repulsion term in the two–site states to be of the order of the inverse distance between two fluorine atoms (6.4 eV) we can thus conclude that the analogous term in the one–site states is of the order of 15–20 eV. For comparison, we recall that the one–site hole repulsion term in doubly core ionized states of hydrocarbons is about 95 eV [35].

The analysis of hole localization exemplified above can very simply be made more quantitative and extended to the whole spectrum by using the described two–hole population analysis [26]. In the case of BF_3 we can thus decompose the 2h

Table 8. Computed DIP (eV) and composition (square leading 2h components) of the dicationic states of BF_3 up to 60 eV. The 2h configurations are indicated by the occupied orbitals of BF_3 from which two electrons are removed. The out–of–plane components discussed in the text are bold faced.

State	DIP	Composition
$^1A_1'$	37.90	0.612 $1a_2'$; 0.221 $3e'$; 0.036 $2e'3e'$
$^3E'$	38.49	0.814 $3e'1a_2'$; 0.034 $2e'1a_2'$
$^3E''$	38.50	0.584 $1e''1a_2'$; 0.240 $3e'1e''$; 0.028 $1a_2''3e'$
$^1E''$	38.54	0.559 $1e''1a_2'$; 0.259 $3e'1e''$; 0.031 $1a_2''3e'$
$^3A_2'$	38.83	**0.870 $1e''$**
$^1E'$	39.60	**0.621 $1e''$** ; **0.245 $1a_2''1e''$**
$^3A_2''$	39.76	0.739 $3e'1e''$; 0.119 $2e'1e''$
$^1A_1''$	39.81	0.439 $1a_2''1a_2'$; 0.428 $3e'1e''$
$^3A_1''$	39.82	0.452 $3e'1e''$; 0.414 $1a_2''1a_2'$
$^1A_2''$	39.84	0.717 $3e'1e''$; 0.139 $2e'1e''$
$^1E'$	40.07	0.500 $3e'$; 0.159 $3e'1a_2'$; 0.111 $2e'1a_2'$; 0.078 $2e'3e'$
$^3A_2'$	40.44	0.704 $3e'$; 0.093 $2a_1'1a_2'$; 0.051 $2e'3e'$; 0.017 $2e'$
$^3E''$	41.01	0.380 $3e'1e''$; 0.292 $1a_2''3e'$; 0.084 $1e''1a_2'$; 0.062 $2e'1a_2''$; 0.040 $2a_1'1e''$
$^1E'$	41.08	0.341 $3e'1a_2'$; 0.242 $2e'1a_2'$; 0.130 $2e'3e'$; 0.081 $3e'$; 0.049 $2a_1'3e'$; 0.024 $2e'$
$^1E''$	41.17	0.358 $1a_2''3e'$; 0.315 $3e'1e''$; 0.072 $1e''1a_2'$; 0.062 $2e'1a_2''$; 0.046 $2a_1'1e''$
$^3E'$	41.41	**0.866 $1a_2''1e''$**
$^3A_1''$	41.99	0.563 $2e'1e''$; 0.188 $1a_2''1a_2'$; 0.112 $3e'1e''$
$^1A_1''$	42.00	0.604 $2e'1e''$; 0.156 $1a_2''1a_2'$; 0.104 $3e'1e''$
$^3E'$	42.03	0.653 $2e'1a_2'$; 0.156 $2e'3e'$; 0.043 $2a_1'3e'$
$^1A_1'$	42.17	**0.428 $1a_2''$** ; **0.395 $1e''$** ; 0.027 $3e'$
$^3E''$	42.46	0.511 $2e'1e''$; 0.176 $1a_2''3e'$; 0.084 $2a_1'1e''$; 0.055 $3e'1e''$; 0.020 $2e'1a_2''$; 0.017 $1e''1a_2'$
$^1E''$	42.49	0.516 $2e'1e''$; 0.123 $1a_2''3e'$; 0.119 $2a_1'1e''$; 0.061 $3e'1e''$; 0.024 $2e'1a_2''$; 0.017 $1e''1a_2'$
$^1A_1'$	42.63	0.370 $3e'$; 0.200 $2e'3e'$; 0.163 $2e'$; 0.060 $1a_2'$; **0.032 $1a_2''$** ; 0.018 $2a_1'$
$^3A_2'$	42.66	0.557 $2e'3e'$; 0.293 $2a_1'1a_2'$; 0.011 $2e'$
$^1A_2'$	42.68	0.573 $2a_1'1a_2'$; 0.290 $2e'3e'$
$^3A_1'$	42.91	0.868 $2e'3e'$
$^3E'$	43.59	0.442 $2a_1'3e'$; 0.387 $2e'3e'$; 0.015 $2e'1a_2'$; 0.013 $2a_1'2e'$
$^1E'$	43.96	0.240 $2a_1'3e'$; 0.219 $2e'3e'$; 0.151 $2a_1'2e'$; 0.124 $2e'1a_2'$; 0.077 $3e'$; 0.045 $2e'$
$^3E''$	43.99	0.420 $2a_1'1e''$; 0.295 $2e'1a_2''$; 0.087 $1a_2''3e'$; 0.027 $3e'1e''$; 0.021 $1e''1a_2'$
$^1E''$	44.06	0.381 $2a_1'1e''$; 0.362 $2e'1a_2''$; 0.068 $1a_2''3e'$; 0.029 $3e'1e''$; 0.020 $1e''1a_2'$
$^3A_2''$	44.38	0.450 $2e'1e''$; 0.362 $2a_1'1a_2''$; 0.044 $3e'1e''$
$^1A_2''$	44.65	0.428 $2a_1'1a_2''$; 0.389 $2e'1e''$; 0.045 $3e'1e''$
$^3A_2'$	44.93	0.365 $2e'$; 0.249 $2a_1'1a_2'$; 0.199 $2e'3e'$; 0.043 $3e'$
$^1E'$	45.42	0.375 $2e'$; 0.204 $2a_1'3e'$; 0.154 $2e'3e'$; 0.071 $2a_1'2e'$; 0.029 $3e'1a_2'$; 0.016 $2e'1a_2'$
$^3E'$	46.57	0.568 $2a_1'2e'$; 0.175 $2a_1'3e'$; 0.067 $2e'3e'$; 0.034 $2e'1a_2'$
$^1A_1'$	47.13	0.420 $2a_1'$; 0.300 $2e'$; 0.126 $2e'3e'$
$^3A_1''$	48.52	0.282 $2e'1e''$; 0.279 $3e'1e''$; 0.246 $1a_2''1a_2'$
$^3E''$	48.87	0.220 $1a_2''3e'$; 0.171 $2e'1a_2''$; 0.168 $2e'1e''$; 0.142 $1e''1a_2'$; 0.097 $3e'1e''$
$^3E''$	51.13	0.290 $2e'1a_2''$; 0.286 $2a_1'1e''$; 0.134 $2e'1e''$; 0.050 $3e'1e''$; 0.045 $1a_2''3e'$
$^1A_1''$	51.16	0.315 $3e'1e''$; 0.253 $1a_2''1a_2'$; 0.242 $2e'1e''$
$^3E'$	51.20	0.260 $2a_1'2e'$; 0.216 $2e'3e'$; 0.183 $2a_1'3e'$; 0.114 $2e'1a_2'$; 0.032 $3e'1a_2'$
$^1E'$	51.22	**0.267 $1a_2''1e''$** ; 0.133 $3e'1a_2'$; 0.126 $2e'1a_2'$; **0.102 $1e''$** ; 0.083 $2e'3e'$; 0.067 $3e'$; 0.029 $2e'$
$^3A_2'$	51.27	0.444 $2e'$; 0.206 $2a_1'1a_2'$; 0.103 $3e'$; 0.051 $2e'3e'$
$^1E''$	51.39	0.212 $1a_2''3e'$; 0.184 $1e''1a_2'$; 0.143 $2e'1a_2''$; 0.139 $2e'1e''$; 0.130 $3e'1e''$
$^3A_2''$	51.44	0.467 $2a_1'1a_2''$; 0.268 $2e'1e''$; 0.072 $3e'1e''$
$^1A_1'$	51.48	0.183 $2e'3e'$; **0.182 $1a_2''$** ; **0.181 $1e''$** ; 0.104 $3e'$; 0.088 $1a_2'$; 0.070 $2e'$
$^1E'$	52.65	**0.289 $1a_2''1e''$** ; **0.118 $1e''$** ; 0.104 $3e'1a_2'$; 0.097 $2e'1a_2'$; 0.066 $3e'$; 0.047 $2a_1'2e'$; 0.046 $2e'$; 0.029 $2e'3e'$
$^1A_1'$	52.84	**0.221 $1e''$** ; 0.208 $2e'3e'$; **0.191 $1a_2''$** ; 0.070 $1a_2'$; 0.060 $3e'$; 0.051 $2a_1'$

continued

Table 8. continued

State	DIP	Composition
$^1A_2'$	53.53	0.536 2e'3e' ; 0.268 2a$_1$'1a$_2$'
$^1E''$	53.83	0.287 2a$_1$'1e'' ; 0.237 2e'1a$_2$'' ; 0.154 2e'1e'' ; 0.059 3e'1e'' ; 0.058 1a$_2$''3e'
$^1E'$	53.96	0.224 2e' ; 0.219 2a$_1$'3e' ; 0.115 2e'1a$_2$' ; 0.105 2a$_1$'2e' ; 0.055 3e'1a$_2$' ; 0.050 3e' ; 0.026 2e'3e'
$^1A_2''$	54.06	0.399 2a$_1$'1a$_2$'' ; 0.309 2e'1e'' ; 0.096 3e'1e''
$^1E'$	57.62	0.395 2a$_1$'2e' ; 0.112 2e'3e' ; 0.105 2a$_1$'3e' ; 0.077 2e' ; **0.019 1a$_2$''1e''** ; 0.012 3e'1a$_2$'
$^1A_1'$	58.03	0.310 2a$_1$' ; 0.282 2e' ; 0.062 2e'3e' ; 0.057 3e' ; **0.020 1e''** ; **0.014 1a$_2$''**

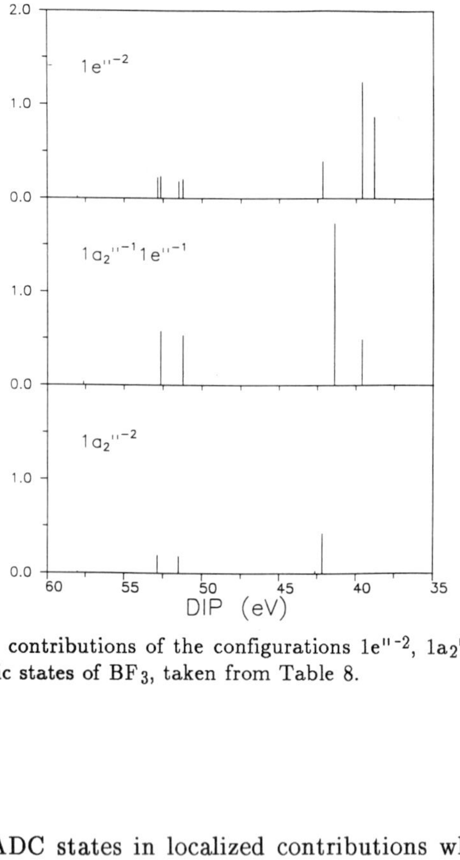

Fig. 5. Bar graph of the contributions of the configurations 1e''$^{-2}$, 1a$_2$''$^{-1}$1e''$^{-1}$ and 1a$_2$''$^{-2}$ to the ADC(2) dicationic states of BF_3, taken from Table 8.

pole strengths of the ADC states in localized contributions which we denote as B^{-2} (two holes on the boron atom), F^{-2} (two holes on the same fluorine atom), $F_1^{-1}F_2^{-1}$ (two holes each on another fluorine), and $B^{-1}F^{-1}$ (one hole on boron and one on a fluorine atom). The corresponding results are reported in Table 9. The analysis confirms fully the occurrence of strong double hole localization phenomena and shows the expected clustering of states with similar hole distribution character. All the states are dominated by the fluorine hole population, the B^{-2} and $B^{-1}F^{-1}$ components being 1 to 2 orders of magnitude smaller and approximately uniform throughout the spectrum. Three distinct groups of states can be identified in the computed energy range, characterized according to their fluorine population: between 37 and 48 eV we

Table 9. DIP (eV) and two-hole population analysis of the Green's function 2h pole strengths for the dicationic states of BF_3.

State	DIP	Population				
		B^{-2}	F^{-2}	$B^{-1}F^{-1}$	$F_1^{-1}F_2^{-1}$	Total
$^1A_1'$	37.90	0.0006	0.0069	0.0240	0.8386	0.8701
$^3E'$	38.49	0.0002	0.0077	0.0463	0.8157	0.8699
$^3E''$	38.50	0.0006	0.0085	0.0388	0.8219	0.8698
$^1E''$	38.54	0.0006	0.0041	0.0404	0.8251	0.8702
$^3A_2'$	38.83	0.0006	0.0000	0.0433	0.8260	0.8698
$^1E'$	39.60	0.0014	0.0045	0.0727	0.7883	0.8668
$^3A_2''$	39.76	0.0012	0.0136	0.0639	0.7864	0.8652
$^1A_1''$	39.81	0.0008	0.0011	0.1060	0.7630	0.8708
$^3A_1''$	39.82	0.0008	0.0035	0.1048	0.7621	0.8711
$^1A_2''$	39.84	0.0013	0.0082	0.0652	0.7902	0.8648
$^1E'$	40.07	0.0015	0.0101	0.0758	0.7784	0.8657
$^3A_2'$	40.44	0.0023	0.0202	0.0851	0.7582	0.8658
$^3E''$	41.01	0.0034	0.0379	0.1032	0.7192	0.8637
$^1E'$	41.08	0.0023	0.0225	0.1371	0.7073	0.8692
$^1E''$	41.17	0.0038	0.0177	0.1114	0.7334	0.8663
$^3E'$	41.41	0.0034	0.0000	0.1478	0.7169	0.8664
$^3A_1''$	41.99	0.0030	0.0154	0.1572	0.6902	0.8658
$^1A_1''$	42.00	0.0032	0.0061	0.1595	0.6978	0.8666
$^3E'$	42.03	0.0017	0.0023	0.1558	0.7063	0.8661
$^1A_1'$	42.17	0.0106	0.0312	0.1385	0.6911	0.8715
$^3E''$	42.46	0.0055	0.0136	0.1684	0.6775	0.8649
$^1E''$	42.49	0.0053	0.0075	0.1669	0.6840	0.8637
$^1A_1'$	42.63	0.0068	0.0329	0.1507	0.6749	0.8652
$^3A_2'$	42.66	0.0050	0.0012	0.1632	0.6940	0.8634
$^1A_2'$	42.68	0.0024	0.0023	0.1452	0.7158	0.8657
$^3A_1'$	42.91	0.0078	0.0000	0.1821	0.6765	0.8650
$^3E'$	43.59	0.0066	0.0079	0.1676	0.6813	0.8634
$^1E'$	43.96	0.0067	0.0235	0.1701	0.6580	0.8583
$^3E''$	43.99	0.0116	0.0291	0.1780	0.6423	0.8609
$^1E''$	44.06	0.0129	0.0149	0.1859	0.6498	0.8635
$^3A_2''$	44.38	0.0107	0.0406	0.1911	0.6156	0.8579
$^1A_2''$	44.65	0.0118	0.0222	0.1950	0.6347	0.8637
$^3A_2'$	44.93	0.0169	0.0416	0.2094	0.5936	0.8615
$^1E'$	45.42	0.0208	0.0204	0.2280	0.5932	0.8623
$^3E'$	46.57	0.0195	0.0457	0.2233	0.5669	0.8554
$^1A_1'$	47.13	0.0223	0.0235	0.2309	0.5790	0.8557
$^3A_1''$	48.52	0.0011	0.6968	0.0928	0.0174	0.8081
$^3E''$	48.87	0.0049	0.6524	0.1142	0.0381	0.8096
$^3E''$	51.13	0.0118	0.5998	0.1816	0.0139	0.8071
$^1A_1''$	51.16	0.0010	0.7119	0.0896	0.0089	0.8113
$^3E'$	51.20	0.0079	0.5741	0.1707	0.0549	0.8076
$^1E'$	51.22	0.0019	0.7056	0.0921	0.0116	0.8112
$^3A_2'$	51.27	0.0112	0.5717	0.1821	0.0421	0.8072
$^1E''$	51.39	0.0041	0.6901	0.1021	0.0155	0.8118
$^3A_2''$	51.44	0.0123	0.5684	0.1921	0.0355	0.8083
$^1A_1'$	51.48	0.0062	0.6822	0.1038	0.0207	0.8130
$^1E'$	52.65	0.0048	0.6946	0.0987	0.0144	0.8125
$^1A_1'$	52.84	0.0085	0.6869	0.1111	0.0081	0.8146
$^1A_2'$	53.53	0.0058	0.6228	0.1743	0.0025	0.8053
$^1E''$	53.83	0.0112	0.5896	0.1889	0.0082	0.7978
$^1E'$	53.96	0.0082	0.5739	0.1841	0.0384	0.8046
$^1A_2''$	54.06	0.0112	0.5764	0.2018	0.0229	0.8123
$^1E'$	57.62	0.0205	0.4814	0.2440	0.0147	0.7605

Table 9. continued

State	DIP	Population				
		B^{-2}	F^{-2}	$B^{-1}F^{-1}$	$F_1^{-1}F_2^{-1}$	Total
$^1A_1'$	58.03	0.0203	0.4848	0.2452	0.0334	0.7838
$^3E'$	62.09	0.0002	0.0015	0.0184	0.0972	0.1173
$^3E'$	62.24	0.0002	0.0021	0.0175	0.0877	0.1075
$^1A_1''$	62.28	0.0015	0.0008	0.0667	0.3028	0.3719
$^1E'$	62.35	0.0019	0.0062	0.0921	0.4249	0.5250
$^3E'$	62.37	0.0011	0.0040	0.0814	0.3806	0.4671
$^1E'$	62.64	0.0007	0.0019	0.0366	0.1685	0.2077
$^1A_2'$	62.64	0.0003	0.0003	0.0162	0.0900	0.1067
$^3A_1''$	62.77	0.0031	0.0008	0.1322	0.5957	0.7317
$^1A_2'$	62.85	0.0006	0.0004	0.0313	0.1805	0.2128
$^1A_1''$	62.86	0.0017	0.0007	0.0733	0.3345	0.4101
$^3E''$	63.09	0.0019	0.0042	0.0638	0.2949	0.3648
$^3E''$	63.15	0.0006	0.0013	0.0242	0.1218	0.1479
$^3A_2'$	63.23	0.0034	0.0049	0.1011	0.5183	0.6277
$^1A_2'$	63.26	0.0012	0.0017	0.0706	0.3446	0.4181
$^3E''$	63.29	0.0006	0.0017	0.0204	0.0869	0.1096
$^1E''$	63.31	0.0015	0.0019	0.0541	0.2637	0.3212
$^3A_1'$	63.43	0.0052	0.0102	0.1278	0.5184	0.6616
$^1E''$	63.53	0.0009	0.0022	0.0309	0.1447	0.1787
$^1E''$	63.54	0.0013	0.0019	0.0365	0.1487	0.1884
$^3E'$	64.10	0.0013	0.0041	0.0330	0.1358	0.1743
$^1A_1'$	64.12	0.0079	0.0098	0.1488	0.5703	0.7369
$^3E'$	64.15	0.0022	0.0139	0.0635	0.2820	0.3616
$^3A_2''$	64.29	0.0018	0.0121	0.0340	0.1368	0.1847
$^3E''$	64.32	0.0020	0.0018	0.0303	0.1070	0.1411
$^1E'$	64.52	0.0052	0.0109	0.1141	0.5050	0.6353
$^1E''$	64.57	0.0103	0.0018	0.1345	0.4638	0.6104
$^3E'$	64.67	0.0013	0.0041	0.0329	0.1296	0.1679
$^3E''$	64.73	0.0019	0.0019	0.0310	0.1233	0.1581
$^3E''$	64.84	0.0013	0.0010	0.0207	0.0778	0.1008
$^1A_2''$	64.91	0.0014	0.0031	0.0198	0.0800	0.1043
$^3E''$	64.91	0.0053	0.0031	0.0726	0.2473	0.3283
$^3A_2''$	64.98	0.0061	0.0308	0.1096	0.3745	0.5210
$^1A_2''$	65.49	0.0023	0.0030	0.0323	0.1112	0.1488
$^1A_2'$	65.55	0.0004	0.0005	0.0412	0.0943	0.1363
$^1A_2''$	65.60	0.0021	0.0025	0.0280	0.0901	0.1227
$^1A_2''$	65.62	0.0056	0.0081	0.0793	0.2880	0.3809
$^3A_2'$	65.68	0.0278	0.0388	0.1379	0.4581	0.6626
$^1A_2'$	65.78	0.0017	0.0015	0.1877	0.4214	0.6122
$^3E'$	65.89	0.0014	0.0183	0.0933	0.2211	0.3341
$^3E'$	65.92	0.0008	0.0133	0.0580	0.1360	0.2081
$^1E'$	66.19	0.0035	0.0043	0.0219	0.0837	0.1133
$^1E'$	66.60	0.0297	0.0099	0.1365	0.4006	0.5766
$^1E'$	67.31	0.0012	0.0081	0.1691	0.3998	0.5782
$^1A_1'$	67.89	0.0131	0.0059	0.0658	0.2079	0.2926
$^1A_1'$	67.98	0.0061	0.0030	0.0312	0.1028	0.1431
$^1A_1'$	68.10	0.0096	0.0054	0.0496	0.1759	0.2406

see states dominated by $F_1^{-1}F_2^{-1}$ character, from 48 to 60 eV by one–site F^{-2}population and finally, beyond 60 eV, the $F_1^{-1}F_2^{-1}$ type states again dominate.

For such an extreme case of double hole localization as BF_3 one cannot of course assume the distribution of total 2h pole strengths of the states to even approximately reproduce the Auger intensity distribution in the fluorine KLL spectrum. The initial core ionized state has one electron vacancy highly localized on one fluorine atom and it is clear that its matrix elements with final states which do not have both holes localized on the same fluorine will be orders of magnitude smaller than for the one–site F^{-2} type states. The two–hole population analysis, providing a consistent measure of the different localization character of the states, suggests immediately to take the F^{-2} pole strengths components as the appropriate weighting factor in the distribution. The gaussian convolution of the vertical DIPs thus obtained (each component with fwhm 1.5 eV) is shown in Fig.6, below the corresponding part of the experimental [33] fluorine Auger spectrum. The figure shows that the procedure gives the desired results, reproducing most details of the Auger profile. The effects of hole localization

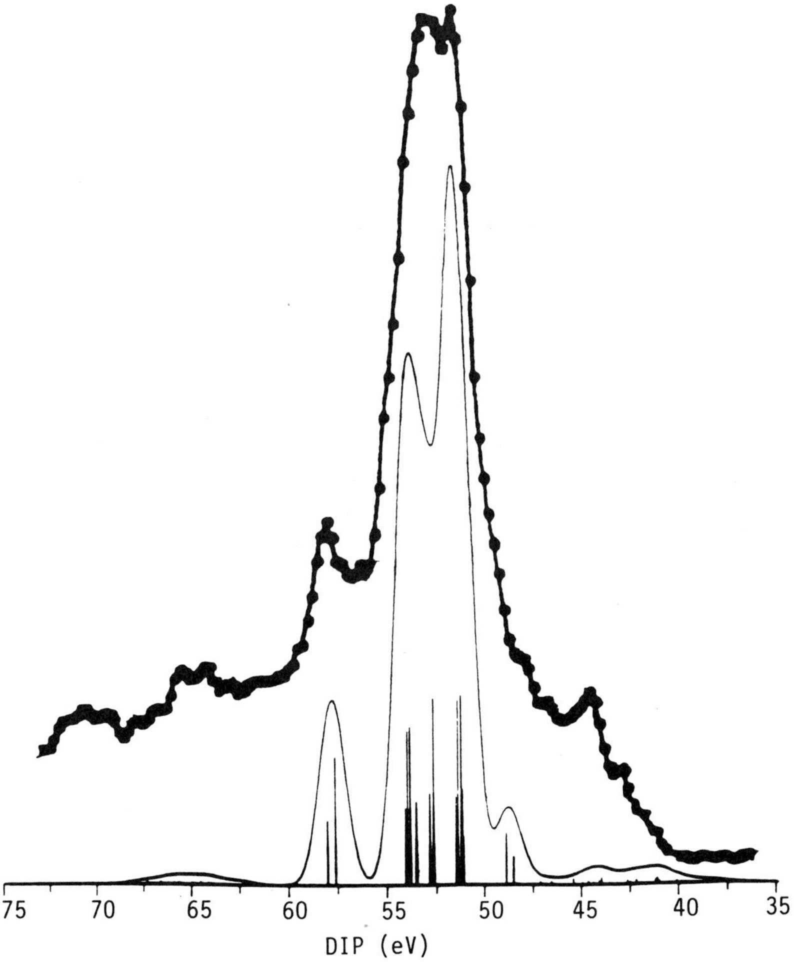

Fig. 6. Experimental (upper) and theoretical (lower) fluorine KLL Auger spectra of BF_3. The theoretical spectrum is a gaussian convolution of the F^{-2} two–hole populations of the ADC(2) pole strengths (shown as vertical bars).

can be visually better appreciated by comparing with Fig.7, where the total 2h pole strength distribution is shown. The selection of the one–site pole strength components correctly reduces the spectral intensity of regions where states dominated by the two–center localized character occur, leaving essentially one intense double peak that matches exactly the experimental lineshape at about 52 eV. Weaker features at both sides are also described in detail. The experimental [33] and theoretical boron KLL spectra, the latter obtained by convolution of the B^{-2} pole strength components, are shown in Fig.8. The boron spectrum appears remarkably more complex than the fluorine one, as expected from the more uniform distribution of the corresponding hole populations. The calculations show clearly the composite nature of the three visible peaks, arising from very many contributions. The relative intensities of the computed peaks show some evident discrepancies with the experiment but it should be considered that the underlying B^{-2} population contributions are very small and, therefore, other rate determining and bandshape factors should play a comparatively large role. For example, the B^{-2} contributions are generally much smaller than the $B^{-1}F^{-1}$ ones. It is evident that the latter character is associated to a strong source of vibrational broadening of the bands which may affect their appearance considerably. The vibrational analysis carried out explicitly in the next example will clearly illustrate this point.

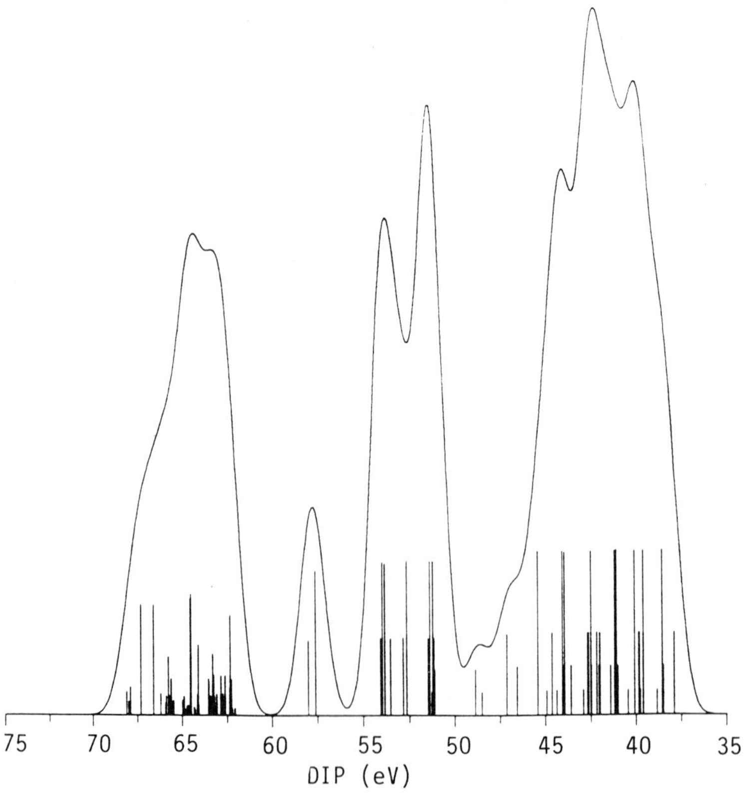

Fig. 7. Gaussian convolution of the ADC(2) total 2h pole strengths for BF_3. Individual state contributions are shown as a bar spectrum.

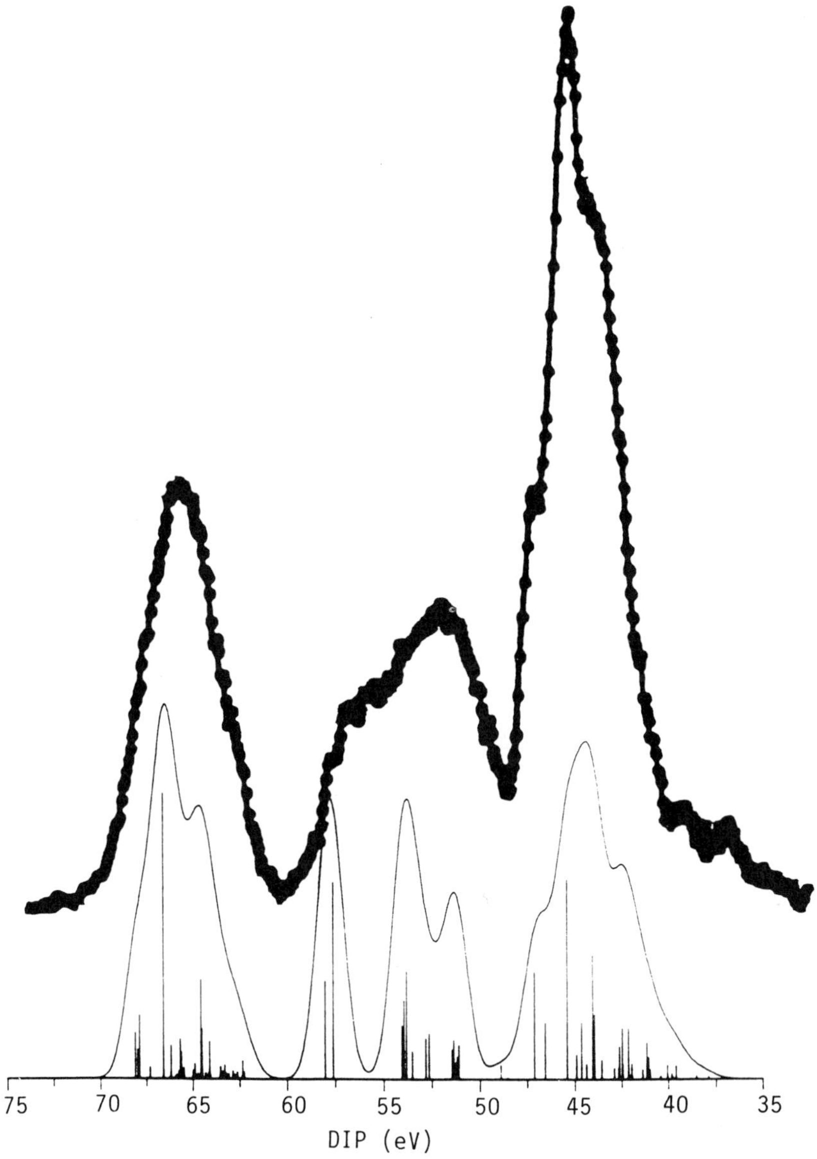

Fig. 8. Experimental (upper) and theoretical (lower) boron KLL Auger spectra of BF_3. The theoretical spectrum is a gaussian convolution of the B^{-2} two–hole populations of the ADC(2) pole strengths (shown as vertical bars).

III.3 – Vibrational effects: carbon monoxide

Despite the small size of the molecule, the carbon and oxygen KLL Auger spectra of CO present many interesting aspects of nontrivial description, and have therefore been repeatedly studied, both experimentally [1a,2,37,56,57] and theoretically [2,16,27b,58–60]. In particular, detailed *ab initio* calculations [2] have shown, by computing Auger vibrational progressions for the lowest lying states, that some very distinctive features of the spectra can only be explained by considering vibrational effects. Striking examples of this are found especially in the carbon spectrum, like the extreme narrowness of the second band, due to the $B^1\Sigma^+$ state and the shape of the first band, which originates from a superposition of the vibrational fine structures of the close lying $X^1\Sigma^+$ and $A^1\Pi$ dicationic states, the latter having a rather broader vibrational envelope than the former. These features constitutes a meaningful and interesting test for our simple method to compute the vibrational envelopes of the Auger bands. Another related point of interest in the calculation of the double ionization spectrum concerns the ordering of the lowest vertical dicationic states of CO [60,61], which are very close lying in energy and give rise to avoided crossings. The reproduction of the correct ordering and local potential energy curve characteristics therefore, requires an adequate account of correlation effects. Finally, with respect to the previous examples of the essentially delocalized holes in the hydrocarbons and the strong localization effects in BF_3, CO represents an intermediate case to verify the applicability of the two–hole population analysis.

Using the ADC(2) method with a triple zeta polarized basis set [62], we have computed [39], and used for generating the Auger spectra, over 140 doubly ionized states of CO having a ionization energy of up to 100 eV. The 50 states with largest 2h weight are reported in Table 10, with their total 2h pole strength and composition. Such a number of states should be compared with the mere 18 independent particle valence dicationic states available for CO, evidencing the general extent of correlation effects in the dication. One notices in Table 10 that, except for the very outermost states, no other main state (and corresponding satellites) can be identified; rather, a substantial breakdown of the two–hole configurations takes place into two or more relevant states. This is true in particular for all the states involving ionization of the inner valence 3σ orbital and, to a lesser extent, for the $4\sigma^{-2}$ configuration. Concerning the computed vertical ordering of the lowest dicationic states, we satisfactorily find that it is in agreement with the interpretation of the Auger spectra [2] and with accurate calculations [2,60,61].

The results of the two–hole population analysis for CO are reported in Table 11, for the same subset of states as in Table 10. We see that in this case many states have a mixed composition, characterized by the fact that both the two–site $C^{-1}O^{-1}$ component and one of the one–site contributions are of comparable magnitude. This is indicative of a partial, but far from complete, delocalization of the two holes which is consistent with the covalent polar character of the bonding in CO. More in particular, we find that large C^{-2} terms (mostly in the very outer valence region) are invariably accompanied by a similar, when not larger, $C^{-1}O^{-1}$ character. For some states, however, a predominance of interatomic localization is observed (i.e., with one hole localized on carbon and one on oxygen), while in other states, particularly those involving 3σ ionization, the two holes appear markedly localized on oxygen alone. In general, we observe that all the states characterized by significant relative values of the one–site population on one atom have a very small relative intra–atomic component on the other atom. This is satisfactorily consistent with the very different appearance of the corresponding Auger spectra of CO.

In the last stage of our calculations, we have computed the derivatives of the double ionization energies of CO at the ground state equilibrium bond distance R_0=1.128 Å [63] (which involved repeating the ADC calculations at two other values of the bond length) and the first and second moments of the Auger vibrational

Table 10. Computed vertical DIPs (eV), total 2h pole strength (PS) and two–hole composition (square coefficients of the leading 2h configurations) for the 50 states of CO^{2+} having largest 2h character. The configurations are indicated by the orbital label of the two holes.

State	Vertical DIP	PS	2h Composition
$^1\Sigma^+$	40.8611	0.834	0.661(5σ) 0.150(4σ5σ) 0.014(4σ) 0.005(3σ5σ) 0.004(1π)
$^3\Pi$	41.1274	0.880	0.876(1π5σ) 0.002(4σ1π) 0.002(3σ1π)
$^1\Pi$	41.6804	0.866	0.859(1π5σ) 0.005(4σ1π) 0.002(3σ1π)
$^3\Sigma^+$	42.9267	0.861	0.853(4σ5σ) 0.006(3σ5σ) 0.002(3σ4σ)
$^1\Sigma$	44.3617	0.771	0.594(4σ5σ) 0.102(5σ) 0.035(4σ) 0.032(1π) 0.009(3σ5σ)
$^3\Sigma^-$	44.9561	0.779	0.779(1π)
$^1\Delta$	46.8040	0.770	0.770(1π)
$^3\Pi$	47.3231	0.764	0.759(4σ1π) 0.005(3σ1π)
$^1\Sigma^+$	48.2805	0.832	0.733(1π) 0.037(4σ) 0.035(4σ5σ) 0.019(5σ) 0.005(3σ) 0.002(3σ4σ)
$^1\Pi$	50.0845	0.711	0.705(4σ1π) 0.006(1π5σ)
$^1\Delta$	51.5148	0.047	0.047(1π)
$^1\Pi$	52.8587	0.074	0.059(4σ1π) 0.013(1π5σ) 0.003(3σ1π)
$^1\Pi$	54.8138	0.039	0.039(4σ1π)
$^1\Sigma^+$	54.8322	0.692	0.582(4σ) 0.044(1π) 0.031(4σ5σ) 0.012(3σ5σ) 0.010(3σ) 0.008(5σ)
$^3\Sigma^+$	55.4322	0.089	0.087(3σ5σ) 0.001(4σ5σ) 0.001(3σ4σ)
$^1\Sigma^+$	57.8854	0.214	0.113(4σ) 0.083(3σ5σ) 0.009(5σ) 0.008(1π) 0.001(4σ5σ)
$^1\Sigma^+$	62.2700	0.229	0.135(3σ5σ) 0.064(5σ) 0.014(4σ5σ) 0.012(3σ4σ) 0.001(1π) 0.001(4σ)
$^3\Sigma^+$	62.3211	0.523	0.511(3σ5σ) 0.009(4σ5σ) 0.002(3σ4σ)
$^1\Sigma^+$	62.4619	0.428	0.352(3σ5σ) 0.037(4σ5σ) 0.017(5σ) 0.017(3σ4σ) 0.004(3σ)
$^3\Pi$	64.5906	0.160	0.158(3σ1π) 0.003(1π5σ)
$^3\Sigma^+$	65.1495	0.158	0.151(3σ5σ) 0.006(3σ4σ) 0.001(4σ5σ)
$^3\Pi$	66.2042	0.531	0.518(3σ1π) 0.014(4σ1π)
$^1\Sigma^+$	66.5391	0.158	0.141(3σ5σ) 0.008(3σ4σ) 0.003(4σ5σ) 0.003(5σ) 0.002(3σ) 0.001(1π)
$^3\Sigma^+$	68.2791	0.487	0.479(3σ4σ) 0.008(3σ5σ)
$^3\Sigma^+$	68.5136	0.222	0.211(3σ4σ) 0.010(3σ5σ)
$^3\Sigma^+$	69.0847	0.034	0.033(3σ4σ) 0.001(3σ5σ)
$^1\Pi$	69.3995	0.205	0.205(3σ1π) 0.001(1π5σ)
$^1\Sigma^+$	69.9944	0.052	0.032(3σ5σ) 0.011(5σ) 0.004(4σ) 0.003(4σ5σ) 0.001(3σ4σ) 0.001(3σ)
$^1\Pi$	70.4208	0.035	0.035(3σ1π)
$^3\Sigma^-$	71.5133	0.040	0.040(1π)
$^1\Sigma^+$	72.9000	0.216	0.204(3σ4σ) 0.005(3σ) 0.004(1π) 0.002(5σ) 0.001(3σ5σ)
$^1\Delta$	72.9555	0.037	0.037(1π)
$^1\Sigma^+$	73.4157	0.464	0.421(3σ4σ) 0.022(3σ5σ) 0.019(3σ) 0.001(4σ) 0.001(1π)
$^1\Pi$	73.7141	0.040	0.029(3σ1π) 0.011(1π5σ)
$^1\Pi$	73.9004	0.143	0.143(3σ1π)
$^1\Pi$	74.3373	0.255	0.254(3σ1π) 0.001(1π5σ) 0.001(4σ1π)
$^1\Sigma^+$	74.6039	0.046	0.039(1π) 0.004(3σ5σ) 0.003(3σ4σ) 0.001(4σ5σ)
$^1\Sigma^+$	74.7382	0.046	0.037(3σ4σ) 0.006(1π) 0.002(5σ) 0.001(3σ5σ)
$^1\Pi$	75.8885	0.066	0.066(3σ1π) 0.001(4σ1π)
$^1\Sigma^+$	79.4410	0.030	0.014(3σ5σ) 0.010(3σ) 0.005(3σ4σ) 0.001(4σ5σ) 0.001(1π)
$^1\Pi$	81.0910	0.036	0.030(4σ1π) 0.004(1π5σ) 0.002(3σ1π)
$^1\Sigma^+$	92.3553	0.193	0.172(3σ) 0.007(3σ5σ) 0.005(3σ4σ) 0.003(4σ5σ) 0.003(4σ) 0.002(1π)
$^1\Sigma^+$	92.7614	0.164	0.140(3σ) 0.012(3σ4σ) 0.007(4σ) 0.001(4σ5σ) 0.001(5σ) 0.001(3σ5σ)
$^1\Sigma^+$	93.0741	0.086	0.081(3σ) 0.002(3σ4σ) 0.001(3σ5σ) 0.001(1π)
$^3\Sigma^-$	93.2986	0.031	0.031(1π)
$^1\Delta$	94.6544	0.033	0.033(1π)
$^1\Sigma^+$	99.4792	0.044	0.033(3σ) 0.006(3σ4σ) 0.004(3σ5σ) 0.001(5σ) 0.001(4σ5σ)
$^1\Sigma^+$	99.5073	0.156	0.144(3σ) 0.005(1π) 0.004(3σ5σ) 0.002(3σ4σ)
$^1\Pi$	100.3142	0.032	0.029(4σ1π) 0.003(1π5σ)
$^1\Sigma^+$	100.5553	0.053	0.045(3σ) 0.004(4σ5σ) 0.002(1π) 0.001(3σ4σ) 0.001(4σ)

Table 11. Two hole atomic population analysis of the 2h components for the dicationic states of CO reported in Table 10.

State	Vertical DIP	Population C^{-2}	O^{-2}	$C^{-1}O^{-1}$
$^1\Sigma^+$	40.8611	0.421	0.001	0.412
$^3\Pi$	41.1274	0.213	0.018	0.649
$^1\Pi$	41.6804	0.212	0.011	0.643
$^3\Sigma^+$	42.9267	0.087	0.000	0.788
$^1\Sigma^+$	44.3617	0.381	0.021	0.370
$^3\Sigma^-$	44.9561	0.048	0.440	0.291
$^1\Delta$	46.8040	0.048	0.435	0.288
$^3\Pi$	47.3231	0.030	0.484	0.250
$^1\Sigma^+$	48.2805	0.080	0.466	0.286
$^1\Pi$	50.0845	0.029	0.446	0.236
$^1\Delta$	51.5148	0.003	0.027	0.018
$^1\Pi$	52.8587	0.008	0.033	0.034
$^1\Pi$	54.8138	0.001	0.025	0.013
$^1\Sigma^+$	54.8322	0.016	0.454	0.222
$^3\Sigma^+$	55.4322	0.022	0.007	0.060
$^1\Sigma^+$	57.8854	0.037	0.082	0.095
$^1\Sigma^+$	62.2700	0.069	0.000	0.164
$^3\Sigma^+$	62.3211	0.143	0.005	0.375
$^1\Sigma^+$	62.4619	0.215	0.000	0.217
$^3\Pi$	64.5906	0.009	0.092	0.059
$^3\Sigma^+$	65.1495	0.027	0.023	0.109
$^3\Pi$	66.2042	0.040	0.280	0.212
$^1\Sigma^+$	66.5391	0.038	0.000	0.122
$^3\Sigma^+$	68.2791	0.013	0.252	0.222
$^3\Sigma^+$	68.5136	0.001	0.140	0.081
$^3\Sigma^+$	69.0847	0.000	0.020	0.014
$^1\Pi$	69.3995	0.012	0.117	0.076
$^1\Sigma^+$	69.9944	0.033	0.008	0.011
$^1\Pi$	70.4208	0.002	0.020	0.013
$^3\Sigma^-$	71.5133	0.003	0.023	0.015
$^1\Sigma^+$	72.9000	0.016	0.158	0.042
$^1\Delta$	72.9555	0.002	0.021	0.014
$^1\Sigma^+$	73.4157	0.016	0.360	0.088
$^1\Pi$	73.7141	0.003	0.020	0.017
$^1\Pi$	73.9004	0.009	0.082	0.053
$^1\Pi$	74.3373	0.017	0.139	0.099
$^1\Sigma^+$	74.6039	0.003	0.024	0.019
$^1\Sigma^+$	74.7382	0.004	0.031	0.010
$^1\Pi$	75.8885	0.005	0.035	0.026
$^1\Sigma^+$	79.4410	0.005	0.012	0.013
$^1\Pi$	81.0910	0.002	0.021	0.013
$^1\Sigma^+$	92.3553	0.016	0.102	0.075
$^1\Sigma^+$	92.7614	0.018	0.080	0.066
$^1\Sigma^+$	93.0741	0.007	0.047	0.033
$^3\Sigma^-$	93.2986	0.002	0.018	0.012
$^1\Delta$	94.6544	0.002	0.019	0.013
$^1\Sigma^+$	99.4792	0.011	0.014	0.020
$^1\Sigma^+$	99.5073	0.011	0.086	0.059
$^1\Pi$	100.3142	0.001	0.021	0.010
$^1\Sigma^+$	100.5553	0.003	0.023	0.026

envelope for each state, according to the expressions derived in Chapter II. In the latter, we have used the experimental [63] ground state frequency, ω, of CO and the derivatives, κ, of the core–hole states computed in ref [2]. The slopes of the most important dicationic states and their resulting vibrational center of gravity and gaussian width for both the carbon and oxygen Auger spectra are listed in Table 12. The centers of gravity are reported as energy shifts with respect to the double ionization energies computed by ADC at R_0. The widths in the table have been computed with the Γ parameter of Eq. (31) (accounting for the core–hole lifetimes and the experimental resolution) set equal to zero, to show the purely vibrational broadening. The table evidences some remarkable features, which are worthwhile to discuss in some detail and are amenable to generalization. Firstly, we notice that all the dicationic state slopes, except for the lowest $^3\Sigma^+$ and the $B^1\Sigma^+$ states, are negative, in obvious agreement with the general expectation of bond lengthening upon removal of two valence electrons. But it is indeed the unusual positive value found for the $B^1\Sigma^+$ state, in agreement with previous calculations [2,12,60], which is particularly relevant in view of its consequences on the carbon Auger spectrum: as is well known, C1s ionization in CO also leads to a shortening of the bond distance, with a computed slope at R_0 of 3.973 eV/bohr [2]. This means that the difference in slope between the C1s hole state and the $B^1\Sigma^+$ final state, and thus the vibrational width of the latter in the carbon Auger, will be particularly small. Notice that, as previously found [2,58] and consistently with its large C^{-2} population (see Table 11), the $B^1\Sigma^+$ state contributes significantly to the carbon spectrum. As observed, the lowest $^3\Sigma^+$ state also has a positive, though smaller, slope. However, since the latter state is a triplet and, in addition, has a computed much smaller C^{-2} population, its expected contribution to the Auger spectrum is marginal.

As Table 12 shows, the computed vibrational widths of the Auger peaks vary substantially from state to state from a fraction of eV to several eV, depending on the relative values of the final state slopes. One can see that, in general, the widths tend to increase in the inner valence part of the spectrum, both as a consequence of the ionization of the more strongly bonding electrons and of the increased density of states. This occurrence should be of general relevance in molecular Auger spectroscopy and, together with the increase in density of states itself and vibronic coupling effects, is the major cause of the substantial broadening of the spectra and lack of sharp peaks as normally observed experimentally in this region.

Inspection of the computed center of gravity shifts in Table 12 leads immediately to some interesting conclusions. The positive κ value of the C1s hole state, in combination with the negative slope computed for the dicationic states (with the two mentioned exceptions), has the consequence that all the computed vibrationally induced shifts of the C Auger bands are positive. By contrast, O1s ionization causes a slight bond length increase and the computed derivative of the O1s hole state at R_0 is –1.380 eV/bohr [2]. This value is smaller in magnitude than all the negative final state slopes (except for the $X^1\Sigma^+$), consequently all but three of the computed center of gravity shifts for the oxygen spectrum are negative. Because of the small κ value for oxygen, we also observe that both the shifts and the widths are smaller in size than for carbon. As a general rule, if we can reasonably expect many of the valence dicationic state slopes to be negative and large compared to the core–hole state slopes, the vibrational center of gravity shifts for those states will have the same sign as κ (see Eqs.(28) and (29)). Therefore, the effect of the shifts on the different appearance of the Auger spectra of a system will tend to be particularly enhanced in those cases, such as carbon monoxide, where different intermediate core–hole states have opposite slopes. Even more relevant is the magnitude of the shifts for the various final states in a given spectrum, which affects the absolute energy position of the peaks and, more important, their separation. One can indeed notice in Table 12 that the shifts can vary substantially over the spectrum and, for large κ values such as that for the carbon core–hole, can reach values of several eV. These quantitative findings

Table 12. Computed center of gravity shift (eV) and gaussian fwhm (eV) for the Auger bands of the states of CO^{2+} reported in Tables 10 and 11. The centers of gravity are obtained by adding the shifts to the shown vertical DIPs (eV). Also shown are the computed bond distance slopes (eV/bohr) at the neutral ground state equilibrium distance.

State	Vertical DIP	Slope	CARBON Shift	CARBON Fwhm	OXYGEN Shift	OXYGEN Fwhm
$^1\Sigma^+$	40.8611	−0.7804	0.5681	1.1839	0.0249	0.0989
$^3\Pi$	41.1274	−5.0714	1.0809	2.2526	−0.1532	0.6090
$^1\Pi$	41.6804	−5.3280	1.1115	2.3166	−0.1639	0.6513
$^3\Sigma^+$	42.9267	0.4444	0.4217	0.8788	0.0757	0.3010
$^1\Sigma^+$	44.3617	2.2407	0.2070	0.4314	0.1503	0.5973
$^3\Sigma^-$	44.9561	−14.8757	2.2525	4.6945	−0.5602	2.2263
$^1\Delta$	46.8040	−14.6825	2.2294	4.6464	−0.5522	2.1945
$^3\Pi$	47.3231	−9.9259	1.6610	3.4617	−0.3547	1.4098
$^1\Sigma^+$	48.2805	−13.9418	2.1409	4.4619	−0.5214	2.0723
$^1\Pi$	50.0845	−11.0212	1.7919	3.7345	−0.4002	1.5905
$^1\Delta$	51.5148	−12.8862	2.0148	4.1990	−0.4776	1.8981
$^1\Pi$	52.8587	−16.3836	2.4327	5.0701	−0.6228	2.4751
$^1\Pi$	54.8138	−3.9630	0.9484	1.9766	−0.1072	0.4261
$^1\Sigma^+$	54.8322	−10.5026	1.7299	3.6054	−0.3787	1.5049
$^3\Sigma^+$	55.4322	−12.2037	1.9332	4.0290	−0.4493	1.7855
$^1\Sigma^+$	57.8854	−10.8122	1.7669	3.6825	−0.3915	1.5560
$^1\Sigma^+$	62.2700	−11.5794	1.8586	3.8735	−0.4234	1.6825
$^3\Sigma^+$	62.3211	−11.6455	1.8665	3.8900	−0.4261	1.6934
$^1\Sigma^+$	62.4619	−11.9339	1.9010	3.9618	−0.4381	1.7410
$^3\Pi$	64.5906	−20.0979	2.8766	5.9952	−0.7770	3.0878
$^3\Sigma^+$	65.1495	−18.0635	2.6335	5.4885	−0.6925	2.7522
$^3\Pi$	66.2042	−17.7063	2.5908	5.3996	−0.6777	2.6933
$^1\Sigma^+$	66.5391	−11.5423	1.8542	3.8643	−0.4218	1.6764
$^3\Sigma^+$	68.2791	−12.0661	1.9168	3.9948	−0.4436	1.7628
$^3\Sigma^+$	68.5136	−9.4630	1.6057	3.3464	−0.3355	1.3334
$^3\Sigma^+$	69.0847	−7.0238	1.3142	2.7389	−0.2343	0.9310
$^1\Pi$	69.3995	−19.8889	2.8516	5.9432	−0.7683	3.0533
$^1\Sigma^+$	69.9944	−3.7196	0.9193	1.9159	−0.0971	0.3859
$^1\Pi$	70.4208	−15.4312	2.3189	4.8329	−0.5833	2.3180
$^3\Sigma^-$	71.5133	−24.8333	3.4425	7.1746	−0.9735	3.8690
$^1\Sigma^+$	72.9000	−16.6243	2.4615	5.1301	−0.6328	2.5148
$^1\Delta$	72.9555	−25.1878	3.4849	7.2629	−0.9883	3.9275
$^1\Sigma^+$	73.4157	−17.0952	2.5178	5.2474	−0.6523	2.5925
$^1\Pi$	73.7141	−13.8095	2.1251	4.4290	−0.5159	2.0504
$^1\Pi$	73.9004	−20.5741	2.9335	6.1138	−0.7967	3.1663
$^1\Pi$	74.3373	−19.9735	2.8618	5.9642	−0.7718	3.0673
$^1\Sigma^+$	74.6039	−19.7989	2.8409	5.9208	−0.7646	3.0385
$^1\Sigma^+$	74.7382	−14.0899	2.1586	4.4988	−0.5276	2.0967
$^1\Pi$	75.8885	−16.1984	2.4106	5.0240	−0.6151	2.4445
$^1\Sigma^+$	79.4410	−17.2222	2.5330	5.2790	−0.6576	2.6134
$^1\Pi$	81.0910	−13.8360	2.1283	4.4356	−0.5170	2.0548
$^1\Sigma^+$	92.3553	−15.6032	2.3395	4.8757	−0.5904	2.3463
$^1\Sigma^+$	92.7614	−17.5556	2.5728	5.3620	−0.6714	2.6684
$^1\Sigma^+$	93.0741	−21.3069	3.0211	6.2963	−0.8272	3.2872
$^3\Sigma^-$	93.2986	−20.7116	2.9500	6.1481	−0.8025	3.1890
$^1\Delta$	94.6544	−20.7169	2.9506	6.1494	−0.8027	3.1899
$^1\Sigma^+$	99.4792	−21.1878	3.0069	6.2667	−0.8222	3.2676
$^1\Sigma^+$	99.5073	−20.9471	2.9781	6.2067	−0.8122	3.2279
$^1\Pi$	100.3142	−22.4127	3.1532	6.5717	−0.8731	3.4697
$^1\Sigma^+$	100.5553	−18.7222	2.7122	5.6526	−0.7199	2.8609

have obvious general implications and should suggest great caution in the usual interpretations of Auger spectra based on vertical electronic transition energies only. Neglect of vibrational effects can lead to *qualitatively* incorrect conclusions, for example in the identification of corresponding peaks in different spectra of the same system (once the appropriate core binding energy is subtracted) as due to the same final states.

To assess more quantitatively the results of our approach, we have finally drawn the theoretical carbon and oxygen Auger spectra of CO, using for each state a gaussian function of area proportional to its degeneracy and to its respective C^{-2} and O^{-2} pole strength population. The origin and fwhm of the gaussians have been taken according to the computed vibrational parameters, but in the widths we have now included, according to Eq.(31), a resolution parameter Γ fixed at 0.2 eV. This quantity accounts for the finite lifetime of the core–hole (0.1 eV) [64] and for the experimental resolution. As usual in our procedure, the low transition rate of the triplet states has been simply accounted for by reducing their computed intensity by 2/3, having again noted that our results are essentially insensitive to fine–tuning this ratio. The resulting spectra are displayed in Fig. 9B and 10B for carbon and oxygen, respectively, together with the experimental spectral profiles taken from ref. [56] and reported on the double ionization energy scale. The striking general agreement between the theoretical and experimental bandshapes in the figures does not need to be emphasized. For comparison purposes, we have also drawn in Fig. 9A and 10A the corresponding distributions of the one–site pole strengths computed without consideration of vibrational effects, i.e., using gaussians centered at the vertical ADC DIPs and of width fixed at an average value of 2 eV. The comparison makes visually evident the effects of the varying vibrational broadening and shift of the bands, especially remarkable, as expected, in the carbon spectrum.

The qualitative assignment of the spectra derivable from our calculations coincides essentially with that of previous studies [2,58]. We would like, however, to illustrate a few more quantitative considerations afforded by our spectra. The accuracy achieved in the carbon spectrum is eyecatching, particularly for the first two bands (at low DIP). The first theoretical band peaks at 41.5 eV, to be compared with the experimentally deduced value of 41.7 eV [2]. In agreement with the vibrational calculations of Correia et al. [2], the band is mainly attributed to a narrow $X^1\Sigma^+$ component which is broadened on the left by the $A^1\Pi$ state. We notice from Table 12 and Fig. 9A that, besides the difference in fwhm between the two components, it is only by accounting for their different vibrational shift, which increases their separation with respect to the vertical DIPs by more than 0.5 eV, that the exact bandshape is reproduced. The shift is also responsible for the accuracy of the computed absolute peak position, as can be seen by comparing with the ADC DIP for $X^1\Sigma^+$ of 40.86 eV. Incidentally, we note that this latter value confirms the observed photoionization appearance threshold for CO^{2+} of 40.75 eV [12]. Similar accuracy is obtained for the second band, attributed to the $B^1\Sigma^+$ state. The fingerprinting sharpness of this band, due to the discussed narrow vibrational envelope, is exactly reproduced, although its absolute energy position is somewhat underestimated. The agreement in the computed width indicates that this latter discrepancy is not due to a too small computed vibrational shift and we should, therefore, attribute it to an underestimation of the vertical DIP which is found in the ADC(2) as well as in other *ab initio* calculations [2,58].

As anticipated, the inner part of the spectrum is much more complex, with many states contributing and intensities generally smearing out. In this situation it is obviously expected that the exact Auger bandshapes not only depend on their vibrational structure, but also on electron correlation terms not fully accounted for in the present calculation (correlation effects are very strong in this energy range and difficult to assess properly) and vibronic coupling effects. Also, the expressions we have derived in Chapter II may become inadequate on a detailed scale, due to avoided crossings among the final states. Very interestingly, however, our calculations show that the straightforward estimation of the major vibrational effects we propose can already be sufficient to improve some theoretical results considerably. Notice, in particular, the prominent feature at 62.5 eV exhibited by the spectrum computed without vibrational effects (Fig. 9A). This is clearly to be associated with the broad experimental band with maximum at about 65 eV but, apart from the evident inadequacy of the fixed width, it appears unsatisfactorily low in energy. As Fig. 9B shows, this discrepancy is almost entirely recovered by accounting for the very large vibrational shift of the band maximum, which is now computed at 64.3 eV.

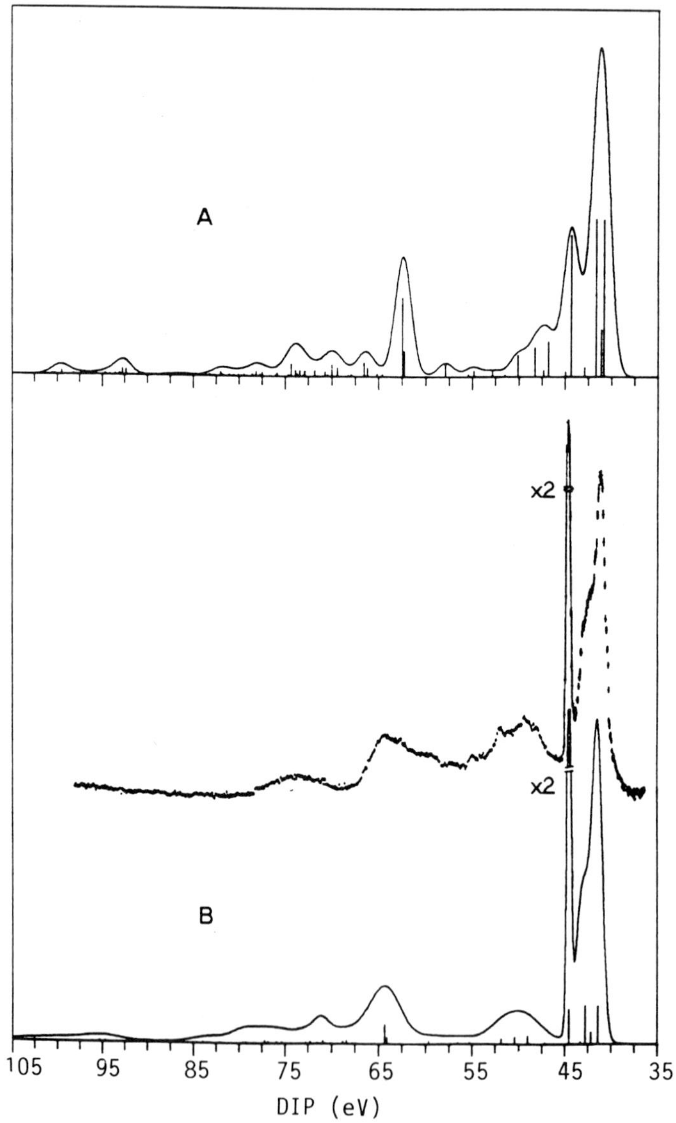

Fig. 9. (A): Gaussian convolution of the C^{-2} populations of the dicationic states of CO at fixed fwhm (2 eV). The individual components are centered at the vertical DIP and shown as a bar spectrum. (B): Experimental (upper) and theoretical (lower) carbon KLL Auger spectra of CO. The computed vibrational width and energy shift of each component have been used in the gaussian convolution.

The main features of the oxygen spectrum are also very well described by our calculations and in this case, as anticipated and can be seen by comparing figures 10A and 10B, the effects of the vibrational broadening and shift are generally far less dramatic than in the carbon spectrum. In the following we compare the computed DIP appearance of the bands with experimental values obtained by subtracting the O1s ionization energy of 542.1 eV [1a] from the Auger kinetic energies extracted from the high resolution spectrum in Fig. 2 of ref. [2]. Exactly as for the carbon spectrum, the first peak is computed at 41.5 eV, to be compared with the observed value of 41.0 eV. Its intensity, mainly due to the $A^1\Pi$ state, appears lower than in the recorded spectrum because of a too small computed O^{-2} weight, especially for the very narrow $X^1\Sigma^+$ component. This may also be the cause of the slight discrepancy in the peak

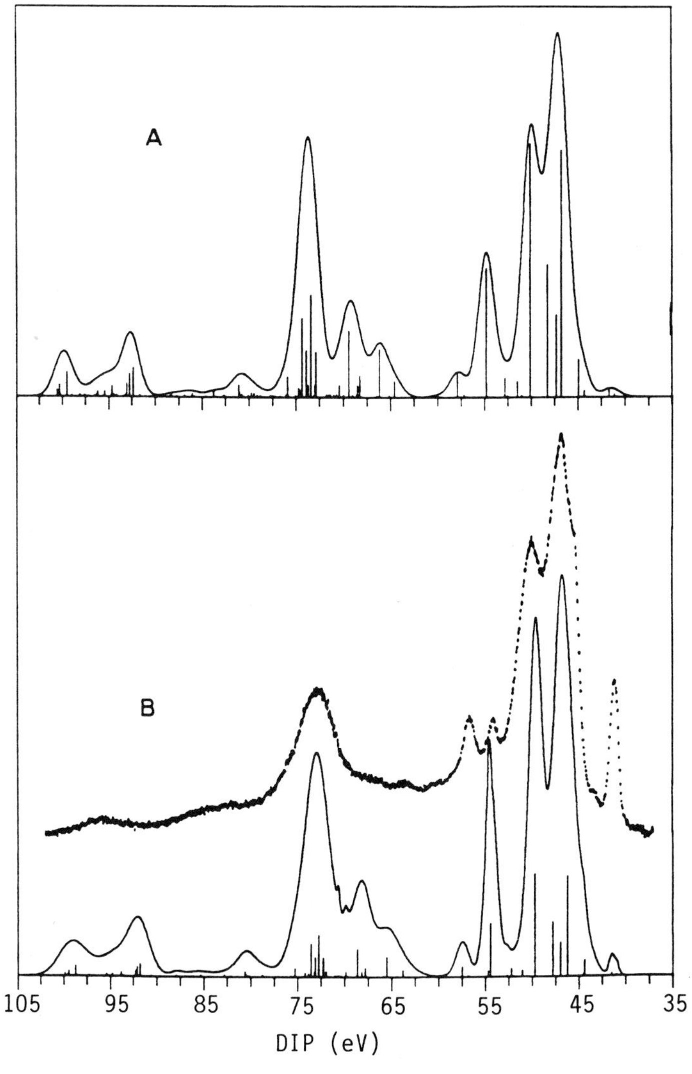

Fig. 10. (A): Gaussian convolution of the O^{-2} populations of the dicationic states of CO at fixed fwhm (2 eV). The individual components are centered at the vertical DIP and shown as a bar spectrum. (B): Experimental (upper) and theoretical (lower) oxygen KLL Auger spectra of CO. The computed vibrational width and energy shift of each component have been used in the gaussian convolution.

position. In this respect, we mention that in the calculations we have carried out using a smaller [4s,2p,1d] basis set [48], we have obtained a somewhat different two–hole population distribution for the lowest four dicationic states, finding higher O^{-2} and correspondingly lower $C^{-1}O^{-1}$ components. In particular, the smaller basis gives an oxygen intensity ratio between $A^1\Pi$ and $X^1\Sigma^+$ of 100:23, in agreement with the value of 100:20 found in ref.[2], and a total intensity for the first peak 3 times larger than in Fig. 10A, yielding an overall better bandshape. This finding reflects apparently the occasional instability of Mulliken–like analyses to the inclusion of more diffuse functions in the basis set. Notice that the correct composition of the first peak, using either basis sets, is again obtained through consideration of the vibrational shift which, contrary to the carbon case, brings the $X^1\Sigma^+$ and $A^1\Pi$ components closer to each other relative to the vertical energies (see Table 12). The two close lying most intense bands of the spectrum contain many significant electronic contributions. Despite this complexity, their profile is very satisfactorily reproduced. In particular, the energy position of the two maxima, computed at 46.8 and 49.7 eV, respectively, matches nearly exactly the values of 46.7 and 49.6 eV extracted from the experiment [2,1a].

The two experimental peaks at 53.5 and 56.2 eV are attributed, as in previous studies [2,58,59], to two states mainly originating from $4\sigma^{-2}$ ionization. It appears clearly, however, that the split of this configuration is unsatisfactorily described at this level of theory, the lowest DIP state having a too larger 2h weigth than the other. The need for an account of correlation effects involving higher final state excitations is evidenced also by the unusual inaccuracy in the energy position of the two peaks, computed at 54.7 and 57.5 eV, respectively.

Our results indicate a very pronounced breakdown of Auger intensity over many states occurring in the energy region above 60 eV, with double ionizations involving one inner valence and one outer valence electrons, or two inner valence electrons. The accuracy of the spectral profile we obtain in this region, as Fig. 10B shows, is probably very close to the limit one can achieve without inclusion of vibronic coupling effects. The most prominent broad feature of the spectrum, is centered at about 72.7 eV in the experimental spectrum and is computed at 72.9 eV. Comparison with Fig. 10A shows again that the good accuracy obtained even in this region is due to consideration of the substantial vibrational shifts and widths. We emphasize again that this finding should be seen in the context of a statistical approach: the high density of states produces effects that may invalidate the approximations used for some *individual* components; nevertheless, the overall appearance of the composite bands is expected to be of good accuracy since the spectral parameters are averaged over many states. Furthermore, a comparison between the computed and experimental broadening of the bands indicates that the intensity breakdown is more pronounced than computed. This is particularly evident above 80 eV, where our calculations still produce some distinct bands whereas an essentially featureless spectrum is observed. In the higher energy range higher excited configurations, as well as more complete basis sets, are needed for a more accurate description.

Acknowledgements

This work has been supported by the "Progetto finalizzato: Materiali speciali per tecnologie avanzate" of CNR (Rome) and by the Deutsche Forschungsgemeinschaft.

References

[1] (a) K.Siegbahn, C.Nordling, G.Johansson, J.Hedman, P.F.Heden, K.Hamrin, U.Gelius, T.Bergmark, L.O.Werme, R.Manne and Y.Baer, *"ESCA Applied to Free Molecules"* (North-Holland, Amsterdam, 1971); (b) M.Thompson, M.D.Baker, A.Christie and J.F.Tyson, *"Auger Electron Spectroscopy"* (Wiley, New York, 1985).

[2] N.Correia, A.Flores-Riveros, H.Ågren, K.Helenelund, L.Asplund and U.Gelius, J.Chem.Phys. **83**, 2035 (1985). See also: T.X.Carroll and T.D.Thomas, J.Chem.Phys. **86**, 5221 (1987).

[3] L.Karlsson, P.Baltzer, S.Svensson and B.Wannberg, Phys.Rev.Lett. **24**, 2473 (1988); S.Svensson, L.Karlsson, P.Baltzer, M.P.Keane and B.Wannberg, Phys.Rev.A **40**, 4369 (1989).

[4] A.Cesar, H.Ågren, A.Naves de Brito, S.Svensson, L.Karlsson, M.P.Keane, B.Wannberg, P.Baltzer, P.G.Fournier and J.Fournier, J.Chem.Phys. **93**, 918 (1990).

[5] W.Eberhardt, E.W.Plummer, I.W.Lyo, R.Reininger, R.Carr, W.K.Ford and D. Sondericker, Aust.J.Phys. **39**, 633 (1986).

[6] D.A.Lapiano-Smith, C.I.Ma, K.T.Wu and D.M.Hanson, J.Chem.Phys. **90**, 2162 (1989).

[7] J.Appell, J.Durup, F.C.Fehsenfeld and P.G.Fournier, J.Phys.B **6**, 197 (1973); P.G. Fournier, J.Fournier, F.Salama, D.Stärck, S.D.Peyerimhoff and J.H.D.Eland, Phys.Rev.A **34**, 1657 (1986).

[8] See, e.g., M.Hamdan, S.Mazumdar, V.R.Marathe, C.Badrinathan, A.G.Brenton and D.Mathur, J.Phys.B **21**, 2571 (1988) and references therein.

[9] R.G.Cooks, T.Ast and J.H.Beynon, Int.J.Mass Spectrom.Ion Phys. **11**, 490 (1973); D.Mathur and C.Bandrinathan J.Phys.B **20**, 1517 (1987) and references therein.

[10] G.Dujardin, S.Leach, O.Dutuit, P.M.Guyon and M.Richard-Viard, Chem.Phys. **88**, 339 (1984); G.Dujardin, L.Hellner, D.Winkoun and M.J.Besnard, Chem.Phys. **105**, 291 (1986).

[11] P.Lablanquie, I.Nenner, P.Millié, P.Morin, J.H.D.Eland, M.-J.Hubin-Franskin and J.Delwiche, J.Chem.Phys. **82**, 2951 (1985); P.Lablanquie, J.H.D.Eland, I.Nenner, P.Morin, J.Delwiche and M.-J.Hubin-Franskin, Phys.Rev.Lett. **58**, 992 (1987); J.H.D.Eland, S.D.Price, J.C.Cheney, P.Lablanquie, I.Nenner and P.G.Fournier, Phil.Trans.R.Soc.Lond.A **324**, 247 (1988).

[12] P.Lablanquie, J.Delwiche, M.-J.Hubin-Franskin, I.Nenner, P.Morin, K.Ito, J.H.D.Eland, J.-M.Robbe, G.Gandara, J.Fournier and P.G.Fournier, Phys.Rev.A **40**, 5673 (1989).

[13] K.Faegri, jr. and H.P.Kelly, Phys.Rev.A **19**, 1649 (1979); M.Higashi, E.Hiroike and T.Nakajima, Chem.Phys. **68**, 377 (1982), Chem.Phys. **85**, 133 (1984); F.P.Larkins and J.A.Richards, Aust.J.Phys. **39**, 809 (1986).

[14] V.Carravetta and H.Ågren, Phys.Rev.A **35**, 1022 (1987); R.Colle and S.Simonucci, Phys.Rev.A **42**, 3913 (1990); K.Zähringer, H.-D.Meyer and L.S.Cederbaum, "Molecular scattering wavefunctions for Auger decay rates: the Auger spectrum of hydrogen fluoride", Phys.Rev.A (submitted for publication).

[15] A.Cesar, H.Ågren and V.Carravetta, Phys.Rev.A **40**, 187 (1989).

[16] H.Ågren, J.Chem.Phys. **75**, 1267 (1981).

[17] (a) L.S.Cederbaum and W.Domcke, Adv.Chem.Phys. **36**, 205 (1977); (b) W.von Niessen, J. Schirmer and L.S.Cederbaum, Comput.Phys.Rep. **1**, 57 (1984); (c) L.S.Cederbaum, Int.J.Quant.Chem.Symp. **24**, 393 (1990).

[18] J.Schirmer and A.Barth, Z.Phys.A **317**, 267 (1984).

[19] A.Tarantelli and L.S.Cederbaum, Phys.Rev.A **39**, 1656 (1989).

[20] A.Tarantelli and L.S.Cederbaum, Phys.Rev.A **39**, 1639 (1989).

[21] J.Schirmer, Phys.Rev.A **43**, 4647 (1991).

[22] F.Tarantelli, A.Tarantelli, A.Sgamellotti, J.Schirmer and L.S.Cederbaum, Chem.Phys.Lett. **117**, 577 (1985); F.Tarantelli, A.Tarantelli, A.Sgamellotti, J.Schirmer and L.S.Cederbaum, J.Chem.Phys. **83**, 4683 (1985); F.Tarantelli, J.Schirmer, A.Sgamellotti and L.S.Cederbaum, Chem.Phys.Lett. **122**, 169 (1985).

[23] F.Tarantelli, A.Sgamellotti, L.S.Cederbaum and J.Schirmer, J.Chem.Phys. **86**, 2201 (1987).

[24] E.Ohrendorf, H.Köppel, L.S.Cederbaum, F.Tarantelli and A.Sgamellotti, J.Chem.Phys. **91**, 1734 (1989).

[25] E.Ohrendorf, F.Tarantelli and L.S.Cederbaum, J.Chem.Phys. **92**, 2984 (1990).

[26] F.Tarantelli, A.Sgamellotti and L.S.Cederbaum, J.Chem.Phys. **94**, 523 (1991).

[27] (a) C.-M.Liegener, Chem.Phys.Lett. **90**, 188 (1982); (b) C.-M.Liegener, Chem.Phys. Letters **106**, 201 (1984); (c) C.-M.Liegener, Chem.Phys. **92**, 97 (1985)

[28] R.L.Graham and D.L.Yeager, J.Chem.Phys. **94**, 2884 (1991).

[29] J.Schirmer, Phys.Rev.A **26**, 2395 (1982); J.Schirmer, L.S.Cederbaum and O.Walter, Phys.Rev.A **28**, 1237 (1983).

[30] A.L.Fetter and J.D.Walecka, *"Quantum Theory of Many–Particle systems"* (McGraw–Hill, New York, 1971).
[31] R.R.Rye and J.E.Houston, J.Chem.Phys. **78**, 4321 (1983).
[32] J.V.Ortiz, J.Chem.Phys. **81**, 5873 (1984).
[33] M.Cini, F.Maracci and R.Platania, J.Electr.Spectr.Rel.Phen. **41**, 37 (1986).
[34] V.Pellizzari, F.Tarantelli, A.Sgamellotti and L.S.Cederbaum, to be published.
[35] L.S.Cederbaum, F.Tarantelli, A.Sgamellotti and J.Schirmer, J.Chem.Phys. **85**, 6513 (1986); **86**, 2168 (1987).
[36] R.R.Rye, D.R.Jennison and J.E.Houston, J.Chem.Phys. **73**, 4867 (1980); T.D. Thomas and P.Weightman, Chem.Phys.Lett. **81**, 325 (1981); P.Weightman, T.D. Thomas and D.R.Jennison, J.Chem.Phys. **78**, 1652 (1983).
[37] J.A.Kelber, D.R. Jennison and R.R.Rye, J.Chem.Phys. **75**, 652 (1981);
[38] W.Eberhardt, E.W.Plummer, C.T.Chen, W.K.Ford, Aust.J.Phys. **39**, 853 (1986).
[39] L.S.Cederbaum, P.Campos, F.Tarantelli and A.Sgamellotti, "Bandshape and vibrational structure in Auger spectra: theory and application to carbon monoxide", J.Chem.Phys. (submitted for publication).
[40] F.Kaspar, W.Domcke and L.S.Cederbaum, Chem.Phys. **44**, 33 (1979).
[41] F.K.Gel'mukhanov, L.N.Mazalov, A.V.Nikolaev, A.V.Kondratenko, V.G.Smirnii, P.I.Wadash and A.P.Sadovskii, Dokl.Akad.Nauk SSSR **225**, 597 (1975); F.K.Gel'mukhanov, L.N.Mazalov, and A.V.Kondratenko, Chem.Phys.Letters **46**, 133 (1977).
[42] H.Köppel, W.Domcke and L.S.Cederbaum, Adv.Chem.Phys. **57**, 59 (1984).
[43] H.C.Longuet–Higgins, U.Öpik, M.H.L.Pryce and R.A.Sack, Proc.Roy.Soc.(London) **A244**, 1 (1958); R.Englman, *"The Jahn–Teller Effect"* (Wiley, New York, 1972).
[44] G.Wentzel, Z.Physik **43**, 521 (1927).
[45] R.Manne and H.Ågren, Chem.Phys. **93**, 201 (1985).
[46] As originally suggested in H.Siegbahn, L.Asplund, P.Kelfve, Chem.Phys.Lett. **35**, 330 (1975).
[47] T.A.Green and D.R.Jennison, Phys.Rev.B **36**, 6112 (1987).
[48] T.H.Dunning, J.Chem.Phys. **53**, 2823 (1970); R.Ahlrichs and P.R.Taylor, J.Chim. Phys. **78**, 315 (1981).
[49] C.Benoit and J.A.Horsley, Mol.Phys. **30**, 557 (1975).
[50] R.R.Rye, T.E.Madey, J.E.Houston and P.H.Holloway, J.Chem.Phys. **69**, 1504 (1978).
[51] D.R.Jennison, Chem.Phys.Letters, **69**, 435 (1980).
[52] F.Tarantelli, A.Sgamellotti and L.S.Cederbaum, to be published.
[53] M.Thompson, P.A.Hewitt and D.S.Wooliscroft, Anal.Chem. **48**, 1336 (1976).
[54] R.Spohr, T.Bergmark, N.Magnusson, L.O.Werme, C.Nordling and K.Siegbahn, Phys.Scr. **2**, 31 (1970).
[55] A.Denis, J.Langlet and J.P.Malrieu, Theor.Chim.Acta, **38**, 49 (1975); L.S.Cederbaum and W.Domcke, J.Chem.Phys. **66**, 5084 (1977).
[56] W.E.Moddeman, T.A.Carlson, M.O.Krause, B.P.Pullen, W.E.Bull and G.K. Schweitzer, J.Chem.Phys. **55**, 2317 (1971).
[57] L.Ungier and T.D.Thomas, J.Chem.Phys. **82**, 3146 (1985).
[58] H.Ågren and H.Siegbahn, Chem.Phys.Letters 72, 498 (1980).
[59] I.H.Hillier and J.Kendrick, Mol.Phys. 31, 849 (1976); D.R.Jennison, J.A.Kelber and R.R.Rye, Chem.Phys. Letters 77, 604 (1981); G.E.Laramore, Phys.Rev.A 29, 23 (1984).
[60] M.Larsson, B.J. Olsson and P.Sigray, Chem.Phys. **139**, 457 (1989), and references therein.
[61] V.R.Marathe and D.Mathur, Chem.Phys.Letters **163**, 189 (1989).
[62] T.H.Dunning, J.Chem.Phys. 55, 716 (1971).
[63] K.P.Huber and G.Herzberg, *"Molecular Spectra and Molecular Structure IV. Constants of Diatomic Molecules"* (Van Nostrand Reinhold, New York, 1979).
[64] M.Tronc, G.C.King, R.C.Bradford and F.H.Read, J.Phys.B 9, L555 (1976).

CALCULATION OF PHOTOIONIZATION CROSS SECTION: AN OVERVIEW

I. Cacelli
Scuola Normale Superiore, Pisa Italy

V. Carravetta and A. Rizzo
I. C. Q. E. M. del C. N. R. ,Pisa Italy

R. Moccia
Dipartimento di Chimica e Chimica Industriale, Pisa Italy

1 Introduction

The electron emission is one of the most efficient decay mechanisms for atoms or molecules excited above the ionization threshold. This makes the knowledge of reliable photoionization cross sections very valuable in a variety of fields. Their importance has grown also for the development of the laser sources and of the synchrotron radiation which have greatly extended both the range of photon energies and the number of photons involved in the processes.

This short review is concerned with the theoretical methods suitable for the calculation of photoionization cross sections due to one- or two-photon absorption. Although several thecniques are considered, particular attention is dedicated to those which rely upon the use of L^2 bases only since they are the most suitable to take advantage of the general codes developed to calculate the electronic structure of atoms and molecules.

The processes that will be taken into consideration are (atomic units, a.u., used throughout)

$$S_{\{0\}} + \omega_1 \longrightarrow S_j^+ + e^-$$
$$S_{\{0\}} + \omega_1 + \omega_2 \longrightarrow S_j^+ + e^-$$

where $S_{\{0\}}$ indicates the atom or the molecule, initially in the state $|0\rangle$, which, after absorbing one or two photons of energy ω_1 (ω_2), yields a parent ion S_j^+ in the state $|j\rangle$, plus a photoelectron with energy ϵ and moment $\vec{k}_e$ ($\epsilon = k_e^2/2$). Although both the states $|0\rangle$ and $|j\rangle$ of the matter and the states of the photons are arbitrary, we will assume $|0\rangle$ to coincide with the ground state.

To not encumber the presentation with too many details the following approximations are adopted throughout:

- Frozen nuclei approximation: all the formulae refer to the Vertical Transition Approximation. In order to allow comparison with gas phase experiments the calculated quantities are orientationally averaged.

Applied Many-Body Methods in Spectroscopy and Electronic Structure
Edited by D. Mukherjee, Plenum Press, New York, 1992

- Dipole approximation and Lowest Order Perturbation Theory (*LOPT*) for the matter radiation interaction.
- For the two-photon processes, photon energies sufficiently far from any excitation energy of the system (off intermediate resonances).

In the following all equations are referred to the orientationally averaged molecular case; the corresponding quantities for the atomic cases, where the orientational average is substituted by an average upon the degenerate components of the initial level and a sum upon the degenerate components of the final level, may be easily derived from them.

2 The cross sections expression

Both the one-photon and the two-photon differential cross sections, once orientationally averaged, are expressible as linear combinations of products of purely geometrical factors times molecular parameters which embody all the relevant characteristics of the molecule [1].

With a_0 and t_0 denoting the a.u. of length and time and α the fine structure constant, the One Photon Differential Ionisation Cross Section may be written as

$$\frac{\partial\sigma^{(1)}_{j,\epsilon,\hat{k}_e\leftarrow\{0\}}(\omega_1,\mu_1)}{\partial\Omega_{\hat{k}_e}} = (2\pi)^2 a_0^2\alpha\omega_1^g \times\frac{1}{12\pi}\sum_J(-1)^J(2J+1)C(1,J,1;-\mu_1,0)P_J(\cos\theta_{\hat{k}_e})F_{j,J}(\epsilon) \tag{2.1}$$

while the Two Photon Differential Ionisation Cross Section is expressible as

$$\frac{\partial\sigma^{(2)}_{j,\epsilon,\hat{k}_e\leftarrow\{0\}}(\omega_1,\mu_1,\omega_2,\mu_2;\beta)}{\partial\Omega_{\hat{k}_e}} = (2\pi)^3 a_0^4 t_0\alpha^2\omega_1^g\omega_2^g \times\sum_{J,M}\sum_L\sum_{j_1,j_2} A_{\mu_1,\mu_2}(L,J,M;j_1,j_2)F_{j,j_1,j_2,J}(\omega_1,\omega_2)Y_{L,0}(\beta,0)Y_{J,M}(\hat{k}_e). \tag{2.2}$$

The resonance condition requires that

$$\epsilon+E_j=\omega_1+E_{\{0\}}$$

for the one photon processes and

$$\epsilon+E_j=\omega_1+\omega_2+E_{\{0\}}$$

for the two-photon processes. The P_J , $Y_{L,M}$ and C are Legendre polynomials, spherical harmonics and Clebsch-Gordan coefficients respectively. The $F_{j,J}$ molecular parameters of eq.(2.1) are defined by

$$F_{j,J}(\epsilon) = \sum_{l_a,l_b}\sum_{m_a,m_b}\sum_{\mu_a,\mu_b} f_J^{(1)}(l_a,m_a,\mu_a,l_b,m_b,\mu_b)\,\mathrm{m}^{(1)\,*}_{j,\epsilon,l_a,m_a,\leftarrow\{0\}}(\mu_a)\mathrm{m}^{(1)}_{j,\epsilon,l_b,m_b,\leftarrow\{0\}}(\mu_b).$$

In the last expression the $f_J^{(1)}$ are products of $3j$ coupling coefficients while the

$$\mathrm{m}^{(1)}_{j,\epsilon,l,m,\leftarrow\{0\}}(\mu) = \langle\Phi_{\{0\}}|\hat{O}_\mu|\Phi^{(-)}_{j,\epsilon,l,m}\rangle \tag{2.3}$$

are the transition moments from the initial state $|\Phi_{\{0\}}\rangle$ to the final state in the continuum $|\Phi^{(-)}_{j,\epsilon,l,m}\rangle$ labelled by its partial wave (l,m) and channel (j) indices, characterized by the energy ϵ of the photoelectron and obeying the boundary conditions appropriate to photoionization.

The asymptotic behaviour of the continuum states, describing a single ionized parent ion plus a photoelectron, may be expressed in terms of the "partial wave channel" (*pwc*) functions $\varphi_{E,j,\epsilon,l,m}$. Assuming that the electrostatic hamiltonian is adequate to our purposes, only states of definite spin multiplicity, equal to that of the initial state, are considered. Thus if Φ^+_{j,M'_S} is the state of the parent ion

$$\phi_j(x_1, x_2, ..., x_{N-1}, \sigma_N) = \sum_{m_S} C(S_j, 1/2, S; M_S - m_S, m_S)\ \Phi^+_{j,M_S-m_S}(x_1, x_2, ..., x_{N-1}, \sigma_N) \mid 1/2, m_S\rangle$$

is the spin coupled ion state, and the global *asymptotic* state corresponding to a specific *pwc*, identified by j, l, m, is given by

$$\varphi_{E,j,\epsilon,l,m}(x_1, ..., x_N) = C\ \mathcal{A}\phi_j(x_1, x_2, ..., x_{N-1}, \sigma_N)u_{j,\epsilon,l,m}(r_N)Y_{l,m}(\hat{r}_N) \tag{2.4}$$

with

$$E = E^+_j + \epsilon\ ,$$

C a suitable normalization constant and $\mathcal{A}$ the antisymmetrizer. It is convenient to indicate by a single letter the *pwc* state labels j, ϵ, l, m: lower case roman for open ($\epsilon > 0$) channels, capital for open and closed ($\epsilon < 0$) channels. For ($r \to \infty$) the one-electron radial orbitals $u_a(r)$ corresponding to positive energies ϵ should be expressible as shifted Coulomb waves, i.e.

$$u_a(r) \underset{r\to\infty}{\Longrightarrow} \cos[\Delta_a(k_a)]\ \{F_{l_a,\epsilon}(k_a r) + tg[\Delta_a(k_a)]G_{l_a,\epsilon}(k_a r)\}$$

where

$$\frac{k_a^2}{2} = \epsilon = E - E_a > 0\ .$$

Thus the asymptotic behaviour of an open *pwc* state will be

$$\varphi_{E_a}(x_1, ..., x_{N-1}, \sigma_N, \vec{r}_N) \underset{r_N\to\infty}{\Longrightarrow} \frac{1}{\sqrt{N}}\phi_{j_a}(x_1, ..., x_{N-1}, \sigma_N) \times Y_{l_a,m_a}(\hat{r}_N)\sqrt{\frac{2}{\pi}}\frac{1}{\sqrt{k_a}}\frac{1}{r_N}\sin\ [\theta_a(r_N) + \Delta_a(k_a)].$$

with

$$\theta_a(r) = k_a r + \frac{1}{k_a}\ln(2k_a r) + \sigma_{l_a}(k_a) - l_a\frac{\pi}{2},$$

$\sigma_{l_a}(k_a)$ the Coulomb wave phase shift and $\Delta_{j_a,l_a,m_a}(k_a)$ being the additional phase shift of the actual wave due to the short-range interaction.

The $\mathrm{m}^{(1)}_{j,\epsilon,l,m,\leftarrow\{0\}}(\mu)$ are computed in the molecular frame and the transition operator $\hat{O}_\mu$, matter counterpart of the interaction hamiltonian, refers to a photon whose polarization, in spherical basis, is denoted by μ. The selection of the transition operator employed (length, velocity etc.) has a bearing upon the value of the exponent g of the photon frequency ω of eqs. (2.1) and (2.2); for instance the Velocity Gauge (VG) requires $g = -1$ while the Length Gauge (LG) requires $g = +1$. $\theta_{\hat{k}_e}$ indicates the angle between the photon polarization and the photoelectron momentum.

In eq.(2.2) $\hat{k}_e$ defines the photoelectron direction in the laboratory frame and β is the angle between the polarization (linear or circular) directions of the two photons which, without loss of generality, have been chosen the first directed along the z axis and the second lying in the xz plane of the laboratory frame. The $A_{\mu_1,\mu_2}(L, J, M; j_1, j_2)$ are purely geometrical factors while the $F_{j,j_1,j_2,J}(\omega_1, \omega_2)$ are molecular parameters whose rather lengthy expression may be found elsewhere [1]. Here we limit ourself to say that, as expected, they are quadratic forms of in general all possible two-photon transition amplitudes computed in the molecular frame

$$\mathrm{m}^{(2)}_{j,\epsilon,l,m\leftarrow\{0\}}(\omega_1,\mu_1,\omega_2,\mu_2) = \langle\Phi_{\{0\}}|\hat{O}^{(2)}_{\mu_1,\mu_2}(\omega_1,\omega_2)|\Phi^{(-)}_{j,\epsilon,l,m}\rangle \ . \tag{2.5}$$

The two-photon transition operator $\hat{O}_{\mu_1,\mu_2}(\omega_1,\omega_2)$ appearing in eq. (2.5), is a complicate object which may be expressed in term of the matter causal resolvent $\hat{G}^{+}_{M}(E_0+\omega)$ as [1]

$$\hat{O}^{(2)}_{\mu_1,\mu_2}(\omega_1,\omega_2) = (1+\mathcal{P}_{1,2})\hat{O}_{\mu_1}\hat{G}^{+}_{M}(E_0+\omega_2)\hat{O}_{\mu_2}. \tag{2.6}$$

The integral cross sections obtained by integrating eqs. (2.1) and (2.2) take the following expressions

$$\sigma^{(1)}_j(\omega_1) = \frac{4\pi^2}{c}\omega_1^g\frac{1}{3}F_{j,0}(\epsilon) = \frac{4\pi^2}{c}\omega_1^g\frac{1}{3}\mathcal{G}^{(1)}_j$$

and

$$\sigma^{(2)}_{j,\mu_1,\mu_2,\beta}(\omega_1,\omega_2) = (2\pi)^3\frac{\omega_1^g\omega_2^g}{c^2}\mathcal{B}^{(2)}_{j,\mu_1,\mu_2,\beta}(\omega_1,\omega_2).$$

where

$$\begin{aligned}\mathcal{B}^{(2)}_{j,\mu_1,\mu_2,\beta}(\omega_1,\omega_2) &= \sum_{l,m}(-1)^{m+\mu_2}\frac{\sqrt{4\pi}}{2l+1}C(1,1,l;\mu_1,m)^2\\ &\times\mathcal{G}^{(2)}_{j,l}(\omega_1,\omega_2)\sum_{\lambda}\frac{1}{\sqrt{2\lambda+1}}C(1,1,\lambda;-\mu_2,\mu_2)C(1,1,\lambda;-m,m)Y_{\lambda,0}(\beta,0)\ .\end{aligned}$$

Thus they require the evaluation of the somehow simpler molecular parameters

$$\mathcal{G}^{(1)}_j = \sum_{L,M,\mu} |\langle\Phi_{\{0\}}\mid\hat{O}_{\mu}\mid\Phi_{j,\epsilon,L,M}\rangle|^2 \tag{2.7}$$

and

$$\mathcal{G}^{(2)}_{j,l}(\omega_1,\omega_2) = \sum_{L,M,m'} |\sum_{\mu'} C(1,1,l;\mu',m'-\mu')\mathrm{m}^{(2)}_{j,\epsilon,L,M\leftarrow\{0\}}(\omega_1,\mu',\omega_2,m'-\mu')|^2 \tag{2.8}$$

which, as it appears from eqs. (2.7) and (2.8), are positive definite and do not depend upon the final *pwc*'s phase shifts $\Delta_{j_a,L,M}(k)$; see eq. (3.5.7). These properties make the $\sigma^{(1)}_j(\omega)$ and, with a slightly more involved procedure, $\sigma^{(2)}_{j,\mu_1,\mu_2,\beta}(\omega_1,\omega_2)$ amenable to a Stieltjes Imaging approach as discussed in sec.3.6.

The above expressions stress the pivotal role played by the transition matrix elements (2.3) and (2.5) involving states in the continuum; they may be calculated by several techniques which are ultimately characterized by the use either of a complete projection on L^2 basis sets or of truly continuum orbitals. Of these two general approaches, while the latter should in principle afford a more accurate evaluation of the properties pertaining to the continuum, the former has some definitive advantage over the traditional scattering methods. Indeed the algorithms that describe on the same basis both bound and continuum orbitals are reduced to a sequence of matrix operations which can be efficiently implemented on modern supercomputers. Moreover the L^2 methods can easily allow different levels of approximation in the calculations which, on the other hand, may be carried out by the well established computer codes already developed to treat bound-state problems. Even at the lowest level of approximation, realistic multi-center non-local potentials like the Hartree-Fock (HF) potential can be easily handled for the description of the electron-molecule(ion) interaction.

The lack of spherical symmetry in molecules curtails the possibility of applying straightforwardly all the methods developed and extensively used to study the photoionization of atoms like the Close Coupling method [2] or the many-body methods for the continuum [3]. Thus there is a need to develop for molecules particular approaches which may take advantage of what has been done for atoms.

Among the methods which resort to the direct calculation of states in the continuum, exploiting variational or other techniques [4, 5], the most extensively applied are: the Iterative Schwinger Variational Method ($ISVM$) [6, 7] , the Multiple Scattering Xα Method ($MS - X\alpha$) [8] and the Linear Algebraic Method (LAM) [9, 10].

As opposed to these techniques that are based on the numerical solution of a set of differential or integral equations, purely L^2 techniques allow to exploit completely the advantages of the quantum chemical algorithms. For instance, the inclusion of coupling between ionization channels is simply obtained by including appropriate configuration states representing different channels in the hamiltonian matrix.

The purely L^2 methods which have been more extensively applied are: the Stieltjes moment theory [11], the Complex Basis Method [12] and the Reactance or K-matrix method [13]. The first method is not directly concerned with the computation of the states in the continuum since its aim is to extract the oscillator strength density in the energy region properly belonging to the continuum, from transition probabilities obtained by an L^2 calculation. Thus it may be applied either to the results yielded by separate calculations of the initial and final states or to those arising from a direct evaluation of the transition density matrices. In fact there are techniques which evaluate directly the transition density matrix based upon the several versions of the Equation of Motion method (EOM), the simplest being the Random Phase Approximation (RPA) [14]. For atomic cases continuum solutions of the RPA equations have already been obtained [15]. As it will be briefly sketched in sec. (3.7), a method based on the K-matrix approach may be adopted to yield RPA transition densities in the molecular continuum employing L^2 basis sets [1].

3 Transition Amplitude in the Continuum; One Photon

In this section we will briefly describe the methods mentioned above, with a little more emphasis on the K-matrix and Stieltjes Imaging techniques on which we have gained more direct experience . [16-24] The discussion will be concerned with the one-photon processes while the peculiarities of the more complicated transition operator (2.6) for the two-photon case will be considered in section (4).

3.1 Multiple Scattering Xα Method

The $MS - X\alpha$ method allows the determination of bound [25, 26] and continuum [8, 26, 27] electronic wavefunctions in the independent particle model.

In the Xα approximation [28] the non-local exchange potential due to each atomic charge distribution is replaced by the corresponding local Xα potential

$$V_{X\alpha}(r) = -3\alpha\,[3\rho(r)/8\pi]^{1/3}$$

where α is a statistical semiempirical parameter whose value approaches 0.7 and ρ is the spherical atomic density charge, which may be fixed or self-consistently determined. Each atom of the molecule is enclosed in a sphere of suitable radius depending on its nuclear charge and the entire molecule is also enclosed in a larger sphere. Inside the tangent atomic spheres the potential is that of the atomic hamiltonian including exchange in the Xα approximation, plus a spherically averaged correction due to the other atoms. Outside the larger sphere the potential is taken to be spherical, while in the interstitial regions the potential is volume averaged. In each space portion so delimited the orbitals are expanded in suitable sets of functions or grid points and, by imposing continuity of the orbital and of its first derivative at the borders of the spheres (where the potential shows a finite discontinuity) a set of secular equations is obtained for a chosen outgoing electron kinetic energy. This constraint allows the evaluation of the K-matrix (see sec. 3.5) among the stationary coulombic waves in the external region, from which suitable incoming waves are finally obtained.

Thus the MS-Xα method is essentially an independent particle approach to photoionization in which an important role is played by the space partition and by the exchange parameter α, as well as by the set of partial waves considered in the expansion of the wavefunction. The commonly employed muffin-tin potential is well known for its deficiencies, i.e. constant potential in the interstitial regions and spherical potential inside the atomic spheres. Some efforts were made in the past to overcome these lacks by introducing overlapping sphere potentials which reduce the volume of the interstitial regions where the assumption of a constant potential is physically incorrect [29, 30]. It as been shown that this brings to a sensible improvement of the computed ionization potentials as well as of the photoionization cross section results [29]. A multipole expansion of the potential inside the atomic spheres and outside the outer sphere eliminates the second deficiency, although it leads to a sensible increase of the computational effort [26]. It is perhaps for this reason that no practical continuum states implementation is present in the literature to our knolewdge.

To facilitate the calculation of the dipole transition moments and to preserve the translational invariance of the transition operators, orthogonality between initial and final single-determinant states is requested. Although the initial and final state orbitals should be in principle determined from different potentials, it is convenient the use of an unique effective local potential so that orthogonality might be automatically achieved [31]. Most authors have opted for an (N-1/2) electrons potential, a reasonable compromise between the initial N and final (N-1) electrons potential. This has been realized by reducing the occupation number of the orbital to be ionized from 1 to 0.5 [29, 32, 33] .

In spite of these deficiencies the $MS - X\alpha$ method has been widely applied to large molecules for which more elaborate techniques are hard to be implemented. Test calculations have shown that other general methods in which the continuum orbitals are described by plane or Coulombic waves are well inferior to the $MS - X\alpha$ method [34] where the continuum orbitals are determined by a realistic, even if approximated, multi-center potential. Although a rather poor quantitative agreement with experiment is generally found, this method is capable of reproducing important features in the photoionization valence [32, 35, 36] and core [33, 37, 38, 39] spectra as shape resonances and Cooper minima. Approach to photoionization of adsorpted diatomics on metal clusters was also reported [40]. Whithout doubts it can be affirmed that the MS-Xα method, when low accuracy is considered satisfactory or when no other method appears to be affordable, is the most versatile technique in use to compute photoionization cross sections and asymmetry parameters for a large class of molecules.

3.2 Iterative Schwinger Variational Method

The $ISVM$, based on an iterative extension of the original Schwinger Variational principle [41, 42], was introduced in the early eighties by McKoy and coworkers [43, 44, 45] as a tool for the numerical evaluation of electron-molecule, electron-molecular ion scattering amplitudes and differential and integral photoionization cross sections in linear molecules. Since then the literature on the *ISVM* has become vast. We restrict our interests to the developments concerning the molecular photoionization, neglecting the wealth of applications to the electron-molecule and electron-molecular ion scattering. In this respect the method has been recently expanded and applied to the calculation of photoionization cross sections and asymmetry parameters of CH_4 [46], H_2O [47] and Cl_2 [48]. We also do not wish to enter the discussion of extensions (like the Multichannel Schwinger Variational Method [49, 50, 51] or the two-potential technique for the treatment of the long range and short range segments of the scattering potential [45]) which to date appear to have been only seldom applied to the calculation of molecular photoionization cross sections.

The details of the method can be found in the original papers. We outline here the main points. Instead of dealing with the Static-Exchange Schrödinger equation

$$\{-\frac{1}{2}\nabla^2 + V_{N-1} - \epsilon\}\Psi_{\vec{k}}^{(-)} = 0 \tag{3.2.1}$$

for the continuum orbitals, the *ISVM* starts with the equivalent integral representation, i.e. the Lippmann-Schwinger equation

$$\Psi_{\vec{k}}^{(-)} = \Psi_{\vec{k}}^{c\,(-)} + G^{c\,(-)}U\Psi_{\vec{k}}^{(-)} \,. \tag{3.2.2}$$

In eqs. 3.2.1 and 3.2.2 V_{N-1} represents the Static Exchange potential of the ion, $\epsilon = k^2/2$ is the energy associated to the continuum orbital, U is the short range portion of the Static Exchange potential ($U = 2V_{N-1} + 2/r$), $G^{c\,(-)}$ is the Coulomb Green's function and $\Psi_{\vec{k}}^{c\,(-)}$ is the Coulomb scattering function. $\Psi_{\vec{k}}^{c\,(-)}$ is often given in terms of its partial wave expansion

$$\Psi_{\vec{k}}^{c\,(-)}(\vec{r}) = (2/\pi)^{1/2}\sum_{l,m} i^l\phi_{klm}^{c\,(-)}(\vec{r})Y_{lm}^*(\Omega_{\hat{k}})$$

where the partial wave Coulomb functions $\phi_{klm}^{c\,(-)}$ have been introduced

$$\phi_{klm}^{c\,(-)}(\vec{r}) = (\frac{1}{kr})e^{-i\sigma_l}F_l(\gamma; kr)Y_{lm}(\Omega_{\hat{r}}) \,.$$

$F_l(\gamma; kr)$ is the regular Coulomb function [52] with $\gamma = -1/k$ while σ_l is the Coulomb phase shift. Notice that the dependence on the nuclear coordinates has been neglected.

An analogous partial wave expansion can be written for the continuum wave function $\Psi_{\vec{k}}^{(-)}(\vec{r})$, representing the ejected electron

$$\Psi_{\vec{k}}^{(-)}(\vec{r}) = (2/\pi)^{1/2}\sum_{l=0}^{l_p}\sum_{m=-l}^{+l} i^l\psi_{klm}^{(-)}(\vec{r})Y_{lm}^*(\Omega_{\hat{k}})$$

with the infinite summation truncated to some convenient upper value l_p. The Lippmann-Schwinger equation 3.2.2 for the partial wave states thus becomes

$$\psi_{klm}^{(-)}(\vec{r}) = \phi_{klm}^{c\,(-)}(\vec{r}) + \langle\vec{r} \mid G^{c\,(-)}U \mid \psi_{klm}^{(-)}\rangle \,. \tag{3.2.3}$$

In the *ISVM* this equation is solved iteratively. The short range potential is approximated by a separable potential of the form

$$U^{S_0} = \sum_{\eta_i,\eta_j \in R} U \mid \eta_i\rangle[U^{-1}]_{ij}\langle\eta_j \mid U = U \mid \boldsymbol{\eta}\rangle\langle\boldsymbol{\eta}U\boldsymbol{\eta}\rangle^{-1}\langle\boldsymbol{\eta} \mid U \tag{3.2.4}$$

with R indicating some initial expansion functions set. Inserting eq. 3.2.4 in eq. 3.2.3 yields

$$\psi_{klm}^{(-)\ S_0}(\vec{r}) = \phi_{klm}^{c\ (-)}(\vec{r}) + \sum_{\eta_i,\eta_j \in R} \langle \vec{r} \mid G^{c\ (-)} U \mid \eta_i \rangle [D^{-1}]_{ij} \langle \eta_j \mid U \mid \phi_{klm}^{c\ (-)} \rangle \qquad (3.2.5)$$

where

$$D_{ij} = \langle \eta_i \mid U - U G^{c\ (-)} U \mid \eta_j \rangle \ .$$

The expansion set R is then augmented by the set of functions

$$S_0 = \{\psi_{kl_1m}^{S_0}, \psi_{kl_2m}^{S_0}, \dots, \psi_{kl_pm}^{S_0}\}$$

obtained as solutions of eq. 3.2.5 . With this augmented set a second set S_1 can be obtained, and the process is continued until the wave functions satisfy some suitable convergence criteria. The set of solutions at the n-th iteration, S_n, is obtained from the set S_{n-1} as

$$\psi_{klm}^{(-)\ S_n}(\vec{r}) = \phi_{klm}^{c\ (-)}(\vec{r}) + \sum_{\chi_i,\chi_j \in R \cup S_{n-1}} \langle \vec{r} \mid G^{c\ (-)} U \mid \chi_i \rangle [D^{-1}]_{ij} \langle \chi_j \mid U \mid \phi_{klm}^{c\ (-)} \rangle \qquad (3.2.6)$$

It has been proved that the converged wave functions are the solutions of the Lippmann-Schwinger equation for the exact potential U [53]. To our knowledge, all the applications of the $ISVM$ to date use the Frozen Core HF approximation for the final state, described as a single determinant with the ionic core orbitals identical to those of the neutral ground state. In order to mantain at all steps during the iteration an orthogonality between $\psi_{klm}^{(-)}(\vec{r})$ and the frozen-core molecular orbitals, a Phillips-Kleinman type of pseudopotential is generally used in the $ISVM$ applications to treat the Static-Exchange interaction. The initial state wave function, which is needed in the computation of the transition amplitudes, has been chosen in some selected applications of the method as to partially include correlation effects, mostly via CI or $MCSCF$-type expansions [44].

In practical implementations the matrix elements arising in solving eq. 3.2.6 were computed by resorting to a single center expansion approach for all the functions involved (continuum orbital, occupied orbitals, potential U, $1/r_{12}$ and $G^{c\ (-)}$). This means that all the functions were expanded about a common center as a sum of products of spherical harmonics times radial functions. The radial integrals were numerically evaluated by putting the radial functions on suitable grids. The angular integrals were then analytical. The use of standing wave boundary conditions allowed the real-value representation of the radial functions.

3.3 Linear Algebraic Method

Another technique originally proposed for describing elastic and inelastic scattering processes in molecules and then easily extended to the calculation of photoionization cross sections and angular distribution parameters is the LAM, developed by Schneider and Collins in the last ten years [54]. It can be considered an hybrid technique in the sense that while the bound orbitals and target states are obtained by calculations using basis sets of L^2 functions, the scattering orbital is determined by a numerical solution of the photoelectron Schrödinger equation.

As well as in the R-matrix theory, the basic assumption is that the space can be divided in a short range region, usually spherical with radius $r = a$, where the ion-electron interaction is described by a non-local, eventually energy dependent, potential and an outer region where the interaction may be approximated by the long range multipole potentials due to the ion charge distribution. If a is chosen large enough, so that the outer region potential is essentially reduced to $1/r$, the short range part of the scattering orbital may be matched to a linear combination of regular and irregular Coulomb waves at the border $r = a$. Adopting the fixed nuclei approximation and the usual partial wave expansion of the continuum orbital, the total scattering wave function is written as

$$|\Psi_\alpha^N\rangle = \sum_\beta |\Phi_\beta^{(N-1)}\phi_{\beta,\alpha}\rangle + \sum_i c_{i,\alpha}|\Psi_i^N\rangle$$

where the first summation (α and β are collections of the (partial-wave) channel quantum numbers) includes all the considered channel states described by the antisymmetrized product of a target state $|\Phi_\beta^{(N-1)}\rangle$ and a scattering orbital $|\phi_{\beta,\alpha}\rangle$ while the second term contains only closed channel states that are necessary in order to describe the short range correlation and polarization effects. Both the $|\Phi_\beta^{(N-1)}\rangle$ and the $|\Psi_i^N\rangle$ states are obtained by quantum chemical calculations, projecting on the same finite set of L^2 internal orbitals. The Schrödinger equation for the photoelectron is derived in the Feshbach approach introducing the projection operators $\hat{P}$ and $\hat{Q}$

$$\hat{P} = \sum_{\alpha,a} |\Phi_\alpha^{(N-1)}\chi_a\rangle\langle\Phi_\alpha^{(N-1)}\chi_a| \qquad \hat{Q} = \sum_j |\Psi_j^N\rangle\langle\Psi_j^N|$$

that are orthogonal because the orbitals χ_a are chosen to be orthogonal to the internal orbitals. The Schrödinger equation for the P-part of the scattering function $|\Psi_\alpha^N\rangle$ is then easily written as

$$(\hat{H}_{PP} - E)\hat{P}|\Psi_\alpha^N\rangle + \hat{H}_{PQ}(E - \hat{H}_{QQ})^{-1}\hat{H}_{QP}\hat{P}|\Psi_\alpha^N\rangle = 0 \tag{3.3.1}$$

and the term

$$\hat{H}_{PQ}(E - \hat{H}_{QQ})^{-1}\hat{H}_{QP} = \hat{V}^{opt}$$

can be seen as a nonlocal energy dependent potential describing the effects of correlation and polarization as well as the interaction among continuum states and autoionizing states. By multiplying eq. (3.3.1) by the target states of the considered channels and by integrating over the coordinates of the bound electrons and the angular coordinates of the photoelectron, the following set of integro-differential equations for the different radial partial waves is obtained

$$[\hat{T}_{kin} + \sum_\beta (V_{\alpha\beta}^{dir} + V_{\alpha\beta}^{exch} + V_{\alpha\beta}^{opt}) - \epsilon_\alpha]|\phi_{\beta,\alpha}\rangle = \sum_j \lambda_{j\alpha}|\phi_j\rangle \tag{3.3.2}$$

where the Lagrange multipliers $\lambda_{j\alpha}$ are introduced to force orthogonality of the continuum orbital to the internal orbitals. All direct interactions $V_{\alpha\beta}^{dir}$ are computed exactly from the density matrices

$$\rho^{\alpha\beta}(\vec{r}) = \sum_{ij} \phi_i^*(\vec{r})\Gamma_{ij}^{\alpha\beta}\phi_j(\vec{r})$$

where the factors $\Gamma_{ij}^{\alpha\beta}$ are obtained by CI calculations for the target wave functions. Both the exchange $V_{\alpha\beta}^{exch}$ and the polarization-correlation terms $V_{\alpha\beta}^{opt}$ are instead computed in a separable approximation projecting on a finite L^2 basis set formed by adding to the internal orbitals some diffuse functions necessary in order to describe adequately the interaction with the continuum orbital. This choice is convenient not only because it allows to employ the results of quantum chemical calculations to approximate the polarization-correlation terms, but also because it permits important simplifications in the solution of the scattering equations 3.3.2. These are converted to a set of coupled integral equations [54] by means, in the case of photoionization, of the Coulomb Green's function $G(r, r')$. The Bloch operator [55] allows to solve the problem in the $r < a$ region. The solution is then extended, by a standard propagation scheme [56], into the asymptotic region where it is matched to a linear combination of regular and irregular Coulomb functions in order to obtain the reactance matrix. In the LAM the set of coupled integral equations [54]

$$\begin{aligned}\phi_{\beta\alpha}(r) &= -\frac{1}{2}G_\beta(r|a)\frac{\partial\phi_{\beta\alpha}}{\partial r}|_a + \sum_\gamma \int G_\beta(r|r')V^{dir}_{\beta\gamma}(r')\phi_{\gamma\alpha}(r')dr' \\ &+ \sum_\gamma \int\int G_\beta(r|\vec{r})V^{exch+opt}_{\beta\gamma}(\vec{r}|r')\phi_{\gamma\alpha}(r')d\vec{r}dr' + \sum_j \lambda_{j\beta}\int G_\beta(r|r')|\phi_j(r')dr'\end{aligned}$$

is reduced, by a discrete quadrature in the finite ($r < a$) region, to algebraic equations that may be solved by standard linear systems routines as well as by the powerful iterative algorithms employed in quantum chemistry to operate on matrices of very large size. As opposed to the electron-scattering process, the calculation of the transition dipole matrix element of the photoionization process requires the knowledge of both the P- and Q- parts of the continuum wave function; this last term is obtained by the equation for the Q space deriving from the Feshbach partition of the Schrödinger equation

$$(\hat{H}_{QQ} - E)\hat{Q}|\Psi^N_\alpha\rangle + \hat{H}_{QP}\hat{P}|\Psi^N_\alpha\rangle = 0$$

Applications of LAM to the calculation of molecular photoionization cross sections and asymmetry parameters have been reported for H_2, N_2, NO and CO_2 in the fixed nuclei, single photoionization channel, HF plus Static-Exchange, frozen core approximations [57]. Interchannel interactions in the scattering equations have been introduced through the effective optical potential in order to describe the influence of autoionizing states on the photoionization cross section of NO [58].

3.4 Complex basis function method

Generalizing the idea behind the complex scaling method [59], originally developed for the treatement of resonances in electron-atom and electron-molecule scattering, Rescigno and McCurdy have recently proposed a technique based on a "local complex distortion" of the energy spectrum [12] for the calculation of photoionization cross sections. As well as other "pure" L^2 methods, such as Stieltjes Imaging (sec. 3.6) and our version of the K-matrix technique (sec. 3.5), it relies entirely on expansions of both bound and continuum wave functions in terms of square integrable basis functions. Differently from the complex coordinates approach which is impractical in calculations on many-electron systems, this method has been succesfully applied for obtaining molecular photoionization cross sections, in the single channel static-exchange approximation [60] and the in the RPA [61], and resonance profiles in atomic photoionization cross sections by large CI calculations [62].

The method is based on the expression of the integral cross section as the imaginary part of a matrix element of the resolvent, or, in other words, of the frequency-dependent polarizability $\alpha(\omega)$

$$\sigma(\omega) = -(4\pi\omega/c)\mathrm{Im}\lim_{\eta\to 0}\langle\Psi_0|\hat{\mu}(E_0+\omega-\hat{H}+i\eta)^{-1}\hat{\mu}|\Psi_0\rangle = \frac{(}{4}\pi\omega/c)\mathrm{Im}\alpha(\omega) \tag{3.4.1}$$

where $|\Phi_0\rangle$ is the initial state wave-function of energy E_0 and $\hat{\mu}$ the dipole operator. An approximation to the matrix element in (3.4.1) may be

$$\lim_{\eta\to 0}\langle\Psi_0|\hat{\mu}(E_0+\omega-\hat{H}+i\eta)^{-1}\hat{\mu}|\Phi_0\rangle \simeq \sum_i^N \frac{\langle\Psi_0|\hat{\mu}|\psi_i\rangle\langle\psi_i|\hat{\mu}|\Psi_0\rangle}{E_0+\omega-E_i} \tag{3.4.2}$$

where E_i and ψ_i are the eigenvalues and eigenfunctions obtained from the diagonalization of the hamiltonian $\hat{H}$ in a finite basis set of square-integrable functions and the inner products are here defined without complex coniugation. If the usual hermitian matrix representation of the hamiltonian is adopted, the approximated matrix element in eq. (3.4.2) will have a row of poles on the real axis, where the limit is taken, instead of the correct branch cut of the continuum spectrum. The prescription of Rescigno and McCurdy is that, in order to move the poles corresponding to a discretized representation of continuum off the real axis, E_i and ψ_i are obtained by diagonalizing a complex symmetric matrix representation of the hamiltonian in a basis including both real and complex L^2 functions. The basis set includes standard Slater or Gaussian functions spanning the molecular region and a limited number of diffuse Slater or Gaussian functions with the exponent multiplied by $e^{-i\theta}$ [60, 61, 62]. Differently from the complex coordinate technique, where the substitution $r \rightarrow r\, e^{i\theta}$ gives rise to the rotation of the full continuous spectrum in the lower half plane, the use of a mixed real-complex basis set allows to induce, by an appropriate choice of the complex valued exponent, only a local distortion in the desired portion of the continuous spectrum. Such technique provides better convergence properties in the calculation of the cross section with respect to the standard complex scaling approach [60]. In fact it avoids the divergence of the continuum matrix element of the resolvent because of the return of the discrete spectrum to the real energy axis for large energies. It should also be stressed that the working equation (3.4.2) of the complex function method may be obtained by a variational expression of the frequency-dependent polarizability in eq. (3.4.1), i.e.

$$[\alpha] = -\langle f|\chi\rangle - \langle\chi'|f\rangle + \langle\chi'|(E-H)|\chi\rangle \tag{3.4.3}$$

with $|f\rangle = \hat{\mu}|\Psi_0\rangle$ and $E = E_0 + \omega$. If $|\chi\rangle$ and $|\chi'\rangle$ are expanded in a finite basis of real and complex functions $|\Psi_i\rangle$, it is easy to verify that, at the stationary point where the first derivatives of the expression in eq. (3.4.3) with respect to the expansion coefficient are equal to zero, eq. (3.4.3) becomes equivalent to eq. (3.4.2). This important property provides a convenient way for optimizing the basis set employed. Due to the representation of resonances as stationary complex eigenvalues close to the real axis, this method allows to resolve, as opposed to the Stieltjes Imaging technique discussed in sec. (3.6), structures of the photoionization cross section that are narrower than the local energy gap of the discretized spectrum. Suggestions have recently been given to extend the complex-basis-function method to the calculation of the photoionization asymmetry parameter and test calculations on the hydrogen atom have been presented [63].

3.5 Reactance K-matrix Method

This method is based upon the energy-variational approach originally pioneered by Fano and Cooper [13] which may be adapted to the usual L^2 approach of quantum chemistry. No orthogonality conditions will be invoked, since the aim is to realize the fastest convergence with the basis set. The method, similarly to the close-coupling employed for atoms, is essentially a *CI* carried out with a basis-set including localized correlating functions and (formally) the asymptotic continua (*pwc*) of the relevant channels. By a judicious choice of these basis-sets, the resulting secular equation is solved, at a given energy E lying in the continuum, by the K-matrix technique, which yields the coupled integral equations [64] for the half-off-shell K-matrix elements. These integral equations are solved by a quadrature upon the grid points supplied by a discretized representation of the "unperturbed" states, which turns the integral equations into a linear system. The matrix elements at arbitrary values of the continuous energy indices are interpolated upon the grid points.

Let us assume that we have at our disposal a complete set of basis functions including the (*pwc*) φ defined by eq.(2.4) (with discrete plus continuous energy indices) and a localized basis, "localized channel", (*lc*), needed to represent bound and quasi bound localized states and to account for the short range correlation of the continuum states. On this basis the Hamiltonian **H** and the overlap **S** matrices has the structure shown in fig.1 where it has been assumed that each channel basis diagonalizes the hamiltonian. In his figure the stars and the horizontal and vertical lines represent the discrete-discrete and discrete-continuum matrix elements, while the dotted regions and the oblique lines represent the continuum-continuum matrix elements.

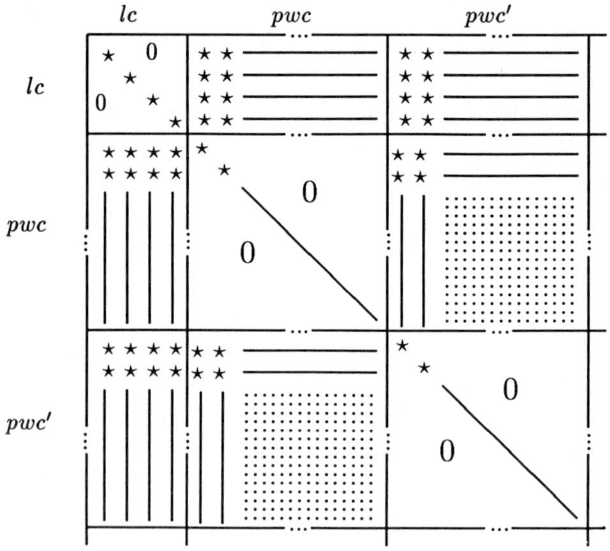

Fig. 1 Structure of the **H**-ES matrix.

The variational coefficients $\mathbf{C}_{\epsilon A,Eb}$ of the eigenvector corresponding to the degenerate component labeled by the *pwc* index b of the energy level E in the continuum (E at least higher than the lowest among the *pcw* thresholds) must satisfy

$$\sum_{A} \sum\!\!\!\!\!\!\int d\epsilon (\mathbf{H}_{\epsilon' D,\epsilon A} - E\,\mathbf{S}_{\epsilon' D,\epsilon A})\mathbf{C}_{\epsilon A,Eb} = 0 \tag{3.5.1}$$

where the $\sum\!\!\!\!\!\!\int d\epsilon$ means summation upon the discrete values plus integration upon the continuous values. According to the K-matrix theory [13] these coefficients may be expressed as

$$\mathbf{C}_{\epsilon A,Eb} = \delta_{Ab}\delta(\epsilon - E) + \frac{\mathrm{P}}{E-\epsilon}\mathbf{K}_{\epsilon A,Eb} \tag{3.5.2}$$

where P indicates the Cauchy principal value, the $\mathbf{K}_{\epsilon A,Eb}$ are the elements of the off-shell reaction matrix and the degeneracy of the eigenvalue E will be given by the number of the available *pwc*'s b open at the energy E considered ($E > E_{t_b}$). The substitution of (3.5.2) into (3.5.1) leads to

$$
\begin{aligned}
\mathbf{K}_{\epsilon' D,Eb} \quad - & \sum_{A(\neq D)} \sum\!\!\!\!\!\!\int d\epsilon (\mathbf{H}_{\epsilon' D,\epsilon A} - E\,\mathbf{S}_{\epsilon' D,\epsilon A}) \frac{\mathrm{P}}{E-\epsilon} \mathbf{K}_{\epsilon A,Eb} \\
= & \ (1-\delta_{Db})(\mathbf{H}_{\epsilon' D,Eb} - E\,\mathbf{S}_{\epsilon' D,Eb})
\end{aligned} \tag{3.5.3}
$$

as the condition which must be satisfied by the $\mathbf{K}$ matrix elements. The normalization of the eigenvectors (3.5.2) may be checked and it is found that

$$
\sum_A \sum\!\!\!\!\!\!\int d\epsilon" \sum_D \sum\!\!\!\!\!\!\int d\epsilon' \mathbf{C}^*_{\epsilon" A,E'b} \mathbf{S}_{\epsilon" A,\epsilon' D} \mathbf{C}_{\epsilon' D,Ef} = \delta(E'-E) \left[\delta_{b,f} + \pi^2 \sum_c \mathbf{K}^*_{Ec,Eb} \mathbf{K}_{Ec,Ef} \right] . \tag{3.5.4}
$$

Thus the set of degenerate eigenvectors $\{\mathbf{U}_{..,Ea}\ \mathbf{U}_{..,Eb}\}$ corresponding to the open available *pwc* is properly normalized to $\delta(E'-E)$ may be expressed as

$$
\{\mathbf{U}(E)\} = \{\mathbf{C}(E)\} \left\{ \left[\mathbf{I} + \pi^2 \mathbf{K}^\dagger(E) \mathbf{K}(E) \right] \right\}^{-1/2} .
$$

In order to have final states $|\ \psi^{(-)}_{Ea}\rangle$ satisfying the boundary conditions appropriated to the photoionization process [65], ingoing waves in all channels outgoing wave only in channel a, we observe that our approximate states should be expressed as

$$
|\ \psi^{(-)}_{Ea}\rangle \simeq \sum_D \sum\!\!\!\!\!\!\int d\epsilon\ |\ \varphi_{\epsilon D}\rangle \mathbf{U}^{(-)}_{\epsilon D,Ea} \tag{3.5.5}
$$

where the vectors $\{\mathbf{U}^{(-)}(E)\}$ are given by

$$
\{\mathbf{U}^{(-)}(E)\} = \{\mathbf{C}(E)\}[\mathbf{I} - i\pi\mathbf{K}(E)]^{-1}
$$

while the vectors

$$
\{\mathbf{U}^{(+)}(E)\} = \{\mathbf{C}(E)\}[\mathbf{I} + i\pi\mathbf{K}(E)]^{-1}
$$

will yield the analogous approximation for the $|\ \psi^{(+)}_E\rangle$ states

$$
|\ \psi^{(+)}_{Eb}\rangle \simeq \sum_D \sum\!\!\!\!\!\!\int d\epsilon\ |\ \varphi_{\epsilon D}\rangle \mathbf{U}^{(+)}_{\epsilon D,Eb} . \tag{3.5.6}
$$

Using eqs. (3.5.4), (3.5.5) and (3.5.6) we obtain for the scattering $\mathcal{S}$ matrix the well known [66, 67] expression

$$
\mathcal{S}_{a,b}(E) = \langle \psi^{(-)}_{Ea}\ |\ \psi^{(+)}_{Eb}\rangle = \left\{ [\mathbf{I} - i\pi\mathbf{K}(E)][\mathbf{I} + i\pi\mathbf{K}(E)]^{-1} \right\}_{a,b}
$$

which shows that its unitarity is ensured by the hermiticity of the $\mathbf{K}(E)$ matrix in spite of the approximations employed. It should be emphasized that this form of the scattering matrix affords a practical way of determining positions and widths of the resonances. To evaluate the differential cross section and the branching ratios we need a proper linear combination of the (3.5.5) which will behave asymptotically like the product of a specific ion state j_a times a coulombic plane wave in the $\hat{k}_a$ direction plus ingoing spherical waves in all channels [65]. These condition are satisfied by

$$
|\ \psi^{(-)}_{E,j_a,\vec{k}_a}\rangle = \sum_{l,m} |\ \Phi^{(-)}_{j_a,\epsilon_a,l,m}\rangle Y^*_{l,m}(\hat{k})
$$

where

$$
|\ \Phi^{(-)}_{j_a,\epsilon_a,l,m}\rangle = \sum_D \sum\!\!\!\!\!\!\int d\epsilon\ |\ \varphi_{\epsilon D}\rangle\ \mathbf{U}^{(-)}_{\epsilon,D,E,j_a,l,m} \imath^l e^{-\imath[\sigma_l(k_a)+\Delta_{j_a,l,m}(k_a)]} . \tag{3.5.7}
$$

The $| \Phi^{(-)}_{j_a,\epsilon_a,l,m}\rangle$ enter in the expressions of the transition matrix elements (2.3) and (2.5). It is expedient to attempt a solution of eq. (3.5.3) by assuming that an adequate grid for the continuous range of the energy indices ϵ' and ϵ will suffice for a satisfactory determination of the $\mathbf{K}_{\epsilon A,Ef}$. In this way it is possible to use an L^2 basis set to represent the u orbitals, yielding a completely discrete representation (which may be taken as the grid values) of the hamiltonian and overlap matrices. By a careful choice of the bases the *pwc*'s should describe essentially only the background part of the continua while the wave packets responsible for the resonance structures of the continuum should be included in the *lc*. This will insure a smooth and regular behaviour of all matrix elements obtained the energy variational calculations that should provide enough grid points in the continuous range of the energy for an accurate interpolation procedure. The same smooth variation of $\mathbf{K}_{\epsilon D,Eb}$ with respect to the first energy index ϵ may be assumed and it is practically verififed a posteriori. Therefore it seems fitting to use a polynomial interpolation to evaluate both $(\mathbf{H}_{\epsilon' D,\epsilon A} - E\ \mathbf{S}_{\epsilon' D,\epsilon A})$ and $\mathbf{K}_{\epsilon A,Eb}$ for arbitrary values of ϵ. Thus if we define

$$(\mathbf{H}_{\epsilon' D,\epsilon A} - E\ \mathbf{S}_{\epsilon' D,\epsilon A})(1-\delta_{DA}) \ = \ \mathbf{V}(E)_{\epsilon' D,\epsilon A}$$

eq. (3.5.3) takes the form

$$\mathbf{K}_{\epsilon' D,Eb} - \sum_A \sum\!\!\!\!\!\!\int d\epsilon\ \mathbf{V}(E)_{\epsilon' D,\epsilon A}\frac{\mathrm{P}}{E-\epsilon}\mathbf{K}_{\epsilon A,Eb} \ = \ \mathbf{V}_{\epsilon' D,Eb} \tag{3.5.8}$$

and the integration for the grid points $\epsilon' D \equiv \epsilon_j D$ may be conveniently carried out, for each channel,

$$\sum\!\!\!\!\!\!\int d\epsilon \mathbf{V}(E)_{\epsilon_j D,\epsilon A}\frac{\mathrm{P}}{E-\epsilon}\mathbf{K}_{\epsilon A,Eb} \ = \ \sum_{J(\epsilon_J<E_{t_A})} \frac{\mathbf{V}_{\epsilon_j D,\epsilon_J A}\mathbf{K}_{\epsilon_J A,Eb}}{E-\epsilon_J}$$
$$+ \ \sum_{i\ (\epsilon_i>E_{t_A})} \int_{\epsilon_i}^{\epsilon_{i+1}} \mathbf{V}(E)_{\epsilon_j D,\epsilon A}\frac{\mathrm{P}}{E-\epsilon}\mathbf{K}_{\epsilon A,Eb}\ .$$

The ϵ_J are the bound states energies of the *pwc* A and the $\epsilon_j, \epsilon_{j+1}...$ are the grid points in the continuum supplied by the variational calculation. The values of $\mathbf{V}_{\epsilon_i D,\epsilon A}$ and $\mathbf{K}_{\epsilon A,Eb}$ in each interval may be approximated by an n-degree polynomial expansion around the midpoint $\overline{\epsilon_i} = (\epsilon_i + \epsilon_{i+1})/2$ obtained for instance by a Lagrange interpolation of the $(n+1)$ closest grid points. Thus if we define the variable

$$x = \epsilon - \overline{\epsilon_i}$$

the interpolated value $F(\epsilon)$ $(\epsilon_i \le \epsilon < \epsilon_{i+1})$ of $f(\epsilon)$ ($\mathbf{V}_{\epsilon_j D,\epsilon A}$ or $\mathbf{K}_{\epsilon A,Eb}$) may be expressed as

$$F(\epsilon) \ = \ \sum_{\nu}\sum_{\mu=0}^{n} \alpha_{\nu,\mu}(i) f(\epsilon_\nu) x^\mu$$

The energy integral for each interval $(\epsilon_i \div \epsilon_{i+1})$ then becomes

$$\int_{\epsilon_i}^{\epsilon_{i+1}} \mathbf{V}(E)_{\epsilon_j D,\epsilon A}\frac{\mathrm{P}}{E-\epsilon}\mathbf{K}_{\epsilon A,Eb} \ = \ \sum_{\nu}\sum_{\nu'} \mathbf{V}(E)_{\epsilon_j D,\epsilon_\nu A}\mathbf{P}^{(i)}_{\nu,\nu'}(E)\mathbf{K}_{\epsilon_{\nu'} A,Eb}$$

where

$$\mathbf{P}^{(i)}_{\nu,\nu'}(E) \ = \ \sum_{\mu}\sum_{\mu'} \alpha_{\nu\mu}(i)\ \alpha_{\nu'\mu'}(i) \int_{\epsilon_i-\overline{\epsilon_i}}^{\epsilon_{i+1}-\overline{\epsilon_i}} \frac{\mathrm{P}}{E+\overline{\epsilon_i}-x} x^{\mu+\mu'} dx$$

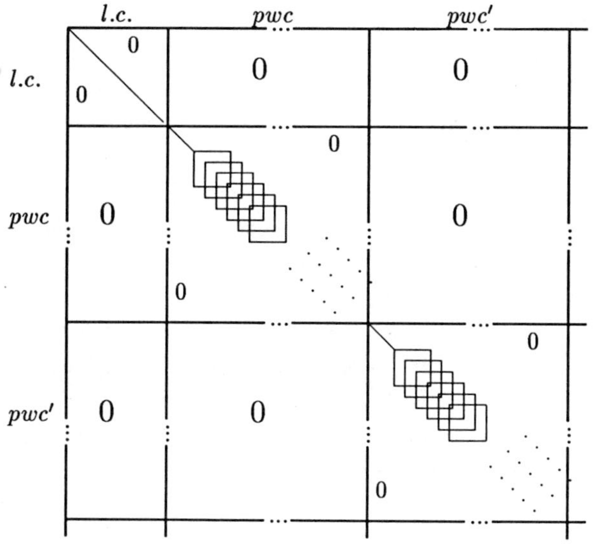

Fig. 2 Structure of the $\mathbf{P}$(E) matrix

may be considered as a block of a block diagonal matrix structured as shown in fig. 2.

By exploiting the foregoing approximations eq. (3.5.8) becomes

$$[\mathbf{I} - \mathbf{V}(E)\mathbf{P}(E)]\mathbf{K}_{..,Eb} = \mathbf{V}(E)_{..,Eb} \, .$$

By this procedure it is possible to obtain approximate expressions for the states (3.5.7) needed to evaluate the transition matrix elements (2.3) and (2.5). More technical details may be found elsewere [1].

The method may be applied at different levels of accuracy. For instance the ion state $|\Phi^+_{j_a}\rangle$ may be approximated as the SCF state of the neutral system with a hole in the orbital j_a while the *pwc* states $\varphi_{E,j_a,\epsilon,l_a,m_a}$ are obtained as antisymmetrized products of the $|\Phi^+_{j_a}\rangle$ times the orbitals $\psi_{\epsilon_{orb.},j_a,l_a,m_a}$ solutions of the Static Exchange Approximation (SEA) singlet hamiltonian

$$\hat{P}_{l_a,m_a}\hat{P}_{virt}(\hat{f} - \hat{\mathcal{J}}_{j_a} + 2\hat{\mathcal{K}}_{j_a})\hat{P}_{virt}\hat{P}_{l_a,m_a}\psi_{\epsilon_{orb.},j_a,l_a,m_a} = \epsilon_{orb.}\psi_{\epsilon_{orb.},j_a,l_a,m_a} \tag{3.5.9}$$

projected upon the l, m spherical harmonic subspace of the virtual orbitals. In the above equation, written for simplicity for a closed shell case and for an Abelian symmetry point group, $\hat{f}$ is the ground state Fock operator, $\hat{P}_{virt}\hat{P}_{l_a,m_a}$ the appropriate projector, $\hat{\mathcal{J}}_{j_a}$ and $\hat{\mathcal{K}}_{j_a}$ the usual Coulomb and exchange operator. The $\psi_{\epsilon_{orb.},j_a,l_a,m_a}$ are supposed to be normalized to $\delta(\epsilon)$ in the continuum and their phase shift is supposed to be known [19].

3.6 Stieltjes imaging

The L^2 methods of quantum chemistry, usually applied to describe bound excited states, can be in principle extended to the calculation of oscillator strengths and excitation energies in the electronic continuum. The finite set of eigenstates with eigenvalues larger then the ion energy, obtained for instance in a CI calculation, can be considered a discretized representation of the electronic continuum .

Among the possible techniques devised for extracting a continous oscillator strength density from a discretized spectrum, the moment method named Stieltjes Imaging (SI) has been by far the most applied. Following the pioneering work of Langhoff and coworkers [11, 68, 69, 70], it has been extensively used for the calculation of integrated one-photon ionization cross sections (for extensive but not exhaustive references, see for instance ref. [71]) and it has been recently extended to the study of other discrete-continuum transition processes [72, 73, 22]. The SI approach is essentially based on the hypothesis that the L^2 discretized spectrum, even if clearly inadequate to describe the structure of a continuum, can anyway give converging approximations of the lowest spectral moments.

Only a short description of the SI technique is briefly sketched here, extensive presentations being available in the literature [11, 68].

Starting from the Kramers - Heisenberg expression of the polarizability as a function of a complex frequency z,

$$\alpha(z) = \int_0^\infty \frac{df(\epsilon)}{\epsilon^2 - z^2} \tag{3.6.1}$$

where

$$\frac{df(\epsilon)}{d\epsilon} = \sum_i^{discrete} f_i \delta(\epsilon - \epsilon_i) + g(\epsilon)$$

is the oscillator strength density containing contributions both from the discrete and the continuous electronic spectrum and ϵ is the excitation energy from the ground state, the cross section $\sigma(\omega)$ is proportional to the imaginary part of the polarizability for real values ω of the frequency or, alternatively, to the oscillator strength density in the continuum

$$\sigma(\omega) = (4\pi\omega/c)Im[\alpha(\omega)] = (2\pi^2/c)g(\omega) \tag{3.6.2}$$

Eq. (3.6.1) is particularly convenient for showing that the polarizability is a Stieltjes integral as far as $(df(\epsilon)/d\epsilon) \geq 0$ in the integration range, which is obviously true for excitations from the ground state. According to the Stieltjes approach, this kind of integral can be convergently approximated, also on the real axis, as

$$\alpha(z) = \int_0^\infty \frac{dF^{(n)}(\epsilon)}{\epsilon^2 - z^2} + R_n(z^2)$$

where $F^{(n)}(\epsilon)$ is an approximate cumulative oscillator strength multistep function

$$F^{(n)}(\epsilon) = \begin{cases} 0 & 0 < \epsilon < \epsilon_1^{(n)} \\ \sum_{j=1}^{l} f_j^{(n)} & \epsilon_l^{(n)} < \epsilon < \epsilon_{l+1}^{(n)} \\ \sum_{j=1}^{n} f_j^{(n)} & \epsilon_n^{(n)} < \epsilon \end{cases} \tag{3.6.3}$$

given in terms of a pseudospectrum $\{\epsilon_j^{(n)}, f_j^{(n)}\}$ defined by the following $2n$ equations

$$S(-2k) = \sum_{j=1}^{n} \frac{f_j^{(n)}}{\left[\epsilon_j^{(n)}\right]^{2k}} \qquad k = 1, 2, ...2n. \tag{3.6.4}$$

Here the $S(-2k)$ are spectral moments of the true oscillator strength function $f(\epsilon)$:

$$S(-j) = \int_0^\infty \frac{df(\epsilon)}{\epsilon^j} .$$

The convergence properties of expression (3.6.3) can be more exactly expressed as [70]

$$F^{(n)}(\epsilon - 0) < F^{(n+1)}(\epsilon - 0) \leq F(\epsilon) \leq F^{(n+1)}(\epsilon + 0) < F^{(n)}(\epsilon + 0)$$

so that, by differentiating the approximate cumulative function $F^{(n)}(\epsilon)$, an approximate oscillator strength density $dF^{(n)}(\epsilon)/d\epsilon$, converging to the correct one on increasing the order n of the pseudospectrum can be obtained. The derivative of $F^{(n)}(\epsilon)$ in the energy continuum $g^{(n)}(\epsilon)$ will be expressed, according to the Stieltjes approach, as

$$g^{(n)}(\epsilon) = \begin{cases} 0 & 0 < \epsilon < \epsilon_1^{(n)} \\ \frac{1}{2}\frac{f_{i+1}^{(n)} + f_i^{(n)}}{\epsilon_{i+1}^{(n)} - \epsilon_i^{(n)}} & \epsilon_i^{(n)} < \epsilon < \epsilon_{i+1}^{(n)} \\ 0 & \epsilon_n^{(n)} < \epsilon \end{cases}$$

and defines, through eq. (3.6.2), a convergent approximation for the cross section. The calculation of $\sigma(\omega)$ is then directly connected to the knowledge of the lowest spectral moments, which, as mentioned before, are supposed to be obtained, with enough accuracy, by L^2 calculations. These calculations will in general produce a spectrum $\{\epsilon_i, f_i\}$ that is strongly basis set dependent in the continuum portion of the spectrum. On the contrary, the pseudospectrum $\{\epsilon_j^{(n)}, f_j^{(n)}\}$, as far as the approximated spectral moments can be considered to converge to the correct ones, is virtually independent of the basis set. It may be recalled that the pseudo excitation energies and oscillator strengths correspond to abscissas and weights of a generalized Gaussian quadrature of order n for the integral in (3.6.1), taking $df(\epsilon)/d\epsilon$ as a weight function. The moment problem expressed by eq. (3.6.4) can be linearized by a Padè approximant to give the polarizability integral

$$\begin{aligned} \alpha(z) &= \sum_{k=1}^{n} S(-2k) z^{2(k-1)} + R_n(z^2) \\ &= \sum_{j}^{n} \frac{f_j^{(n)}}{(\epsilon_j^{(n)})^2 - z^2} + R_n(z^2) = \frac{P_{n-1}(z^2)}{Q_n(z^2)} + R_n(z^2) \end{aligned} \qquad (3.6.5)$$

with

$$P_{n-1}(z^2) = \sum_{i=0}^{n-1} a_i^{(n)} z^{2i} \qquad\qquad Q_n(z^2) = 1 + \sum_{i=1}^{n} b_i^{(n)} z^{2i} \; .$$

The pseudo spectrum $\{\epsilon_j^{(n)}, f_j^{(n)}\}$ is then obtained from the roots and residues of the Padè approximant. The coefficients $a_i^{(n)}$ and $b_i^{(n)}$ can be obtained by solving a set of linear equations

$$\sum_{L=1}^{n} S(-2M + 2L - 2) b_L^{(n)} = -S(-2M - 2) \qquad n \leq M \leq (2n - 1) \qquad (3.6.6)$$

$$a_M^{(n)} = \sum_{L=0}^{M} S(-2M + 2L - 2) b_L^{(n)} \qquad 0 \leq M \leq (n - 1) \qquad (3.6.7)$$

easily derived from eq. (3.6.5). A non trivial solution of eq. (3.6.6) can always be found in so far as the $S(-2k)$'s are moments of a non decreasing distribution, but the numerical implementation of eqs. (3.6.6) (3.6.7) is not convenient if a very large number of moments is employed. In fact the spectral moments are highly redundant and in order to avoid numerical problem they must be computed with high accuracy. Several suggestions can be found in the literature for overcoming this problem [69, 71, 74]. All the proposed algorithms are essentially based on the use of more stable quantities, instead of the spectral moments, to compute the Padè approximants; nevertheless it can be said that none of these techniques is completely trouble free if pseudospectra of high order are computed and the use of double precision in the computer implementation of the algorithms is higly recommended.

The SI technique here reviewed for the calculation of photoionization cross sections can be in principle generalized to any discretized "spectrum" formed by "energy" values and "intensities" as far as these are positive quantities. It has been effectively applied to the calculation of Auger intensities [72] , shake-off intensities [73] as well as two-photon ionization cross sections [22].

As shown previously, the two-photon cross section may be written [1] in terms of positive definite quantities $\mathcal{G}_j^{(2)}(\omega_1,\omega_2)$ (2.8), that can be regarded as proportional to partial two-photon oscillator strengths, explicitely depending in this case, not only on the energies and polarization of the absorbed photons, but also on the final state energy. By an L^2 calculation we can compute these partial oscillator strengths, for fixed values of the two-photon parameters, for a discretized set of final states and form with them and the corresponding final state energy values a "spectrum" to which the SI procedure previously described can be easily applied. The SI will actually generate only a set of derivative values, i.e. of two-photon oscillator strength densities values at energies generally different from $E_a + \omega_1 + \omega_2$ and a simple interpolation will be necessary in order to match the resonance condition.

Essentially the same procedure can be followed to extimate by the Golden Rule

$$\Gamma_\epsilon = 2\pi|\langle\Phi_d|H - E_d|\Phi_\epsilon\rangle|^2 \tag{3.6.8}$$

the linewidth Γ_{E_d} of a resonance structure in the photoionization spectrum due for instance, to the presence of a discrete state $|\Phi_d\rangle$ belonging to a closed channel, embedded and interacting with the continuum states $|\Phi_\epsilon\rangle$ of open channels. The positive quantity Γ_ϵ in eq. (3.6.8) can be computed for the discrete set of continuum energies ϵ_j obtained by an L^2 calculation. The SI procedure applied to the "spectrum" $\{\epsilon_j, \Gamma_{\epsilon_j}\}$ will produce Stieltjes derivative values that, interpolated at the energy E_d will offer a reasonable approximation to the resonance linewidth. This method has been effectively applied to the calculation, within Wentzel's ansatz approximation, of molecular Auger spectra [72].

A limit of the SI technique can be seen in its "energy resolution capability". The pseudo spectrum, optimized in the sense of being a "quadrature spectrum", will offer, as previously discussed, a uniform and basis set independent representation of the continuum, but it cannot improve the energy resolution of the input spectrum, that is, of course, proportional to the number of L^2 states for unit energy interval. This is strictly connected to the fact that only a small number of the computed low order spectral moments will satisfactorily approximate the correct values and will consequentely be employed in the SI calculation.

The so called Stieltjes-Tchebycheff procedure [69, 70], that differs from the plain SI technique in the sense that one of the pseudo spectrum energy may be varied continuosly, has been considered a natural extension of SI in order to compute continuos values of the Stieltjes derivative. Unfortunately, as verified by us as well as by other authors [75], the Stieltjes-Tchebycheff algorithm is affected by numerical instabilities that easily introduce artificial "modulations" of the computed strength density making this technique, at least in the present numerical implementations, pratically useless for obtaining clear information about the presence of resonance structure in the spectrum.

3.7 Random Phase Approximation for Continuum

Although there are in the literature some examples of the *RPA* equations solved in the continuum [15] they are restricted to atoms where the spherical symmetry affords considerable simplification of the problem. It is possible however, by using a technique similar to that used to discretize the K-matrix equation, to extract the matrix elements for transitions to the degenerate continua from the results of an L^2 finite orbital basis *RPA* calculation.

In this context it must be noted, as already mentioned in the sec. 3.4, that a finite L^2 basis set version of the *RPA* employing a mixture of real and complex basis functions has been proposed and applied to compute the integral photoionization cross section of N_2 [76].

Here the presentation is limited to the *RPA* for singlet closed shell cases and for Abelian molecular point symmetry groups; the extension to other cases may be realized by applying the same general procedures outlined below. Let a, b, c .... indicate the set of *pwc* indices $j\ l\ m$ (i.e. $a \equiv j_a\, l_a\, m_a$) where j stands for an occupied *SCF* orbital and l, m specify the assigned partial wave of an orbital projected upon the virtual orbital set of the *SCF* ground state problem. The appropriate excitation *RPA* operators are thus defined as

$$\hat{T}^{\dagger}_{Eb} = \sum_{D} \not\!\!\sum d\epsilon \left(\mathbf{X}_{\epsilon D,Eb}\, \hat{a}^{\dagger}_{\epsilon D} \hat{a}_{j_D} + \mathbf{Y}_{\epsilon D,Eb}\, \hat{a}^{\dagger}_{j_D} \hat{a}_{\epsilon D} \right)$$

where the index b labels the degenerate components of the final level corresponding to the excitation energy E and the singlet spin coupling scheme is understood. The coefficients $\mathbf{X}$ and $\mathbf{Y}$ are determined by the *RPA* equation, with discrete plus continuous indices,

$$\begin{pmatrix} \mathbf{M} & \mathbf{Q} \\ \mathbf{Q}^* & \mathbf{M}^* \end{pmatrix} \begin{pmatrix} \mathbf{X}_{.,Eb} \\ \mathbf{Y}_{.,Eb} \end{pmatrix} = E \begin{pmatrix} \mathbf{I} & \mathbf{0} \\ \mathbf{0} & -\mathbf{I} \end{pmatrix} \begin{pmatrix} \mathbf{X}_{.,Eb} \\ \mathbf{Y}_{.,Eb} \end{pmatrix}. \tag{3.7.1}$$

The orbital basis for each *pwc* is determined by solving the *SEA* equation (3.5.9). It should be emphasized that in spite of the lack of orthogonality between the excited orbitals related to different occupied orbital indices j, the *RPA* eq. (3.7.1) will retain the same structure. By this choice the matrix elements of the *RPA* eq. (3.7.1) may be written as

$$\mathbf{M}_{\epsilon' d',\epsilon d} = \delta_{d',d}\delta_{\epsilon',\epsilon}\epsilon + \mathbf{V}_{\epsilon' d',\epsilon d} \tag{3.7.2}$$

where now ϵ represent the "unperturbed" excitation energy given as the difference between the orbital energy $\epsilon_{orb.}$ and the ionization threshold, which is now the energy ϵ_j of the occupied orbital of the corresponding *pwc*, and

$$\begin{aligned} \mathbf{V}_{\epsilon' d',\epsilon d} &= \delta_{j_{d'},j_d}(1 - \delta_{l_d m_d, l_{d'} m_{d'}})\langle \epsilon' d' | \hat{f} | \epsilon d \rangle \\ &\quad + (1 - \delta_{d,d'}) \left\{ 2\, [\epsilon' d'\ j_d | j_{d'}\ \epsilon d] - [\epsilon' d'\ \epsilon d | j_{d'}\ j_d] \right\} \end{aligned} \tag{3.7.3a}$$

$$\mathbf{Q}_{\epsilon' d',\epsilon d} = 2\, [j_{d'}\ \epsilon' d' | j_d\ \epsilon d] - [j_{d'}\ \epsilon d | j_d\ \epsilon' d'] \,. \tag{3.7.3b}$$

For energies corresponding to excitation in the continuum eq. (3.7.1) may be solved using a procedure similar to the K-matrix technique described in sec. (3.5), i.e. by adopting the following expressions for the $\mathbf{X}$ and $\mathbf{Y}$ *RPA* coefficients

$$\mathbf{X}_{\epsilon\ d,E\ b} = \delta_{d,b}\delta(\epsilon - E) + \frac{\mathrm{P}}{E - \epsilon}\mathbf{K}_{\epsilon d,Eb} \tag{3.7.4a}$$

$$\mathbf{Y}_{\epsilon d,Eb} = \mathbf{L}_{\epsilon d,Eb}. \tag{3.7.4b}$$

It is easy to verify that, using (3.7.2),(3.7.3a) and (3.7.3b), these last expressions, once substituted in eq. (3.7.1), yield the following pair of coupled equations

$$\mathbf{K}_{\epsilon' d',Eb} - \sum_d \sum\!\!\!\!\!\!\int d\,\epsilon \left[\mathbf{V}_{\epsilon' d',\epsilon d}\frac{\mathrm{P}}{E-\epsilon}\mathbf{K}_{\epsilon d,Eb} + \mathbf{Q}_{\epsilon' d',\epsilon d}\mathbf{L}_{\epsilon d,Eb}\right] = \mathbf{V}_{\epsilon' d',Eb} \tag{3.7.5a}$$

$$-(\epsilon' + E)\mathbf{L}_{\epsilon' d',Eb} - \sum_d \sum\!\!\!\!\!\!\int d\epsilon \left[\mathbf{Q}^*_{\epsilon' d',\epsilon d}\frac{\mathrm{P}}{E-\epsilon}\mathbf{K}_{\epsilon d,Eb} + \mathbf{V}^*_{\epsilon' d',\epsilon d}\mathbf{L}_{\epsilon' d',Eb}\right] = \mathbf{Q}_{\epsilon' d',Eb}. \tag{3.7.5b}$$

With an appropriate choice of the basis, which must include a large number of diffuse functions, the elements of the matrices $\mathbf{M}$ and $\mathbf{Q}$ may be regular and reasonably smooth functions of the energy indices. Thus also the dependence upon the first energy index of the $\mathbf{K}$ and $\mathbf{L}$ matrices is expected to be smooth. It is then justifiable to interpolate them at arbitrary values of the energy from a discretized representation of the solutions of eq. (3.5.9) obtainable by a large finite L^2 basis set.

Applying the same procedures utilized in sec. (3.5), to eqs. (3.7.5a), (3.7.5b) the determination of the required $\mathbf{K}$ and $\mathbf{L}$ is reduced to the solution of the finite dimensional linear equation system

$$\begin{pmatrix} 1 - \mathbf{VP}(E) & -\mathbf{QR} \\ -\mathbf{Q}^*\mathbf{P}(E) & -(1E + \boldsymbol{\epsilon}) - \mathbf{V}^*\mathbf{R} \end{pmatrix} \begin{pmatrix} \mathbf{K}_{.,Eb} \\ \mathbf{L}_{.,Eb} \end{pmatrix} = \begin{pmatrix} \mathbf{V}_{.,Eb} \\ \mathbf{Q}_{.,Eb} \end{pmatrix} \tag{3.7.6}$$

where $\mathbf{P}(E)$ is the "integration" matrix previously defined, $\mathbf{R}$ is a similar matrix but for the absence of the $\mathrm{P}/(E-\epsilon)$ term and $\boldsymbol{\epsilon}$ is the diagonal matrix of the "unperturbed" excitation energies.

The discretized vectors $\mathbf{K}_{.,Eb}$ and $\mathbf{L}_{.,Eb}$, solutions of eq. (3.7.6), may then be used to obtain the transition matrix elements to the states (2.4) in the continuum

$$\begin{aligned}\langle\Psi_{\{0\}}|\hat{O}^{(1)}|\varphi_{Ea}\rangle \simeq \\ \langle j_b|\hat{O}^{(1)}|Eb\rangle + \sum_d \sum_{\mu,\nu} \left[\langle j_d|\hat{O}^{(1)}|\epsilon_\mu d\rangle \mathbf{P}^d_{\mu,\nu}(E)\mathbf{K}_{\epsilon_\nu d,Eb} + \langle j_d|\hat{O}^{(1)}|\epsilon_\mu d\rangle^* \mathbf{R}^d_{\mu,\nu}\mathbf{L}_{\epsilon_\nu d,Eb}\right] .\end{aligned}$$

It may be shown that the vectors (3.7.4a) and (3.7.4b) yielded by eq. (3.7.6) are not normalized to unit energy range. It is not difficult to obtain the transformation to the new **X'** and **Y'** coefficients obeing the appropriate normalization condition

$$\sum_d \sum\!\!\!\!\!\!\int d\epsilon \left(\mathbf{X}'^*_{\epsilon d\,Ea}\mathbf{X}'_{\epsilon d,bE'} - \mathbf{Y}'^*_{\epsilon d,Ea}\mathbf{Y}'_{\epsilon d,E'b}\right) = \delta(E-E')\delta_{a,b} ,$$

which, with a little algebra and taking into account the boundary conditions mentioned in the introdution yields the RPA version of the transition matrix elements (2.3)

$$\langle\Psi_{\{0\}}|\hat{O}^{(1)}_\mu|\Phi^{(-)}_{Ea}\rangle = \sum_d \langle\Psi_{\{0\}}|\hat{O}^{(1)}_\mu|\varphi_{Ed}\rangle\,[1 - i\pi\mathbf{K}(E)]^{-1}_{d,a}\, i^{l_a} e^{i[\sigma_{l_a}+\Delta_a(k_a)]} .$$

required to compute the differential and the integral cross sections for a one photon absorption process.

4 Two-Photon Transition Operators

The two-photon transition operator (2.6) appearing in the matrix elements (2.5) requires in principle the knowledge of the resolvent $\hat{G}^+_M(E_0 + \omega)$.

By employing its representation upon the eigenstates of the hamiltonian and denoting, with a simplified notation, the initial and final states with $|i\rangle$ and $|f\rangle$ respectively, the matrix element (2.5) may be cast in the form

$$\mathrm{m}^{(2)}_{f\leftarrow i}(1,2) = (1+\mathcal{P}_{12}) \sum\!\!\!\!\!\!\int_c \frac{\langle f|\hat{O}_1|c\rangle\langle c|\hat{O}_2|i\rangle}{\omega_2 - (E_c - E_i) + i0^+} \; . \tag{4.1}$$

When any photon energy satisfies the resonance condition $\omega = E_c - E_i$, i.e. when it coincides with an excitation energy of the system, one speaks of "resonant processes", characterized in the experiments by an extraordinary increase of the cross section. This occurence will not be discussed here except for mentioning that for a resonant process a more adequate approximation for matrix element (2.5) in the case of two equal photons may be written as [1]

$$\mathrm{m}^{("2")}_{f\leftarrow i}(1,1) = \sum\!\!\!\!\!\!\int_c \frac{\langle f|\hat{O}_1|c\rangle\langle c|\hat{O}_1|i\rangle}{\omega_1 + E_i - E_c - \Delta_c + i\Gamma_c/2}$$

where Δ_c and Γ_c indicate the shift and the width of the level c caused by the interaction with the radiation and the superscript (2) has been put between commas since this quantity does not correspond to a well-defined order in perturbation theory.

The computational efforts required by many-body systems like atoms or molecules are so formidable that exact results have been obtained only for the Hydrogen atom [77]. Furthermore, even in this one-body case and under the *LOPT*, dipolar and non-relativistic approximations, the final formulae are so complicated that noticeable efforts are still needed to work out both exact and approximate simplifications [78, 79]. It is obvious that for more complicate systems it is mandatory to introduce further approximations. Here we limit the presentation to a few of the more used techniques.

4.1 Truncated Summation Method

In principle, a straightforward approximation for (4.1) is provided by a quadrature of the resolvent upon a set of unit-normalized states. These states ϕ_l and their eigenvalues ϵ_l are most commonly obtained by diagonalizing the Hamiltonian upon suitable basis sets; the extension and the quality of the set $\{\phi_l\}$ needed to achieve reliable results depends critically on the system and on the photon energies involved and general rules for arbitrary situations cannot be stated. When both the initial and the final states are low-energy ones, their matrix elements $\langle c|\hat{O}|i\rangle$, $\langle f|\hat{O}|c\rangle$ with the highly excited states are small and therefore the approximations

$$\hat{G}^+_M(E) = \sum\!\!\!\!\!\!\int_c \frac{|c\rangle\langle c|}{E - E_c + i0^+} \approx \sum_l \frac{|\phi_l\rangle\langle\phi_l|}{E - \epsilon_l} \tag{4.1.1}$$

$$\mathrm{m}^{(2)}_{f\leftarrow i}(1,2) = (1+\mathcal{P}_{12}) \sum\!\!\!\!\!\!\int_c \frac{\langle f|\hat{O}_1|c\rangle\langle c|\hat{O}_2|i\rangle}{\omega_2 - (E_c - E_i)} \approx (1+\mathcal{P}_{12}) \sum_l \frac{\langle\phi_f|\hat{O}_1|\phi_l\rangle\langle\phi_l|\hat{O}_2|\phi_i\rangle}{\omega_2 - (\epsilon_l - \epsilon_i)} \tag{4.1.2}$$

require a set $\{\phi_l\}$ of limited dimension but generally of high quality. In particular, this set must include accurate representations for the lowest states (especially those with $E_c < E_f$ which may yield intermediate resonances), while the contribution from the continuum may be safely approximated by a sum over a small number of broad wavepackets [80]. Since the cross section depends critically on the position of the intermediate excitation energies, their theoretical values are often replaced by the experimental counterparts, or the states are calculated employing phenomenological model potentials which fix the energy differences. Since

$$\langle f|\hat{O}_1^p|c\rangle\langle c|\hat{O}_2^p|i\rangle = -(E_f - E_c)(E_c - E_i)\langle f|\hat{O}_1^r|c\rangle\langle c|\hat{O}_2^r|i\rangle \tag{4.1.3}$$

where $\hat{O}^r$ and $\hat{O}^p$ stand for the LG and the VG form of the interaction operator, one expects that the summation for $\mathrm{m}^{(2)}_{f\leftarrow i}(1,2)$ (4.1.2) should converge more rapidly in the LG than in the VG. In these cases, therefore, the LG results are expected to be more reliable than the VG ones, but this is not necessarily true. On the other hand, when the final state $|f\rangle$ is an highly excited Rydberg state or lies in the continuum many intermediate states may give sizeable contributions to $\mathrm{m}^{(2)}_{f\leftarrow i}(1,2)$ and therefore the set $\{\phi_l\}$ must contain satisfactory approximations for many excited states. The VG appears preferable in these cases, since the LG operator may give spurious results for the matrix elements between the diffuse final state and the wavepackets which represent the intermediate continuum. Even worse, when the variational basis contains the diffuse functions needed to represent Rydberg and continuum states, the low-energy ϕ_l states may acquire small spurious tails which affect their LG matrix elements with a diffuse final state. This may be particularly serious when these matrix elements are small and the photon energies are nearly resonant. Finally, $E_i + \omega_{1(2)}$ may fall in the continuum spectrum, in which case the variational states must allow to interpolate accurately the matrix elements at energies near E. Needless to say, the VG appears the best gauge for these calculations. The truncated summation technique has a wide applicability because many methods are available in quantum chemistry to approximate atomic and molecular states [24]. Let us note that the methods commonly employed in quantum chemistry give symmetry-adapted functions and that therefore degenerate components may be easily treated by efficient group-theoretical techniques. This ensures that the approximate representations of the resolvent are totalsymmetric ones and that the selection rules are obeyed.

4.2 Dalgarno and Lewis Method

A skillful technique for calculating (2.5) has been developed by Schwartz and Tiemann [81] from a work of Dalgarno and Lewis [82]. The matrix element $\mathrm{m}^{(2)}_{f\leftarrow i}(1,2)$ may be thought as $\langle f|\hat{O}_1|\theta_2\rangle + \langle f|\hat{O}_2|\theta_1\rangle$ where

$$|\theta_j\rangle = \hat{G}^+_M(E_i + \omega_j)\hat{O}_j|i\rangle \qquad j = 1,2 \tag{4.2.1}$$

may be determined by solving the inhomogeneous differential equation

$$(\hat{H}_M - E_i - \omega_j)|\theta_j\rangle = \hat{O}_j|i\rangle \tag{4.2.2}$$

under suitable boundary conditions. If these equations are approximatively solved by projection upon a finite basis, the formulae of the truncated summation method are obtained. Differently from this method, aimed at a representation of the resolvent, the Dalgarno and Lewis technique deals with the perturbed states $|\theta_j\rangle$, which have a more direct physical significance; it may therefore allow to develop approximations which better exploit the physical characteristics of the system. For instance, in the two-photon detachment of H^- [83] this method allowed to exploit the fact that the adiabatic approximation in hyperspherical coordinate is in this case a "single particle" description more accurate than the HF one. Here again, the calculations are very cumbersome unless a single degree of freedom may be singled out from the system. Nevertheless, the two-photon ionization of Argon, including some correlation effects, has been recently calculated by a variant of this method in which the intermediate state- final state one-particle density matrices instead of the actual intermediate state are directly determined [84]. Finally we mention that an approximate and modified version of the Dalgarno and Lewis method has been employed for the vibronic motions in H_2 and OH [85].

4.3 Many-Body Perturbation Methods

The many-body perturbation treatment of the multiphotonic transitions is a variant of the standard method in which both the matter-radiation interaction $\hat{O}$ and the fluctuation potential $\hat{V}_f = \hat{H} - \hat{H}^{(0)}$ act as perturbers. The model Hamiltonian $\hat{H}^{(0)}$ is generally the HF one, but a few other choices may be also employed. Let us discuss the first terms for the absorption of two identical photons, starting from the HF unperturbed picture (i.e. $\hat{H}^{(0)} = \hat{H}_{HF}$) as done recently for Argon [86]. We denote by $T^{(n,k)}_{f\leftarrow i}$ the contribution of order n in the photon interaction (for $n = 1$ the many-body formulae for the one-photon case are obtained) and k in the fluctuation potential; obviously the $LOPT$ result $\mathrm{m}^{(2)}_{f\leftarrow i}(1,2)$ between the true states is (in the limit of complete one-electron basis) the sum of all the contributions of order 2 in $\hat{O}$ and of any order in $\hat{V}_f$:

$$\mathrm{m}^{(2)}_{f\leftarrow i}(1,2) = \sum_{k=0}^{+\infty} T^{(2,k)}_{f\leftarrow i} \tag{4.3.1}$$

The lowest-order contribution to the transition amplitude is

$$T^{(2,0)}_{f\leftarrow i} = \langle \Phi^{(0)}_f | \hat{O}\hat{Q}[E^{(0)}_i + \omega - \hat{H}^{(0)}]^{-1}\hat{O}|\Phi^{(0)}_i\rangle$$

where $|\Phi^{(0)}_j\rangle$ are the HF states and $\hat{Q} = 1 - |\Phi^{(0)}_i\rangle\langle\Phi^{(0)}_i|$. This equation gives the $LOPT$ result for the unperturbed model.

The third-order contribution (second order in the radiation interaction and first order in the fluctuation potential) is

$$\begin{aligned} T^{(2,1)}_{f\leftarrow i} &= \langle\Phi^{(0)}_f|\hat{O}\frac{\hat{Q}}{E^{(0)}_i + \omega - \hat{H}^{(0)}}\hat{O}\frac{\hat{Q}}{E^{(0)}_i - \hat{H}^{(0)}}\hat{V}_f|\Phi^{(0)}_i\rangle \\ &+ \langle\Phi^{(0)}_f|\hat{O}\frac{\hat{Q}}{E^{(0)}_i + \omega - \hat{H}^{(0)}}\hat{V}_f\frac{\hat{Q}}{E^{(0)}_i + \omega - \hat{H}^{(0)}}\hat{O}|\Phi^{(0)}_i\rangle \\ &+ \langle\Phi^{(0)}_f|\hat{V}_f\frac{\hat{Q}}{E^{(0)}_i + 2\omega - \hat{H}^{(0)}}\hat{O}\frac{\hat{Q}}{E^{(0)}_i + \omega - \hat{H}^{(0)}}\hat{O}|\Phi^{(0)}_i\rangle. \end{aligned}$$

As usual, this perturbative expansion may be represented diagrammatically as reported in the fig.3 .

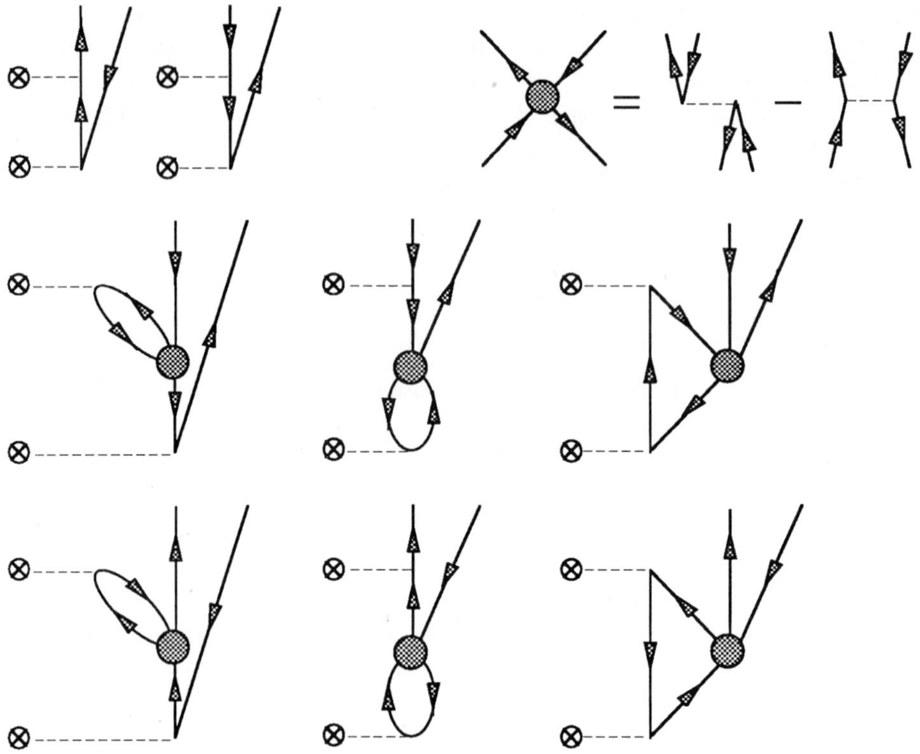

Fig. 3. Some of the lowest-order Feynman diagrams contributing to the two-photon transition amplitude in the double-perturbation treatment. Downward lines correspond to occupied HF orbitals and upward lines to virtual ones. $\otimes - - - - -$ represent the photon interaction, while the big dot defined in the right side of the first row is the symmetrized (Coulomb-Exchange) electron interaction. The two diagrams in the left first row are the only contributions of second order. The six diagrams of the second and third rows correspond to the second addendum of $T^{(2,1)}_{f\leftarrow i}$; equation (4.3.2) of text.

Two kinds of contributions are possible at fourth-order, namely $T^{(4,0)}_{f\leftarrow i}$ and $T^{(2,2)}_{f\leftarrow i}$; their structure should be clear from the two previous cases. $T^{(4,0)}_{f\leftarrow i}$ represents the lowest radiative correction to the $LOPT$; it may be represented by $8 = 2^{4-1}$ diagrams like those for $T^{(2,0)}_{f\leftarrow i}$, corresponding to all the up/down possibilities on the internal lines. Many more diagrams correspond to $T^{(2,2)}_{f\leftarrow i}$; Pindzola and Kelly [87] have estimated the most important in the two-photon ionization of Argon and found their contribution negligible; these authors give also an approximate expression for $T^{(4,0)}_{f\leftarrow i}$. Another approach, which may be analized in terms of many-body theory diagrams, has been proposed by Wendin and coworkers [88, 89]. The basic idea is to replace the matter-radiation interaction $\hat{V}(x;\omega)$ (LG only!), expressed in the coordinate and energy representation, by an effective screened one $\hat{V}_{eff}(x;\omega)$ which satisfies the integral equation

$$\hat{V}_{eff}(x;\omega) = \hat{V}(x;\omega) + \int dx' \int dx'' \; v(x-x')R^{(1)}(x',x'';\omega)\hat{V}_{eff}(x'';\omega)$$

where $v(x' - x'')$ is the electron-electron Coulomb interaction and $R^{(1)}$ is the linear response function. This rather intuitive approach, which demands an appropriate definition of $R^{(1)}$, has been used to evaluate, in the one-electron approximation, different types of processes for some atoms [89].

4.4 Method of the Response Function

Using a field-theretical approach the two-photon transition matrix element may be written as

$$\mathrm{m}^{(2)}_{j,\epsilon,l,m\leftarrow\{0\}}(\omega_1,\mu_1,\omega_2,\mu_2) = (1+\mathcal{P}_{1,2})\int dx'_a dx_a dx'_b dx_b \hat{O}_{\mu_1}(x'_a,x_a)\hat{O}_{\mu_2}(x'_b,x_b) \times \sum_M \frac{\langle\Phi_0\hat{\psi}^\dagger(x'_a)\hat{\psi}(x_a)\Phi_M\rangle\langle\Phi_M\hat{\psi}^\dagger(x'_b)\hat{\psi}(x_b)\Phi^{(-)}_{j,\epsilon,l,m}\rangle}{\omega_2 - E_M + E_0}$$

$\hat{\psi}^\dagger(x)$ being the field operator generating a particle in the spin-space point x. This expression, which involves the transition density matrices

$$X^0_M(x_a,x'_a) = \langle\Phi_0\hat{\psi}^\dagger(x'_a)\hat{\psi}(x_a)\Phi_M\rangle \tag{4.4.1}$$

$$X^M_{j,\epsilon,l,m}(x_b,x'_b) = \langle\Phi_M\hat{\psi}^\dagger(x'_b)\hat{\psi}(x_b)\Phi^{(-)}_{j,\epsilon,l,m}\rangle \tag{4.4.2}$$

is suitable for the introduction of the widely diffuse Response techniques, i.e. of those techniques involving the study of the response of a reference state to a general time-dependent external perturbation.

The interested reader will find a more detailed discussion on this topic in a previous review[1]. Here we simply observe that the Linear Response function (often indicated as Retarded Polarization Propagator or Two-time Green's Function) [90] furnishes generally satisfying approximations to the transition density matrices $X^0_M(x_a,x'_a) = \langle\Phi_0\hat{\psi}^\dagger(x'_a)\hat{\psi}(x_a)\Phi_M\rangle$ through its residues and to the excitation energies $\omega_M = E_M - E_0$ through its poles. In the Linear Response theory the excitation energies ω_M of the system are determined by the equation

$$\langle\Phi_0 \mid [\delta\hat{O}_M,\hat{H},\hat{O}^\dagger_M] \mid \Phi_0\rangle = \omega_M\langle\Phi_0 \mid [\delta\hat{O}_M,\hat{O}^\dagger_M] \mid \Phi_0\rangle \tag{4.4.3}$$

which can be obtained when studying the EOM of the excitation operator $\hat{O}^\dagger_M$. In eq. (4.4.3) $\delta\hat{O}_M$ is a variation of $\hat{O}_M$ and

$$[\hat{A},\hat{B},\hat{C}] = \frac{1}{2}[\hat{A},[\hat{B},\hat{C}]] + \frac{1}{2}[[\hat{A},\hat{B}],\hat{C}].$$

The transition moments between the reference state Φ_0 and and excited state Φ_M are then given by the following expression

$$\langle\Phi_0 \mid \hat{O}_\mu \mid \Phi_M\rangle = \int dx_a dx'_a \hat{O}_\mu(x'_a,x_a)X^0_M(x_a,x'_a) = \langle\Phi_0 \mid [\hat{O}_\mu,\hat{O}^\dagger_M] \mid \Phi_0\rangle\ .$$

Explicit expressions of the transition matrix elements obtained by selecting $(MC)SCF$ approximations for the reference exact state and suitable forms of the excitation operator $\hat{O}^\dagger_M$ have been published[1]. The excited state - excited state transition density matrices $X^M_{j,\epsilon,l,m}(x_b,x'_b)$ of eq. (4.4.2) on the other hand may be computed by using the reference state transition density matrices $X^0_M(x_a,x'_a)$ obtained in a (Multi Configurational) Linear Response calculation. By exploiting the formal properties of the excitation operators $\hat{O}_M$ and the form of the solution of the EOM equation (4.4.3), it is possible to write [91]

$$\langle\Phi_M \mid \hat{O}_\mu \mid \Phi_{M'}\rangle = \int dx_b dx'_b \hat{O}_\mu(x'_b,x_b)X^M_{M'}(x_b,x'_b)$$

$$= \langle\Phi_0 \mid [\hat{O}_M,\hat{O}_\mu,\hat{O}^\dagger_{M'}] \mid \Phi_0\rangle + \frac{1}{2}\langle\Phi_0 \mid \{[\hat{O}_M,\hat{O}^\dagger_{M'}],\hat{O}_\mu\} \mid \Phi_0\rangle$$

where the braces indicate an anticommutator. Once again the reader should refer to Ref. [1] for explicit expressions in terms of the transition density matrix elements $\mathbf{X}_{pq,M}$, $\mathbf{Y}_{pq,M}$, $\mathbf{X}_{n,M}$ and $\mathbf{Y}_{n,M}$.

References

[1] I. Cacelli, V. Carravetta, R. Moccia, and A. Rizzo. *Computer Phys. Rep.*, in press (1991).

[2] R. J. W. Henry and L. Lipsky. *Phys. Rev.*, 153, 51 (1967).

[3] H. P. Kelly. *Chem. Phys. Lett.*, 20, 547 (1973).

[4] B. H. Brandsen. *Atomic Collision Theory.* W. A. Benjamin, Inc. , New York (1970).

[5] R. K. Nesbet. *Variational Methods in Electron-Atom Scattering Theory.* Plenum Press, New York (1980).

[6] R. R. Lucchese and V. McKoy. *Phys. Rev. A*, 21, 112 (1980).

[7] R. R. Lucchese, G. Raseev, and V. McKoy. *Phys. Rev. A*, 25, 2572 (1982).

[8] D. Dill and J. L. Dehmer. *J. Chem. Phys.*, 61, 692 (1974).

[9] L. A. Collins and B. I. Schneider. *Phys. Rev. A*, 24, 2387 (1981).

[10] L. A. Collins and B. I. Schneider. *Phys. Rev. A*, 29, 1695 (1984).

[11] P. W. Langhoff. In B. J. Dalton, S. M. Grimes, J.P. Vary, and S. A. Williams, editors, *Theory and Application of Moment Methods in Many-Fermions Systems*, page 191. Plenum, New York (1980).

[12] T.N. Rescigno and C.W. McCurdy. *Phys. Rev. A*, 31, 624 (1985).

[13] U. Fano and J. W. Cooper. *Rev. Mod. Phys.*, 40, 441 (1968).

[14] D. J. Rowe. *Rev. Mod. Phys.*, 40, 153 (1968).

[15] M. Ya Amusia and N. A. Cherepkov. *Case Studies in Atomic Physics*, 5, 47 (1975).

[16] I. Cacelli, R. Moccia, and V. Carravetta. *Chem. Phys.*, 90, 313 (1984).

[17] I. Cacelli, V. Carravetta, and R. Moccia. *J. Phys. B: At. Mol. Phys.*, 18, 1375 (1985).

[18] I. Cacelli, V. Carravetta, and R. Moccia. *Mol. Phys.*, 59, 385 (1986).

[19] I. Cacelli, V. Carravetta, and R. Moccia. *J. Chem. Phys.*, 85, 7038 (1986).

[20] I. Cacelli, V. Carravetta, R. Moccia, and A. Rizzo. *J. Phys. Chem.*, 92, 979 (1988).

[21] I. Cacelli, V. Carravetta, and R. Moccia. *Chem. Phys.*, 120, 51 (1988).

[22] I. Cacelli, V. Carravetta, R. Moccia, and A. Rizzo. *J. Chem. Phys.*, 89, 7301 (1988).

[23] I. Cacelli, V. Carravetta, A. Rizzo, and R. Moccia. *Chem. Phys. Lett.*, 155, 210 (1989).

[24] I. Cacelli, V. Carravetta, A. Rizzo, and R. Moccia. *J. Chem. Phys.*, 92, 2883 (1990).

[25] K. H. Johnson. *Adv. Quantum Chem.*, 7, 143 (1973).

[26] J. Siegel, D. Dill, and J. L. Dehmer. *J. Chem. Phys.*, 64, 3204 (1976).

[27] J. W. Davenport. *Phys. Rev. Letters*, 36, 945 (1976).

[28] J. C. Slater. *Adv. Quantum Chem.*, 6, 1 (1972).

[29] F. A. Grimm, T. A. Carlson, W. B. Dress, P. Agron, J. O. Thomson, and J. W. Davenport. *J. Chem. Phys.*, 72, 3041 (1980).

[30] M. Roche, D. R. Salahub, and R. P. Messmer. *J. Electron Spectr.*, 19, 273 (1980).

[31] A. Görling and N. Rösch. *J. Chem. Phys.*, 93, 5563 (1990).

[32] T. A. Carlson, P. Gerard, M. O. Krause, F. A. Grimm, and B. P. Pullen. *J. Chem. Phys.*, 86, 6918 (1987).

[33] J. S. Tse, Z. F. Liu, J. D. Bozek, and G. M. Bancroft. *Phys. Rev. A*, 39, 1791 (1989).

[34] W. Thiel. *Chem. Phys.*, 57, 227 (1981).

[35] D. Mehaffy, P. R. Keller, T. A. Carlson, and F. A. Grimm. *J. Electron Spectr.*, 28, 239 (1983).

[36] J. D. Bozek, G. M. Bancroft, J. N. Cutler, K. H. Tan, B. W. Yates, and J. S. Tse. *Chem. Phys.*, 132, 257 (1989).

[37] J. R. Swanson, D. Dill, and J. L. Dehmer. *J. Chem. Phys.*, 75, 619 (1981).

[38] P. M. Dittman, D. Dill, and J. L. Dehmer. *Chem. Phys.*, 78, 405 (1983).

[39] J. S. Tse. *Chem. Phys. Lett.*, 163, 392 (1989).

[40] A. Schichl, D. Menzel, and N. Rösch. *Chem. Phys. Lett.*, 105, 285 (1984).

[41] J. Schwinger. *Phys. Rev.*, 72, 742 (1947).

[42] J. M. Blatt and J. D. Jackson. *Phys. Rev.*, 76, 18 (1949).

[43] R. R. Lucchese and V. McKoy. *Phys. Rev. A*, 21, 112 (1980).

[44] R. R. Lucchese, G. Raseev, and V. McKoy. *Phys. Rev. A*, 25, 2572 (1982).

[45] R. R. Lucchese, K. Takatsuka, and V. McKoy. *Phys. Rep.*, 131, 147 (1986).

[46] M. Braunstein, V. McKoy, L. E. Machado, L. M. Brescansin, and M. A. P. Lima. *J. Chem. Phys.*, 89, 2998 (1988).

[47] L. E. Machado, L. M. Brescansin, M. A. P. Lima, M. Braunstein, and V. McKoy. *J. Chem. Phys.*, 92, 2362 (1990).

[48] M. Braunstein and V. McKoy. *J. Chem. Phys.*, 92, 4887 (1990).

[49] M. A. P. Lima, T. L. Gibson, K. Takatsuka, and V. McKoy. *Phys. Rev. A*, 30, 1741 (1984).

[50] K. Takatsuka and V. McKoy. *Phys. Rev. A*, 30, 1734 (1984).

[51] M. A. P. Lima and V. McKoy. *Phys. Rev. A*, 38, 501 (1988).

[52] M. Abramowitz. *Handbook of Mathematical Functions.* North-Holland Publishing Company (1969).

[53] R. R. Lucchese, D. K. Watson, and V. McKoy. *Phys. Rev. A*, 22, 421 (1980).

[54] B.I. Schneider and L.A. Collins. *Computer Physics Reports*, 10, 49 (1989).

[55] C. Bloch. *Nucl. Phys.*, 4, 503 (1957).

[56] B.I. Schneider and R.B. Walker. *J. Chem. Phys.*, 70, 2466 (1979).

[57] L.A. Collins and B.I. Schneider. *Phys. Rev. A*, 29, 1695 (1984).

[58] D.L. Lynch, B.I. Schneider, and L.A. Collins. *Phys. Rev. A*, 38, 4927 (1988).

[59] C. W. McCurdy. In D. G. Truhlar, editor, *Resonances in Electron-Molecule Scattering, van der Waals Complexes, and Reactive Chemical Dynamics*, page 17. American Chemical Society 263, Washington D. C. (1984).

[60] Chin hui Yu, Russel M. Pitzer, and C. W. McCurdy. *Phys. Rev. A*, 32, 2134 (1985).

[61] S. Yabushita, C. W. McCurdy, and T. N. Rescigno. *Phys. Rev. A*, 36, 3146 (1987).

[62] T.N. Rescigno. *Phys. Rev. A*, 31, 607 (1985).

[63] C.W. McCurdy and T.N. Rescigno. *Phys. Rev. A*, 35, 657 (1987).

[64] P. Dhez and D.L. Ederer. *J. Phys. B: At. Mol. Phys.*, 6, L59 (1973).

[65] G. Breit and H. A. Bethe. *Phys. Rev.*, 93, 888 (1954).

[66] A. F. Starace. *The theory of atomic photoionisation, The Encyclopedia of Physics Vol.XXXI.* Springer Verlag , Berlin (1982).

[67] C. J. Joachain. *Quantum Collision Theory.* North-Holland Publishing Co. NewYork (1975).

[68] P.W. Langhoff and C.T. Corcoran. *J. Chem. Phys.*, 61, 146 (1974).

[69] P.W. Langhoff, C.T. Corcoran, J.S. Sims, F. Weinhold, and R.M. Glover. *Phys. Rev. A*, 14, 1042 (1976).

[70] T. N. Rescigno, B. V. McKoy, and B. Schneider, editors. Plenum, New York (1979).

[71] F. Muller-Plathe and G. Dierksen. *Phys. Rev. A*, 40, 696 (1989).

[72] V. Carravetta and H. Ågren. *Phys. Rev. A*, 35, 1022 (1987).

[73] H.Ågren and V.Carravetta. *J. Chem. Phys.*, 87, 370 (1987).

[74] R.K. Nesbet. *Phys. Rev. A*, 14, 1065 (1976).

[75] J. J. Delaney, V. R. Saunders, and I. H. Hillier. *J. Phys. B: At. Mol. Phys.*, 14, 819 (1981).

[76] S. Yabushita, C. W. McCurdy, and T. N. Rescigno. *Phys. Rev. A*, 36, 3146 (1987).

[77] E. Karule. *J. Phys. B: At. Mol. Phys.*, 4, L67 (1971).

[78] A. Costescu, I. Brandus, and N. Mezincescu. *J. Phys. B: At. Mol. Phys.*, 18, L11 (1985).

[79] V. Florescu, S. Patrascu, and O. Stoican. *Phys. Rev. A*, 36, 2155 (1987).

[80] R. Moccia and P. Spizzo. *J. Phys. B: At. Mol. Opt. Phys.*, 21, 1145 (1988).

[81] C. Schwartz and J.I. Tiemann. *Ann. Phys. (N.Y.)*, 6, 178 (1959).

[82] A. Dalgarno and J.T. Lewis. *Proc. Roy. Soc.*, page 70 (1955).

[83] M.G.J. Fink and P. Zoller. *J. Phys. B: At. Mol. Phys.*, 18, L373 (1985).

[84] A.F. Starace and Tsin-Fu Jiang. *Phys. Rev. A*, 36, 1705 (1987).

[85] G.L. Bendazzoli, S. Evangelisti, and P. Palmieri. *Nuovo Cimento D*, 9, 45 (1987).

[86] C. Pan, B. Gao, and A.F. Starace. *Phys. Rev. A*, 41, 6271 (1990).

[87] M.S. Pindzola and H.P. Kelly. *Phys. Rev. A*, 11, 1543 (1975).

[88] A. L' Huillier, L. Jonsson, and G. Wendin. *Phys. Rev. A*, 33, 3938 (1986).

[89] A. L' Huillier and G. Wendin. *Phys. Rev. A*, 36, 4747 and 5632 (1987).

[90] J. Oddershede, P. Jørgensen, and D. L. Yeager. *Comp. Phys. Rep.*, 2, 33 (1984).

[91] P. Jørgensen, P. Swanstrøm, D. L. Yeager, and J. Olsen. *Int. J. Quantum Chem.*, 23, 959 (1983).

MULTICONFIGURATIONAL GREEN'S FUNCTION (PROPAGATOR) TECHNIQUES FOR EXCITATION ENERGIES, IONIZATION POTENTIALS, AND ELECTRON AFFINITIES: AN OVERVIEW

Danny L. Yeager

Chemistry Department
Texas A&M University
College Station, Texas 77843-3255
USA

ABSTRACT

With Green's function or propagator techniques electronic excitation energies, ionization potentials, electron affinities, electronic transition probabilities, and other response properties can be accurately calculated. These techniques have several significant advantages over more conventional *ab initio* approaches to electronic energy difference calculations such as ΔMCSCF and ΔCI. Included among these are: 1) Energy differences such as electronic excitation energies and ionization potentials are calculated directly rather than as differences of two (fairly large) total electronic energies and hence, "correlation corrections" are more easily "balanced" than with most other approaches; 2) For excitation, ionization, and attachment energies, propagator techniques mimic Δ full CI at a small fraction of the computer time and cost; 3) The polarization propagator is the correct linear response of a system to an external perturbation (e.g. an electromagnetic field) so response properties such as frequency dependent polarizabilities are calculated reliably and accurately; and 4) The length, velocity, and acceleration forms of the oscillator strength are equal in the limit of a complete basis set of orbitals.

Propagator techniques which are based on an initial multiconfigurational (MC) reference state and which do not use perturbation theory methods are often preferable to perturbational propagator approaches. This is because with these MC propagator techniques even highly correlated and open shell atoms and molecules can be easily and accurately handled. The single particle multiconfigurational Green's function approach I will discuss here gives excellent inner and outer valence shake-up and principal ionization potentials (IPs) and electron affinities (EAs). It is known as the multiconfigurational spin tensor electron propa-

Applied Many-Body Methods in Spectroscopy and Electronic Structure
Edited by D. Mukherjee, Plenum Press, New York, 1992

gator method (MCSTEP). Another of our multiconfigurational Green's function approaches is known as the multiconfigurational time dependent Hartree Fock (MCTDHF) or multiconfigurational linear response (MCLR). With MCTDHF/MCLR accurate atomic and molecular electronic excitation energies, oscillator strengths, polarizabilities, and other response properties are obtained. We have also recently proposed and developed the multiconfigurational particle-particle propagator (MCP2P) for directly and accurately determining double ionization potentials and double electron affinities. In addition to theoretical developments I will present results of several calculations which demonstrate MCSTEP, MCTDHF/MCLR, and MCP2P.

I. INTRODUCTION

Properties involving energy differences or properties involving the response of a system to an external perturbation are most frequently determined in a laboratory. Determining energy differences such as electronic excitation energies, electron affinities, or ionization potentials using most *ab initio* methods involves calculating an accurate initial state wavefunction and total energy and several accurate final state wavefunctions and corresponding total electronic energies. Then to obtain energy differences a fairly large total electronic initial state energy is subtracted from the other (fairly large) total electronic energies. Compared to the total electronic energy of a system, electronic excitation energies, ionization potentials, and electron affinities are fairly small. Hence, procedures which obtain energy differences by explicit subtraction between two large numbers to get a smaller number can lead to significant errors, especially if the initial and final states are fundamentally different, e.g. electron affinities, excitations to Rydberg states, etc. For example, it is well known that it is often difficult to properly balance "correlation corrections" in Δ large scale CI between neutral initial and ionic final states. In contrast, with Green's function/propagator methods the focus is on obtaining these energy differences directly. Thus, in addition to anticipated savings in computational time (since separate calculations are not made for the initial and final states); it is expected that these Green's function/propagator approaches will give, in general, more accurate energy differences.

A multitude of propagator or Green's function methods have been developed. The nature of the physical property that we wish to describe dictates which propagator has to be considered. If we are interested in electron addition or removal processes the electron propagator or one (or single) particle Green's function is considered. The poles of the electron propagator are the electron affinities (EAs) and ionization potentials (IPs) of the atomic or molecular system. For describing electronic excitation spectra and linear response properties like the dynamic dipole polarizability it is necessary to consider the polarization (part of the two-particle) Green's function or the polarization propagator. Other propagators are relevant for the description of processes like double ionization in Auger spectroscopy and two-photon absorption.

Methods for Green's function/propagator methods for electronic structure have usually focused on perturbational approaches.[1-12] That is, the resulting equations are solved consistently through a certain order or approximately through a certain order in the electron-electron interaction (the perturbation). While this can frequently give excellent agreement with experiment (e.g. when the equations are at least third order for IPs and EAs and at least second order for excitation energies), perturbational approaches are sometimes unreliable and/or converge slowly. Perturbational approaches usually work well only when the initial (or reference) state is closed shell and when initial state correlation effects are not large (i.e. when initial state non-dynamical correlation is small). Many important problems of chemical and physical interest involve such initial states.

It is expected that perturbational approaches to Green's function calculations will have difficulty when the initial state is open shell or when its inherently multiconfigurational. That's in part because consistently reliable and general perturbation theory approaches have not been developed for electronic wavefunctions for these cases. For example, open shell MBPT usually uses a UHF zero order state which leads to some spin contamination. We have developed and studied several methods to correctly handle open shell and/or highly multiconfigurational initial states with Green's function methods which do not use perturbation theory. Several aspects of these approaches will be discussed below.

In the following I will first present the theory and some results for our one-electron MC propagator methods (MCEP, MCSTEP, and RMCSTEP). More theoretical details will be given for the one-electron case since many of the theoretical details for the other propagators are analogous. In Section III excitation energy/linear response methods (MCTDHF/MCLR) will be discussed and in Section IV MC propagator methods for double ionization potentials (DIPs) and double electron affinities (DEAs) will be presented. The focus in these sections will be on theoretical principals and demonstrative calculations. Several recent research efforts have been devoted to developing and programming large scale techniques which can handle up to several hundred thousand or more configurations in Green's function approaches.[13-15] These large scale methods will not be discussed in this paper. The final section will be a brief summary and conclusions.

II. THE ELECTRON PROPAGATOR OR THE SINGLE PARTICLE GREEN'S FUNCTION FOR DIRECTLY DETERMINING IONIZATION POTENTIALS AND ELECTRON AFFINITIES

A. The single-particle Green's function

We here discuss the theory of the single-particle Green's function or, equivalently, the electron propagator for the direct determination of ionization potentials and electron affinities.[1,2,16]

1. The spectral resolution of the electron propagator. The electron propagator for an exact N-electron state of total spin S_0 and spin projection M_0, denoted $|\psi_0^{NS_0M_0}\rangle$, is defined as

$$G_{r\sigma,p\sigma'}(E) = \langle\psi_0^{NS_0M_0}|\, a^{+}_{r\sigma}\,(E + E_0^{NS_0} - H)^{-1}\, a_{p\sigma'}\, |\psi_0^{NS_0M_0}\rangle$$

$$+ \langle\psi_0^{NS_0M_0}|\, a_{p\sigma'}\,(E - E_0^{NS_0} + H)^{-1}\, a^{+}_{r\sigma}\,|\psi_0^{NS_0M_0}\rangle, \tag{1}$$

where

$$H|\psi_0^{NS_0M_0}\rangle = E_0^{NS_0}|\psi_0^{NS_0M_0}\rangle, \tag{2}$$

$$S^2|\psi_0^{NS_0M_0}\rangle = S_0(S_0 + 1)|\psi_0^{NS_0M_0}\rangle, \tag{3}$$

$$S_z|\psi_0^{NS_0M_0}\rangle = M_0|\psi_0^{NS_0M_0}\rangle. \tag{4}$$

H is the nonrelativistic electronic Hamiltonian in the Born-Oppenheimer approximation

$$H = \sum_{ij\sigma} h_{ij} a^{+}_{i\sigma} a_{j\sigma} + (1/2) \sum_{ijkl\sigma\sigma'} \langle ij|kl\rangle\, a^{+}_{i\sigma} a^{+}_{j\sigma'} a_{l\sigma'} a_{k\sigma} \tag{5}$$

The creation and annihilation operators in H are summed over a formally complete orthonormal basis of spatial orbitals $\{\phi_i\}$ and the spin indices, σ, σ' can have the value of α or β spin.

The spectral resolution of Eq. (1) is

$$G_{r\sigma,p\sigma'} = \sum_f \frac{\langle\psi_0^{NS_0M_0}|a^{+}_{r\sigma}|\psi_f^{(N-1)S_fM_f}\rangle \langle\psi_f^{(N-1)S_fM_f}|a_{p\sigma'}|\psi_0^{NS_0M_0}\rangle}{E + E_0^{NS_0} - E_f^{(N-1)S_f}}$$

$$+ \sum_f \frac{\langle\psi_0^{NS_0M_0}|a_{p\sigma'}|\psi_f^{(N+1)S_fM_f}\rangle \langle\psi_f^{(N+1)S_fM_f}|a^{+}_{r\sigma}|\psi_0^{NS_0M_0}\rangle}{E - E_0^{NS_0} + E_f^{(N+1)S_f}} \tag{6}$$

where the summations in the first and second terms contain a complete set of exact $N - 1$ and $N + 1$ electronic eigenstates, respectively, and

$$H\,|\psi_f^{(N\pm1)S_fM_f}\rangle = E_f^{(N\pm1)S_f}|\psi_f^{(N\pm1)S_fM_f}\rangle, \tag{7}$$

$$S^2|\psi_f^{(N\pm1)S_fM_f}\rangle = S_f(S_f + 1)|\psi_f^{(N\pm1)S_fM_f}\rangle, \tag{8a}$$

$$S_z | \psi_f^{(N\pm1)S_fM_f} \rangle = M_f | \psi_f^{(N\pm1)S_fM_f} \rangle. \tag{8b}$$

It can easily be seen from Eq. (6) that the exact ionization potentials and electron affinities of the initial N-electron system are the poles of the Green's function.

2. Superoperator notation. Using superoperator notation, the electron propagator can be written in a form more convenient for calculation. Arranging the set of annihilation operations $\{a_i\}$ as a superrow vector, **a**, Eq. (1) can be rewritten as[2,17,18]

$$G(E) = (\mathbf{a} | (E\hat{I} - \hat{H})^{-1} | \mathbf{a}). \tag{9}$$

In Eq. (9), the superoperators $\hat{I}$ and $\hat{H}$ are characterized by their effect on an arbitrary operator Op:

$$\hat{I}\,\mathrm{Op} = \mathrm{Op}, \tag{10}$$

$$\hat{H}\,\mathrm{Op} = [\mathrm{H},\mathrm{Op}], \tag{11}$$

and the binary product is defined as

$$(\mathrm{Op}_i | \mathrm{Op}_j) = \langle \psi_0^{NS_0M_0} | \{\mathrm{Op}_i^+, \mathrm{Op}_j\} | \psi_0^{NS_0M_0} \rangle. \tag{12}$$

In Eq. (12), {,} denotes the anticommutator. Taking the inner projection[19] of the superoperator resolvent gives

$$G(E) = (\mathbf{a} | \mathbf{h})(\mathbf{h} | E\hat{I} - \hat{H} | \mathbf{h})^{-1}(\mathbf{h} | \mathbf{a}). \tag{13}$$

Equation (13) can be easily verified by twice inserting unity[20]:

$$\hat{I} = | \mathbf{h})(\mathbf{h} | \mathbf{h})^{-1}(\mathbf{h} | \tag{14}$$

and utilizing the identity

$$(\mathbf{h} | \mathbf{h}) = (\mathbf{h} | E\hat{I} - \hat{H} | \mathbf{h})(\mathbf{h} | \mathbf{h})^{-1}(\mathbf{h} | (E\hat{I} - \hat{H})^{-1} | \mathbf{h}). \tag{15}$$

If the operator manifold **h** is complete,[21,22] Eqs. (9) and (13) are identical.

Since the IPs and EAs occur at the poles of the electron propagator, Eq. (13) shows that the IPs and EAs can be determined as eigenvalues of the matrix

$$(\mathbf{h} | E\hat{I} - \hat{H} | \mathbf{h}). \tag{16}$$

Approximations that are introduced into the one-electron propagator bring the eigenvalue equation [Eq. (16)] to a computationally tractable form.

Equation (16) is the same as the equations-of-motion for ionization potentials and electron affinities[6,8]

$$\langle \psi_0^{NS_0M_0} | \{\delta O_f^+,[H,O_f]\} | \psi_0^{NS_0M_0} \rangle$$

$$= \omega_f \langle \psi_0^{NS_0M_0} | \{\delta O_f^+, O_f\} | \psi_0^{NS_0M_0} \rangle \qquad (17)$$

when the ionization potential/electron affinity operator O_f^+ which is identified with the ionic state $| \psi_f^{(N\pm1)S_f M_f} \rangle$ is expanded with the operator manifold $\mathbf{h}$. ω_f is the ionization potential or electron affinity. In practical calculations with either Eq. (16) or Eq. (17), the symmetric double anticommutator is usually used:

$$\{A,B,C\} = (1/2)(\{A,[B,C]\} + \{[A,B],C\}). \qquad (18)$$

In Eq. (16) the operator manifold is complete. Of course, in practical calculations a truncated operator manifold is used. Dalgaard[21] and Manne[22] give one choice for a complete operator manifold which includes single electron removal with all possible excitations and single electron addition with all possible excitations from a specified single determinant. The operators in this manifold act on a state which is not orthogonal to the specified single determinant to give a complete N ± 1 basis of states.

While the operator manifold of Dalgaard and Manne is formally complete, it is not the most convenient for calculations where the initial (reference) state is open shell. This can be easily seen by examining the effect of simple electron removal, $a_{i\sigma}$, from an open shell state, with e.g. S=1. By simple rules for angular momentum addition, the resulting N - 1 states for this example will generally be a mixture of a doublet and a quartet since no explicit coupling has been introduced. That is, for open shell reference states' ionization potentials and electron affinities pure ion states will not be obtained in a method which utilizes a truncated (rather than a complete and, hence, infinite) operator manifold.

We have handled this problem in two ways:

1. In the multiconfigurational electron propagator method (MCEP)[23] we used modified creation and destruction operators {u, d, u$^+$, d$^+$} instead of {a, a$^+$}. These operators were first introduced by Pickup and Mukhopadhyay[24] and are designed to generate ion states with pure spin symmetry. They involve S_+ and S_- in combination with creation and destruction operators and depend on the spin symmetry of the state they are acting on. Hence, they are somewhat awkward to use.

2. In the multiconfigurational spin tensor electron propagator method (MCSTEP)[25] we explicitly use tensor operators and tensor states to assure proper spin coupling. This approach is more elegant than the

MCEP approach and gives essentially identical results. This method is easier to extend and generalize.

B. Approximate electron propagator calculations and MCSTEP

In all practical electronic Green's function calculations, the exact initial state is approximated and the operator manifold is truncated.

With both MCEP and MCSTEP the initial state is multiconfigurational, usually chosen to be an MCSCF state. With MCSTEP (but not MCEP) the reference state is a tensor state (see below) which contains all M_S components for a given S.

The operator manifold for both MCEP and MCSTEP is $\{a_i, a_i|\Gamma\rangle\langle 0|, |0\rangle\langle\Gamma|a_i\}$ (with appropriate modifications with either method to assure that the final states are eigenfunctions of S^2 and S_z). The states $\{|\Gamma\rangle\}$ are states in the MCSCF orthogonal complement space to the initial state $|0\rangle$. The IP and EA transfer operators, $\{a_i|\Gamma\rangle\langle 0|\}$ and $\{|0\rangle\langle\Gamma|a_i\}$, respectively, are included analogously as in MC linear response (and MCSCF optimization) where equivalent N electron transfer operators occur naturally.[26,27] The IP and EA transfer operators are necessary to obtain good ionization potentials and electron affinities.

The (truncated) operator manifold of Dalgaard and Manne is also not particularly convenient or useful for the MC reference state even if the initial state is closed shell since a large number of operators would be necessary to duplicate the important effects of $\{a_i|\Gamma\rangle\langle 0|\}$ and $\{|0\rangle\langle\Gamma|a_i\}$.

The use of tensor states and tensor operators in the MCSTEP approach allows for the use of tensor algebra[28,29] in the derivation and subsequent evaluation of the MCSTEP equations.

In the MCSTEP technique, we employ a multiconfigurational self-consistent field (MCSCF) state as one component of the reference state. An MCSCF state is used so that atoms and molecules with highly correlated initial and final states can be handled accurately. A complete active space (CAS) MCSCF is usually used only in order to simplify the evaluation of certain matrix elements.[25] We rapidly and reliably obtain the MCSCF state by use of our Newton-Raphson complete second order procedures[30-32] with guaranteed convergence.[33] Upon convergence the characteristics of the stationary point are determined to assure that all five criteria are fulfilled for it to be a proper representation of the exact eigenstate of the Hamiltonian.[34-36] We also check to be certain that the obtained stationary point is desired[36] as well as proper. The MCSCF state, $|NS_0M_0\rangle$, is one component of an N-electron tensor state, $|NS_0\rangle\rangle$. The MCSCF reference tensor state (with $2S_0 + 1$ components) is composed of individual states that are eigenfunctions of S^2 and S_z and are interconnected by S_+ and S_-.

The operator manifold for MCSTEP consists of tensor operators. In order to ensure that N ± 1 electron states of pure spin symmetry are obtained even for open-shell reference states for which the initial spin is nonzero these operators are angular momentum coupled to the components of $|NS_0\rangle\rangle$. The MCSTEP operator manifold is

$$h = \{\mathbf{h}^{(1)},\mathbf{h}^{(2)},\mathbf{h}^{(3)}\} = \{a_r(\tfrac{1}{2}),\ T_{rc}(N-1,k),T_{rc}(N+1,k)\}, \tag{19}$$

where $\{a_r(\frac{1}{2})\}$ are the spin-tensor destruction operators of spin rank 1/2 and $\{T_{rc}(N \pm 1,k)\}$ are spin-tensor transfer-type operators of rank k. These latter tensor operators are formed by explicitly tensor coupling the single electron removal operators on the ket part of the tensor state transfer operators $\{|NS_c\rangle\rangle \times \langle\langle NS_0|\}$ to give $\{T_{rc}(N-1,k)\}$ and single electron removal operators on the bra part of the tensor state transfer operators $\{|NS_0\rangle\rangle \times \langle\langle NS_c|\}$ to give $\{T_{rc}(N+1,k)\}$. $|NS_cM_c\rangle$ is the Mc component of one of the tensor states, $|NS_c\rangle\rangle$, in the orthogonal complement space of the MCSCF initial state (note that $\mathbf{h}^{(2)}$ and $\mathbf{h}^{(3)}$ refer to IP and EA transfer-type operators, respectively, and <u>not</u> to operators of the form $a_i^+ a_j$ and $a_i^+ a_j a_k$).

In MCSTEP we use a modified version of Eq. (16) where the operator manifold of Eq. (19) is explicitly vector coupled to the different spin components in an MCSCF reference tensor state with spin S_0 to form a new N − 1 electron state with spin S_f. Subsequent Racah recoupling[29] results in the generalized matrix eigenvalue equation

$$\mathbf{M}X_f = \omega_f \mathbf{N}X_f, \tag{20}$$

where

$$M_{r,p} = \sum_{\Gamma}(-1)^{S_0-\Gamma-S_f-\gamma_r}\, W(\gamma_r\gamma_p S_0 S_0;\Gamma S_f)(2\Gamma+1)^{1/2}$$

$$\times \langle\langle NS_0 || \{h_r^+(\bar{\gamma}_r),H,h_p(\gamma_p)\}^{\Gamma} || NS_0\rangle\rangle, \tag{21}$$

$$N_{r,p} = \sum_{\Gamma}(-1)^{S_0-\Gamma-S_f-\gamma_r}\, W(\gamma_r\gamma_p S_0 S_0;\Gamma S_f)(2\Gamma+1)^{1/2}$$

$$\times \langle\langle NS_0 || \{h_r^+(\bar{\gamma}_r),h_p(\gamma_p)\}^{\Gamma} || NS_0\rangle\rangle, \tag{22}$$

and ω_f is an IP or EA to the final ion state $|N \pm 1\ S_f\rangle\rangle$ which has spin S_f. W is the usual Racah coefficient and $h_p(\gamma_p)$ and $h_r^+(\bar{\gamma}_r)$ are tensor operators from Eq. (19) with ranks γ_p and γ_r, respectively.

C. Characterization

1. Optimal MCSCF orbitals. The orbitals used in the MCSTEP calculation are the converged MCSCF orbitals of the reference state. The occupied and partially occupied orbitals directly obtained from an MCSCF calculation are optimal for defining and correlating the N-electron MCSCF state but may not be optimal for describing the ionic state as well. We note that CAS MCSCF orbitals are separated into three sets: an inactive set where each orbital is always doubly occupied in each configuration of the CAS MCSCF, an active set where an orbital is partially occupied overall (i.e., doubly occupied, singly occupied or unoccupied in each determinant in $|NS_0\rangle\rangle$), and an unoccupied (virtual) set where each orbital is empty in each configuration of the CAS for any $|NS_0\rangle\rangle$ component.

In performing an MCSTEP calculation we first set up Eq. (22) only with $\mathbf{h}^{(1)}$ (simple electron destruction) tensor operators. The three blocks in Eq. (22) for which r and p both represent inactive (doubly occupied) orbitals, both active (partially occupied) orbitals, and both unoccupied (virtual) orbitals are then separately diagonalized. A unitary transformation within each of the three orbital sets is thus obtained. This transformation does not allow the inactive, active, or unoccupied orbitals to mix with each other (but, of course, it does allow orbital mixing within each block). MCSCF orbitals can be arbitrarily rotated (mixed) with their own kind (e.g., inactive with inactive) without moving to another stationary point on the MCSCF hypersurface.[25] Hence the MCSCF reference state energy is unchanged. These orbitals for S=0 are the same as those later proposed and used by Andersson, Malmqvist, Roos, Sadlej, and Wolinski.[37]

We define an MCSTEP calculation as one in which the operator manifold of Eq. (19) and a complete active space MCSCF initial state is used. Therefore, we utilize the unitary transformation matrix to transform the original MCSCF orbitals to ones that diagonalize the $\mathbf{h}^{(1)} - \mathbf{h}^{(1)}$ block of the M matrix separately in the inactive-inactive, active-active, and unoccupied-unoccupied blocks. The state formed with these transformed orbitals and the original MCSCF configuration state functions (with new state expansion coefficients which diagonalize the CAS MCSCF CI using the new orbitals) has the same electronic energy and properties as the original MCSCF state.[25] However, the orbitals better describe simple electron addition and removal similar to Koopmans' theorem. This rotation of the MCSCF orbitals is analogous to the rotations which can be used in closed-shell SCF calculation among doubly occupied orbitals and separately among the virtual orbitals to obtain canonical Hartree-Fock orbitals and virtual orbitals which have Koopmans' theorem orbital energies. In fact, in the limit of a single determinant initial state the orbitals obtained from our rotation procedure are the canonical Hartree-Fock orbitals and virtual orbitals. (However, with MCSTEP we do not use or define a Fock operator or H_0 and do not use perturbation theory so the difficulties previously discussed for open-shell and highly correlated atoms and molecules do not arise.)

This orbital transformation procedure is useful in obtaining

MCSTEP contributions from primarily only one operator rather than from several operators in the $\mathbf{h}^{(1)}$ manifold when the IP is a principal IP and thus, serves as an aid in characterization.

2. Renormalization for Characterization. (In this section I will ignore, for clarity, all reference to tensor coupling in MCSTEP or {u, d, u^+, d^+} operators in MCEP.) The exact final N±1 electron states, $|\Psi_f^{(N-1)S_f M_f}\rangle$ and $\langle\Psi_f^{(N+1)S_f M_f}|$, are given by[21,38,39]

$$O_f|\Psi_0^{NS_0M_0}\rangle = |\Psi_f^{(N-1)S_f M_f}\rangle \text{ for IPs} \tag{23a}$$

and

$$\langle\Psi_0^{NS_0M_0}|O_f = \langle\Psi_f^{(N+1)S_f M_f}| \text{ for EAs} \tag{23b}$$

where

$$O_f = \sum_{i=1}^{\infty} X_{if}\, h_i \tag{24}$$

and X_f is the generalized eigenvector of the exact generalized eigenvalue problem Eq. (20). The MCSTEP equations (similarly to those of the polarization propagator) may also be derived making use of Eqs. (23) and the 'killer' condition[21,38]

$$O_f^+|\Psi_0^{NS_0M_0}\rangle = 0 \text{ for IPs} \tag{25a}$$

and

$$\langle\Psi_0^{NS_0M_0}|O_f^+ = 0 \text{ for EAs.} \tag{25b}$$

The commutator norm (i.e. superoperator binary product) of the approximate O_f operator with respect to the initial (MCSCF or CI) approximate state $|0\rangle$ is

$$\langle 0|\{O_f^+,O_f\}|0\rangle = \langle 0|O_f^+\, O_f|0\rangle + \langle 0|O_f\, O_f^+|0\rangle = 1. \tag{26}$$

Using Eqs. (25) the second term is approximately zero (in the limit of a complete operator manifold and exact initial state it is exactly zero) for an IP eigenvector, and the first term is approximately zero for an EA eigenvector. Using Eq. (23) the first term is approximately unity for a IP operator and the second term is approximately unity for a EA operator. This may be used to determine whether the obtained, approximate eigenoperator O_f corresponds to a IP or EA type process.

To obtain a proper statistical interpretation for characterization of these N±1 electron states we require

$$<(N\text{-}1)S_fM_f \,|\, (N-1)S_fM_f> \equiv <0\,|\,O_f^{+}O_f\,|\,0> = 1 \quad \text{(IPs)} \tag{27a}$$

and

$$<(N+1)S_fM_f \,|\, (N+1)S_fM_f> \equiv <0\,|\,O_fO_f^{+}\,|\,0> = 1 \quad \text{(EAs)} \tag{27b}$$

Thus we can re-normalize the obtained, approximate eigenoperator O_f using the the norm defined in Eqs. [27]. (Remember that $<0|O_f^{+}O_f|0>$ is exactly one for the exact eigenvector (Eq. [24]) but is only close to but not exactly unity for an approximate IP eigenvector and $<0|O_fO_f^{+}|0>$ is only close to but not exactly unity for an approximate EA eigenvector.) The final states $|(N\pm1)S_fM_f>$ are expanded in terms of the N±1 electron configurations, and the dominant configurations are used in the characterization of these states.

D. Calculations

Ionization potential calculations using usual electron propagator (or, equivalently, Equations-of-Motion (EOM) or single particle Green's function) methods which use perturbation theory to solve the equations through a certain order are reliable for low-lying, vertical, principal (i.e., IPs which can primarily be described by simple removal of an electron from an orbital) valence ionizations from closed shell initial state atoms and molecules with relatively small amounts of correlation.[1-12] For example, vertical, principal, outer valence ionization potentials may be reliably and accurately determined for many closed-shell atoms and molecules with third-order perturbative-type single particle Green's function (PTSPGF) methods and extensions in which some higher order terms are taken into account. With these methods vertical, principal, outer valence IPs for closed-shell systems without large correlation effects can be determined to about ±0.2 eV for first row (i.e., Li – Ne) diatomics and triatomics provided both d and f functions are included in slightly larger than double zeta (s,p) basis sets and some higher order (greater than third order) terms are taken into account or estimated. Without some estimation or accounting of higher order terms or without inclusion of f functions, errors may be ±0.1 eV larger. These methods are usually more reliable for calculating these kinds of IPs than other, non-Green's function methods such as Δ large scale configuration interaction (ΔCI), Δ coupled cluster (ΔCC) and Δ many-body perturbation theory (ΔMBPT).

Although these perturbative Green's function methods have been very successful for vertical, outer valence, principal IPs in closed shell atoms and molecules there are many cases for which PTSPGF techniques have serious problems. When correlation effects become important, as expected, PTSPGF methods are not reliable, e.g. for inner valence IPs, or when several configurations are of importance for describing the reference (initial) state (such as e.g. in Be). Shake-up transitions (in which the primary processes involve electron removal and simultaneous electronic excitation) are handled only with difficulty with PTSPGF and usually at a lower level of approximation than the principal IPs. For shake-up ionizations these calculations (with extensions of usual theory) are often of only qualitative accuracy. More accurate shake-up IP calculations using typical closed-shell electron propagator techniques require very complicated extensions of theories and computational methods. Because of difficulties in consistently choosing an adequate zero order Hamiltonian for perturbation theory and frequent breakdown in the perturbation series, open-shell perturbative-type Green's function methods[40-42] have not been particularly successful or generalizable. Furthermore, if the initial open shell state is not totally symmetric, because of the use of simple creation $\{a_r^+\}$ and destruction $\{a_r\}$ operators and simple combinations of these operators in the usual Green's function techniques, the ionization processes described are to ion states which frequently are not eigenfunctions of S^2 and S_z.[43]

Nichols, Yeager and Jørgensen[23] proposed, coded, and initially tested a non-perturbative Green's function procedure called the multiconfigurational electron propagator (MCEP). The MCEP solves the problems encountered in PTSPGF calculations when applied to highly correlated reference states or to open-shell atoms and molecules. Initial calculations with MCEP were performed for the vertical IPs below 50 eV in N_2 and O_2[23] and later for F_2.[44] These MCEP results are very accurate giving vertical IPs below 25 eV ± 0.3 eV away from experiment using valence <4s3p1d> or <5s4p1d> basis sets and giving qualitatively correct energies for all IPs below 50 eV.[23,44]

In Table I the vertical IPs in O_2 are listed, along with Δ multireference configuration interaction (CI) and experiment. It is apparent that MCEP gives accurate and reliable IPs in O_2 below 30 eV even for IPs, e.g. to the b $^4\Sigma_g^-$ and B $^2\Sigma_g^-$ states where multireference CI results are significantly in error. MCSTEP O_2 calculations gave essentially identical results.[45]

Comparison calculations were obtained for the valence ionization potentials in F_2 with MCEP and other large scale techniques within the same basis set, i.e. ΔSCF, ΔMCSCF, PTSPGF, and Δ large scale multireference CI.[44] In Table II the lowest three vertical, principal IPs in F_2 are listed comparing several different large scale methods.

Koopman's Theorem F_2 IPs (the negatives of the X $^1\Sigma_g^+$ canonical SCF orbital energies) are in the wrong order and differ by as much as 3 eV from experiment. ΔSCF IPs are better but are still in the wrong order. ΔMCSCF IPs are in the correct order but are too low (by as much as 1.9 eV).

Table I. The observable ionization potentials of O_2 below 30 eV.[a]

Ionic state	MCEP[b]	Multireference CI[c]	Experiment PES[d]
$X^2\Pi_g$	12.23	12.11	12.3
$a^4\Pi_u$	16.70	16.68	16.8
$A^2\Pi_u$	17.77	17.79	17.7
$b^4\Sigma_g^-$	18.54	17.83	18.4
$B^2\Sigma_g^-$	20.86	19.69	20.7
$^2\Pi_u$	24.56	---	24.0
$C^4\Sigma_u^-$	24.87	---	24.6
$^2\Sigma_u^-$	28.41	---	27.8
$^4\Sigma_u^-$	29.42	---	---

[a]All results in eV.

[b]Ref. [23]. MCSTEP results are essentially identical (see Ref. [45]).

[c]Ref. [46].

[d]Ref. [47].

Table II. Comparison of the outer valence, vertical (R=2.68 a.u.) potentials of F_2 in eV using a <4s3p1d> basis set (Ref. [44]).

State	Koopman's theorem	ΔSCF	ΔMCSCF[a]	Δ Large scale multi-reference CI[b]	PTSPGF (third-order)[c]	MCEP[d]	Exp[e]
$\tilde{X}^2\Pi_g(1\pi_g)^{-1}$	18.13	16.63	14.47	15.29	15.43	16.03	15.83
$\tilde{A}^2\Pi_u(1\pi_u)^{-1}$	21.98	20.58	17.56	18.40	18.56	19.04	18.90
$\tilde{B}^2\Sigma_g^+(3\sigma_g)^{-1}$	20.14	18.63	19.15	20.69	20.93	21.39	21.10

[a]Multiconfigurational self-consistent field. The $X^1\Sigma_g^+$ MCSCF included all configurations of the correct symmetry involving 14 electrons in all the valence orbitals ($2\sigma_g$, $2\sigma_u$, $3\sigma_g$, $1\pi_u$, $1\pi_g$, $3\sigma_u$). The ion MCSCF's included all configurations of the appropriate symmetry with 13 electrons in all the valence orbitals.

[b]Configuration interaction. Each CI used the orbitals from the corresponding MCSCF calculation and the corresponding MCSCF configurations plus all single and double excitations of the appropriate symmetry from the CAS orbitals into the unoccupied MCSCF virtual orbitals.

[c]Third-order perturbation-type Green's function (standard EOM). For details see Ref. [6].

[d]Multiconfigurational electron propagator.

[e]Ref. [48].

For the F_2 Δ large scale multireference CI calculations, separate CI calculations were performed on both the ground and ionic states. First, separate multiconfigurational self consistent field (MCSCF) calculations were performed on the X $^1\Sigma_g^+$ and each of the ion states. The X $^1\Sigma_g^+$ MCSCF included all configurations of the correct symmetry involving 14 electrons in all the valence (CAS) orbitals function ($2\sigma_g$, $2\sigma_u$, $3\sigma_g$, $1\pi_u$, $1\pi_g$, $3\sigma_u$). The ion MCSCFs included all configurations of the appropriate symmetry with 13 electrons in all the valence orbitals. This procedure gives 8, 20, 20, and 18 configurations for the $X^1\Sigma_g^+$, $\tilde{X}^2\Pi_g$, $\tilde{A}^2\Pi_u$, and $\tilde{B}^2\Sigma_g^+$ states, respectively. Each CI used the orbitals from the corresponding MCSCF calculation and the corresponding MCSCF configurations plus all single double excitations of the appropriate symmetry from the CAS (i.e. valence) orbitals into the unoccupied MCSCF virtual orbitals. This procedure gives 27,040; 90,201; 90,201, and 103,596 configurations for the $X^1\Sigma_g^+$, $\tilde{X}^2\Pi_g$, $\tilde{A}^2\Pi_u$, and $\tilde{B}^2\Sigma_g^+$ states, respectively. This is probably the most widely used MRCI procedure.

Considerable errors can persist in Δ very large scale multireference CI even for the lowest IPs (see Table II). The lowest three Δ very large scale multireference CI IPs differ from experiment by 0.41 - 0.54 eV (all too low). This error demonstrates the difficulty "balancing" correlation corrections for ionization potentials and excitation energies in even Δ large scale CI. The ΔSCF IPs are too large. The all valence CAS MCSCF procedure has more configurations for each ion state than for the ground state. Thus, the ΔMCSCF calculations over–compensate by "correlating" the ionic states more than the neutral and the IPs are all too low. The ΔMRCI better balances correlation corrections. However, there are considerably more configurations for each ionic state than for the ground state. Thus, most likely, the ionic states are again "over-corrected" compared to the neutral and the IPs are consequently somewhat low. It should be re-emphasized that this MRCI procedure used for obtaining orbitals and configurations is probably the method in most wide-spread use.

Similar fairly large errors have been reported with Δ many body perturbation theory (MP4) (see e.g. Refs. [49,50]) and Δ coupled cluster (see e.g. Refs. [51,52]). It is also extremely difficult or impossible to calculate all of the IPs 0-50 eV in e.g. F_2 using Δ very large scale multireference CI or other wavefunction based techniques (whereas it is straightforward and not difficult with MCEP and MCSTEP).

The best results in Table II compared with experiment are given by MCEP where the calculated IPs are within 0.3 eV of experiment. Given the size of the basis set used, i.e. <4s3p1d>, this is as good as can reasonably be expected. PTSPGF IPs are also fairly good giving values within -0.4 eV of experiment. However, as noted above, PTSPGF approaches are not reliable for other systems which are open shell or highly correlated. In addition with PTSPGF inner valence IPs are determined only with difficulty and not very reliably.

We have recently used MCSTEP to predict the low-lying IPs in NH_2.[53] These results are presented in Table III. The NH_2 UV photoelectron spectrum was obtained by Dunlavey et al.[54] It is cluttered and difficult to interpret due to the presence of contaminants, reactants, and reaction by-products. Our results are an aid in analyzing and interpreting this spectrum. For example, we predict the presence PES peaks at 16.86 eV, 18.00 eV, and 18.26 eV which should be observed in addition to the reported observed peaks we calculate at 11.77 eV, 12.22 eV, and 14.03 eV.

Although MCEP and MCSTEP give accurate low lying IPs higher lying IPs are qualitatively but not quantiatively correct. For example, the MCEP IPs to the $^4\Sigma_g^-$ and $^2\Sigma_g^-$ states of O_2 calculated at 42.00 eV and 42.30 eV, respectively, are apparently 2-3 eV from experiment.[23] This is because the operators included in MCEP and MCSTEP do not allow for electron removal + simultaneous excitation of the remaining electrons to diffuse ion states. We have modified MCSTEP to take these processes into account. Our new method is known as the repartitioned MCSTEP (RMCSTEP).[25]

RMCSTEP IPs for Be are given in Table IV. The RMCSTEP IPs to the lowest eight ion states are within ±0.07 eV except for the IP to the ^2D state which is 0.14 eV from experiment. This (still relatively small) error for the IP to the ^2D state compared with the other IPs is due to the choice of basis set. Except for the lowest two IPs, these RMCSTEP IPs involve primarily simple electron removal + excitation to diffuse ion states. We are currently testing RMCSTEP for the high-lying IPs of other atoms and molecules of chemical and physical interest.

We have recently applied the multiconfigurational electron propagator for the first time to calculate electron affinities.[56] The calculated electron affinities of Li, Na, and K differ from experiment on average by 0.003 eV, giving the best theoretical values reported to date. These results are given in Table V.

III. MULTICONFIGURATIONAL LINEAR RESPONSE (ALSO KNOWN AS POLARIZATION PROPAGATOR, POLARIZATION GREEN'S FUNCTION, RANDOM PHASE APPROXIMATION, AND TIME DEPENDENT HARTREE-FOCK) TECHNIQUES

A. Introduction

Propagators were introduced in statistical physics in order to describe the response of a system to an external perturbation.[63] Polarization propagators are closely related to response properties. The propagator which describes the linear response is the so-called retarded polarization propagator. For example, the frequency dependent polarizability is correctly determined by the linear response of a system to a time

Table III. Vertical IPs of NH_2 (in eV).[a]

	<4s3p1d/2s1p>		6-311G+(2d,p)	6-311G+(2df,pd)		
Ion State	(Basis I) MCSTEP	(Basis I) MR-SDCI	(Basis II) MCSTEP	(Basis III) MCSTEP	EXP[b]	Other Calc[b]
$\tilde{X}\,^3B_1$	11.572	11.511	11.763	11.767	12.00	11.37
$\tilde{A}\,^1A_1$	12.101	12.089	12.213	12.218	12.45	12.10
$^1B_1$	13.981		14.030	14.031	14.27	13.71
$^3A_2$	16.783		16.864	16.859		16.35
$^1A_1$	17.815		17.974	17.995		---
$^1A_2$	18.155		18.254	18.256		17.66
$^3B_2$	19.230		19.381	19.385		---
	.125[d]		.006[d]			

[a]The calculations are carried out at the experimental X^2B_1 equilibrium geometry.

[b]Ref. [54]. The calculational results are Δ single configuration singles + doubles CI.

[c]Ref. [43]. 2ph-extended TDA (a PTSPGF approach). This result is an IP obtained by determing the lowest EA of the closed shell $^1A_1$ positive ion state. Since PTGF methods are generally unreliable for open shell initial states, no other of the low-lying IPs of NH_2 can be accurately determined by this method.

[d]Average absolute deviation from MCSTEP using Basis III.

Table IV. MCSTEP and Re-partitioned MCSTEP ionization potentials of Be atom using the complete active space orbitals (2s, 2p, 3s, 3p, 3d) (Ref. [25]).[a]

State	MCSTEP IP	Re-partitioned MCSTEP IP	Exp[b]
X^2S	9.31	9.28	9.32
^2P	13.42	13.33	13.28
^2S	23.08	20.21	20.26
^2P	---	21.25	21.28
^2D	26.09	21.62	21.48
^2S	---	23.58	23.64
^2P	---	23.99	24.05

[a]All results in eV. The basis set is composed a <5s5p1d> contracted, Cartesian Gaussian valence basis augmented with 2 diffuse s, 2 diffuse p and 2 diffuse d functions.

[b]Ref. [55].

Table V. The electron affinities of the alkali atoms Li, Na, and K (in eV).

Atom	Expt.[a]	MCSTEP[b]	Kaldor[c] MR-CCSD	Partridge *et al.*[d] MR-SDCI	Christensen-Dalsgaard[e] hyperspherical	Ortiz[f] PTGF
Li	0.618	0.619	0.610		0.596	0.549
Na	0.548	0.544	0.552	0.541	0.539	0.482
K	0.501	0.496	0.507	0.492	0.499	0.497
Average[g]		0.003	0.006	0.008	0.011	0.046

[a]Ref. [57].

[b]Ref. [56].

[c]Refs. [58 and 59]. Multireference coupled cluster.

[d]Ref. [60]. Multireference CI.

[e]Ref. [61].

[f]Ref. [62.] Perturbational-type single-particle Green's function (electron propagator) method. These results are full third order with the most important fourth-order terms included.

[g]Average absolute deviation from experiment.

varying electric field. The poles of the polarization propagator/Green's function give the excitation energies of the system.

Initial application of polarization propagator methods was performed both within the framework of *ab initio* and semiempirical molecular orbital theory using the Random Phase Approximation (RPA)[3,64,65] (also known as the time dependent Hartree Fock (TDHF)) or self-consistent extensions of RPA.[3] Most subsequent applications of polarization propagator methods concentrated on extensions of the RPA/TDHF approximation which are based on a perturbation expansion of the polarization propagator in which RPA/TDHF is the first order approximation. These methods (and closely related techniques) we will refer to as perturbational-type polarization Green's function (PTPGF) approaches. While PTPGF methods have been generally useful, successful and reliable for closed shell atoms and molecules with a relatively small amounts of initial state correlation they have the same problems inherent in perturbation approaches to single particle Green's function techniques, i.e. it is difficult or impossible to reliably calculate excitation energies, oscillator strengths, polarizabilities, and other response properties when non-dynamical effects are important or when the initial state is open shell.

Yeager and Jørgensen first proposed[66] and developed the multiconfigurational time dependent Hartree Fock (MCTDHF/MCLR) approach.[67] (This is equivalent to the multiconfigurational linear response, multiconfigurational polarization Green's function, and multiconfigurational polarization propagator methods). Other developments with multiconfigurational polarization propagators have been made by Banerjee, Kenney, and Simons[68]; McWeeny[69]; and Dalgaard[70].

The Green's function we examine for excitation energies and response properties is:

$$G_{r_\sigma s_{\sigma'}, p_{\sigma''} q_{\sigma'''}}(E) = -\langle\Psi_0^{NS_0M_0}|\, a_{r_\sigma}^+ a_{s_{\sigma'}} (E+E_0^{NS_0}-H)^{-1} a_{p_{\sigma''}}^+ a_{q_{\sigma'''}} \,|\Psi_0^{NS_0M_0}\rangle$$

$$+ \langle\Psi_0^{NS_0M_0}|\, a_{p_{\sigma''}}^+ a_{q_{\sigma'''}} (E-E_0^{NS_0}+H)^{-1} a_{r_\sigma}^+ a_{s_{\sigma'}} \,|\Psi_0^{NS_0M_0}\rangle \qquad (28)$$

In practical MC calculations the initial state is chosen to be an MCSCF state and the operator manifold is $\{a_i^+ a_j, |\Gamma\rangle\langle 0|\}$. These are consistent with the operator set used for MCSCF optimization. The resulting generalized matrix eigenvalue equation is:

$$\begin{pmatrix} \mathbf{A} & \mathbf{B} \\ \mathbf{B} & \mathbf{A} \end{pmatrix} \begin{pmatrix} Y_f \\ Z_f \end{pmatrix} = \omega_f \begin{pmatrix} \mathbf{S} & \Delta \\ -\Delta & -\mathbf{S} \end{pmatrix} \begin{pmatrix} Y_f \\ Z_f \end{pmatrix} \qquad (29)$$

where ω_f refers to an excitation energy. The matrix elements are given elsewhere.[2]

It can easily be shown that the RPA/TDHF is equivalent to the proper linear response of an SCF single-determinant wavefunction to an external perturbation, such as an electromagnetic field.[2,9] Similarly, MCTDHF/MCLR is equivalent to the correct linear response of a multiconfigurational self consistent field (MCSCF) wavefunction.[2,9] It has been shown as well that correct response properties are more difficult to attain using wavefunctions, such as CI or coupled cluster wavefunctions, for which all variational parameters are not fully optimized[2] (i.e. in those approaches the orbitals are not optimized for the configurations included in the wavefunction and the Hellmann-Feynman Theorem is not satisfied).

In the limit of a complete basis set of orbitals, length, velocity, and acceleration forms of the oscillator strength are equal in RPA/TDHF or in MCTDHF/MCLR. Furthermore with these the Thomas-Reiche-Kuhn (TRK) sum rule is satisfied for a complete basis set. For the practical MCTDHF/MCLR calculations reported below the length, velocity, and acceleration forms of the oscillator strength are almost exactly equal and the TRK sum rule is almost exactly fulfilled. This is not true for most other *ab initio* methods.[2,9,66,67]

B. Calculations

We have recently applied MCTDHF/MCLR to Be atom for twenty-five low lying singlet and triplet excitations and compared the results to Δ full CI.[71] Be atom provides an excellent demonstrative case for multiconfigurational based Green's function methods because of the fairly strong mixing (i.e. non-dynamical correlation) of the $2p^2$ configuration (~10%) with the $2s^2$ (~90%) in the ground state. These MCTDHF/MCLR results for excitations to triplet states are shown in Table VI (similar results are obtained for singlet states). These excitation energies mimic Δ full CI usually to within a few thousandths of eV. However, each full CI energy required approximately 1-2 CPU days on an FPS-164. All the MCTDHF/MCLR excitation energies were obtained on a VAX 11/780 in under two hours CPU time total.

To further demonstrate the efficacy and accuracy of MCTDHF/MCLR we present comparison average deviations from full CI for the lowest five singly excited states of ^3S and ^3P symmetry and lowest four singly excited states of ^1S and ^1P symmetry with other newly developed MC methods.[73,74] The same basis set was used in all these calculations. These results are given in Table VII. The MR-MBPT[73] and CCDPPA[74] calculations for the low-lying doubly excited states $2p^2$ $^3P^e$, $2p^2$ ^1D, and $2p^2$ ^1S were not presented in the original MR-MBPT and CCPPA publications, possibly because of the difficulty in accurately calculating doubly excited states with other MC methods. No $^{1,3}D$ results were reported for MR-MBPT. CCSDPPA results were slightly worse than CCDPPA values.[74]

Transition moments and both frequency dependent and frequency independent polarizabilities have also been determined for Be using the same <9s9p5d> contracted Cartesian Gaussian basis set.[75] The lowest

Table VI. $X^1S \rightarrow {}^3S$, ${}^3P$, and ${}^3D$ excitation energies of Be [71].[a]

State	ΔMCSCF (2s3s2p3p3d)[b]	MCTDHF/MCLR (2s3s2p3p3d)[c]	Δfull CI	Expt[b]
$2s2p\,^3P^o$	2.739	2.732	2.733	2.725
$2s3s\,^3S$	6.434	6.439	6.444	6.457
$2s3p\,^3P^o$	7.290	7.303	7.295	7.303
$2p^2\,^3P^e$	7.457	7.524	7.423	7.401[e]
$2s3d\,^3D$	8.479	7.746	7.741	7.694
$2s4s\,^3S$		7.992	7.985	7.998
$2s4p\,^3P^o$		8.281	8.272	8.283
$2s4d\,^3D$		8.457	8.449	8.423
$2s5s\,^3S$		8.569	8.560	8.556
$2s5p\,^3P^o$		8.694	8.686	8.688
$2s6s\,^3S$		8.894	8.886	8.823
$2s6p\,^3P^o$		8.967	8.957	8.893

[a]All excitation energies are reported in eV. The basis set is a <9s9p5d> set of contracted Cartesian Gaussian functions.

[b]The complete active space (CAS) of two electrons in the (2s3s2p3p3d) orbitals were used for both the excited state and the ground state MCSCF calculations.

[c]The orbitals in parentheses refer to the CAS used in the MCSCF reference state in the MCTDHF/MCLR calculations.

[d]Ref. [55].

[e]Ref. [72].

Table VII. Average absolute error (eV) for excitation in Be using new MC methods compared with Δ full CI.[a,b)]

	MCTDHF/MCLR[c)]			MR-MBPT[d)]		CCDPPA[e)]
	(2s2p) CAS	(2-3s, 2-3p) CAS	(2-3s, 2-3p, 3d) CAS	Fifth order (MR-MBPT(5))	MR-MBPT(5)+ [2,2] Pade' approximate	
^1S	0.182	0.001	0.008	0.032	0.018	0.08
^3S	0.145	0.005	0.007	0.036	0.016	0.06
^1P	0.144	0.016	0.007	0.032	0.016	0.03
^3P	0.143	0.002	0.007	0.036	0.017	0.09
^1D	0.204	0.026	0.022	f)	f)	0.12
^3D[f)]	0.175	0.004	0.006	f)	f)	0.04
$2p^2\ ^3P^e, ^1D$	0.212	0.142	0.070	g)	g)	g)

a)Lowest five singly excited state of each triplet S,P symmetry. Lowest four singly excited state of singlet S,P symmetry. The basis set is the same (as in Table VI) for all methods.

b)All results in eV.

c)Multiconfigurational time dependent Hartree-Fock (also known as the multiconfigurational linear response). Ref. [71].

d)Multiconfigurational mady body perturbation theory. Fifth order results and fifth order plus Pade' approximate correction. Ref. [73].

e)Coupled cluster doubles polarization propagator approximation. CCSDPPA results were slightly worse. Ref. [74].

f)Lowest two singly excited states of singlet, triplet D symmetry. No $^{1,3}D$ results were reported with MR-MBPT.

g)No doubly excited states were reported for MR-MBPT or CCPA. Full CI results were not obtained for $2p^2\ ^1S$.

transition moment is calculated at 1.39 au with MCTDHF/MCLR compared with 1.34 ± 0.05 experimentally.[76] A comparison list of frequency independent polarizabilities is given in Table VIII. Since the MCTDHF/MCLR is the correct linear response for an MCSCF wavefunction, we anticipate that our best value of 37.64 au is the most accurate calculated to date. In addition, with MCTDHF/MCLR it is easy to calculate frequency dependent polarizabilities.[67,75]

Several other MCTDHF/MCLR calculations have been done on several other systems of chemical and physical interest. These include: O_2[67,90,91], CH_2[92], CO[93], N_2[94,95], CH^+[15,96], C_2H_2[97], Ne[98,99], He[99] and Ar[99] and several other systems.

Some other interesting and important uses of multiconfigurational linear response are for geometry optimization.[2,100-102]

Another is for characterizing MCSCF stationary points,[34-36] i.e. for determining if an obtained MCSCF stationary points if a good representation of the exact, desired state. For example, if one is trying to obtain the Nth state the MCSCF stationary point should give N–1 negative MCTDHF/MCLR transition energies and no instabilities. Unfortunately, most MCSCF stationary points are not checked for this necessary characteristic. No current, forefront MCSCF code checks this property on convergence. Hence, many of these MCSCF stationary points obtained with these programs may be incorrect.

IV. THE MULTICONFIGURATIONAL PARTICLE-PARTICLE PROPAGATOR (MCP2P) METHOD FOR DIRECTLY DETERMINING VERTICAL DOUBLE IONIZATION POTENTIALS AND ELECTRON AFFINITIES

We have recently presented and developed the multiconfigurational particle-particle propagator (MCP2P) method for directly determining vertical double ionization potentials (DIPs) and double electron affinities (DEAs).[103] The particle-particle propagator for the exact N-electron state is

$$G_{r_\sigma s_{\sigma'}, p_{\sigma''} q_{\sigma'''}}(E) = -\langle\Psi_0^{NS_0M_0} | a_{r_\sigma}^+ a_{s_{\sigma'}}^+ (E+E_0^{NS_0}-H)^{-1} a_{p_{\sigma''}} a_{q_{\sigma'''}} | \Psi_0^{NS_0M_0}\rangle$$
$$+ \langle\Psi_0^{NS_0M_0} | a_{p_{\sigma''}} a_{q_{\sigma'''}} (E-E_0^{NS_0}+H)^{-1} a_{r_\sigma}^+ a_{s_{\sigma'}}^+ | \Psi_0^{NS_0M_0}\rangle \qquad (30)$$

We choose an MCSCF state as our approximation to the exact initial state. For the operator manifold we choose

$$\{a_i a_j,\ |\lambda^{(N-2)}\rangle\langle 0|,\ |0\rangle\langle\lambda^{(N+2)}|\}$$

where $|\lambda^{(N\pm2)}\rangle$ represents all independent N±2 electron configurations in the reference CAS. New large scale methods were developed and used for the MCP2P as well.[103]

Table VIII. Frequency independent X^1S polarizabilities of Be [Ref. 75].

Method	α(o) (au)
Ref. [69]	
TDHF/SCLR	45.64
MCTDHF/MCLR (2s2p)	36.40
MCTDHF/MCLR (2s2p3s3p)	37.72
MCTDHF/MCLR (2s2p3s3p3d)	37.64
HRPA	37.50
Previous Calculations	
TDHF/SCLR[b]	45.88
MCTDHF/MCLR (2s2p)[b]	36.38
CI (S + D)[c]	37.6
CI (S + D)[c]	37.3
RHF[d]	45.63
PNO-CI[d]	39.44
PNO-CEPA[d]	37.84
NHF[e]	45.615
MCSCF (base)[f]	45.55
MCSCF (corr)[f]	36.51
Per. Var. (CI)[g]	37.05
Per. Var. (CI)[g]	38.14
Pseudopotential[h]	36.7
MBPT[i]	46.77
SOPPA[j]	43.54
SD-MBPT [1/1][k]	35.92
Exact Lower Bound[l]	35.75
Density Functional Linear Response[m]	43.0
Time Dependent Kohn Sham[n]	39.1

[a]CAS is in parentheses. Basis set is the same as in Table VI. HRPA results are EOM with double excitations and double deexcitations neglected.
[b]Ref. [77].
[c]Ref. [78].
[d]Ref. [79].
[e]Ref. [80].
[f]Ref. [81].
[g]Ref. [82].
[h]Ref. [83].
[i]Ref. [84].
[j]Ref. [85]; 50 function STO calculation.
[k]Ref. [86].
[l]Ref. [87].
[m]Ref. [88].
[n]Ref. [89].

In the first calculations with this method (obtained using a moderate-sized CAS and moderate-sized basis set) the average absolute deviation from Δ full configuration interaction within the same basis set for DIPs to the lowest (in energy) five $^{1,3}S$ Be^{2+} states is 0.04 eV.

V. SUMMARY

Multiconfigurational based Green's function or propagator techniques provide accurate and reliable excitation energies, ionization potentials, and electron affinities. They consistently out perform Δ large scale multireference CI and many other large scale *ab initio* methods both in computational efficiency and accuracy compared with experiment. When Δ full CI can be obtained, our results to date show that these multiconfigurational Green's function approaches mimic Δ full CI at a fraction of the computer time.

ACKNOWLEDGEMENTS

I would like to acknowledge the Robert A. Welch Foundation (Grant A-770) for research support and to thank Rebecca K. Thomson for typing this manuscript.

REFERENCES

1. J. Linderberg and Y. Öhrn, "Propagators in Quantum Chemistry," Academic Press, London (1973).
2. P. Jørgensen and J. Simons, "Second Quantization-Based Methods in Quantum Chemistry," Academic Press, New York (1981).
3. C. W. McCurdy, T. Rescigno, D. L. Yeager and V. McKoy, in: "Methods of Electronic Structures," H. F. Schaefer III, ed., Plenum, New York (1977).
4. L. S. Cederbaum, G. Hohlneicher, and W. von Niessen, Chem. Phys. Lett. 18:503 (1973).
5. L. S. Cederbaum, W. Domcke, J. Schirmer, and W. von Niessen, Adv. Chem. Phys. 65:115 (1986).
6. M. F. Herman, K. F. Freed, and D. L. Yeager, Adv. Chem. Phys. 48:1 (1981).
7. J. Oddershede, Adv. Chem. Phys. LXIX:201 (1987).
8. J. Simons, Annual Rev. Phys. Chem. 28:15 (1977).
9. J. Oddershede, P. Jørgensen, and D. Yeager, Comp. Phys. Rep. 2:33 (1984).
10. J. Oddershede, Adv. Quantum Chem. 11:257 (1978).
11. G. Born and Y. Öhrn, Adv. Quantum Chem. 13:1 (1981).
12. J. V. Ortiz, Chem. Phys. Lett. 136:387 (1987).
13. J. Olsen, B. J. Roos, P. Jørgensen, and H. J. Aa. Jensen, J. Chem. Phys. 89:2185 (1989).
14. J. Olsen, H. J. Aa. Jensen and P. Jørgensen, J. Comp. Phys. 74:265 (1988).
15. J. Olsen, D. Yeager, and P. Jørgensen, J. Chem. Phys. 91:381 (1989).

16. D. N. Zubarev, "Nonequilibrium Statistical Thermodynamics," Consultants Bureau, New York (1974).
17. O. Goscinski and B. Lukman, Chem. Phys. Lett. 7:573 (1970).
18. B. T. Pickup and O. Goscinski, Mol. Phys. 26:1013 (1973).
19. P.-O. Löwdin, Phys. Rev. 139:357 (1965).
20. J. Simons, J. Chem. Phys. 64:4541 (1976).
21. E. Dalgaard, Int. J. Quantum Chem. XV:169 (1979).
22. R. Manne, Chem. Phys. Lett. 45:470 (1977).
23. J. Nichols, D. Yeager, and P. Jørgensen, J. Chem. Phys. 80:293 (1984).
24. B. T. Pickup and A. Mukopadhyay, Chem. Phys. Lett. 79:109 (1981).
25. J. T. Golab and D. L. Yeager, J. Chem. Phys. 87:2925 (1987).
26. D. L. Yeager and P. Jørgensen, Chem. Phys. Lett. 65:77 (1979).
27. E. Dalgaard, J. Chem. Phys. 72:816 (1980).
28. D. J. Rowe and C. Ngo-Trong, Rev. Mod. Phys. 47:471 (1975).
29. M. E. Rose, "Elementary Theory of Angular Momentum," John Wiley, New York (1957).
30. D. L. Yeager and P. Jørgensen, J. Chem. Phys. 71:755 (1979).
31. D. Yeager and P. Jørgensen, Mol. Phys. 39:587 (1980).
32. J. Olsen, D. L. Yeager, and P. Jørgensen, Adv. Chem. Phys. 54:1 (1983).
33. P. Jørgensen, P. Swanstrøm, and D. L. Yeager, J. Chem. Phys. 78:347 (1982).
34. J. Golab, D. Yeager, and P. Jørgensen, Chem. Phys. 78:175 (1983).
35. J. Golab, D. Yeager, and P. Jørgensen, Chem. Phys. 93:83 (1985).
36. A. Rizzo and D. Yeager, J. Chem. Phys. 93:8011 (1990).
37. K. Andersson, P. Malmqvist, B. Roos, A. Sadlej, and K. Wolinski, J. Phys. Chem. 94:5483 (1990).
38. D. J. Rowe, "Nuclear Collective Motion," Methuen, London (1970).
39. M. F. Herman and K. F. Freed, Chem. Phys. 36:383 (1979).
40. P. Albertsen and P. Jørgensen, J. Chem. Phys. 70:3254 (1979); *ibid* 71:4652 (1979).
41. G. D. Purvis and Y. Öhrn, J. Chem. Phys. 62:2045 (1975).
42. L. S. Cederbaum and J. Schirmer, Z. Physik 271:221 (1974); J. Schirmer, L. S. Cederbaum, and W. von Niessen, Chem. Phys. 56:285 (1981).
43. P. Tomasello, W. von Niessen, J. Schirmer, and L. S. Cederbaum, J. Electron Spec. 40:193 (1986).
44. J. Golab, B. Thies, D. Yeager, and J. Nichols, J. Chem. Phys. 84:284 (1986); erratum, J. Chem. Phys. 87:778 (1987).
45. J. Golab, D. Yeager, and P. Swanstrøm, Chem. Phys. 110:339 (1986).
46. K. Kimura, S. Katsumata, Y. Achiba, T. Yamazaki, and S. Iwata, "Handbook of HeI Photoelectron Spectra of Fundamental Organic Molecules," John Wiley, New York (1981).
47. J. A. R. Samson, J. L. Gardner, and G. N. Haddad, J. Electron Spectrosc. 12:28 (1977).
48. H. van Lonkhuzen and C. A. de Lange, Chem. Phys. 89:313 (1984).
49. J. Baker, R. Nobes, and L. Radom. J. Comp. Chem. 7:349 (1986).
50. J. V. Ortiz, Chem. Phys. Lett. 136:387 (1987).
51. H. Nakasuji, Int. J. Quantum Chem. S17:241 (1983).
52. A. Haque and U. Kaldor, Int. J. Quantum Chem. 34:425 (1986).
53. R. Graham, J. Golab, and D. Yeager, J. Chem. Phys. 88:2572 (1988).

54. S. J. Dunlavey, J. M. Dyke, N. Jonathan, and A. Morris, Mol. Phys. 39:1121 (1980); J. M. Dyke, N. Jonathan, and A. Morris, Int. Rev. Phys. Chem. 2:3 (1982).
55. S. Bashkin and J. A. Stoner Jr., "Atomic Energy Levels and Grotrian Diagrams I," North Holland, Amsterdam (1975).
56. R. Graham, D. Yeager, and A. Rizzo, J. Chem. Phys. 91:5451 (1989); erratum, J. Chem. Phys. 92:6336 (1990).
57. H. Hotop and W. C. Lineberger, J. Phys. Chem. Ref. Data 14:731 (1985).
58. U. Kaldor, J. Chem. Phys. 87:4693 (1987).
59. U. Kaldor, J. Chem. Phys. 88:5248 (1988).
60. H. Partridge, D. A. Dixon, S. P. Walch, C. W. Bauschlicher Jr., and J. L. Gole, J. Chem. Phys. 79:1859 (1983).
61. B. L. Christensen-Dalsgaard, J. Phys. B 21:2359 (1988).
62. J. V. Ortiz, J. Chem. Phys. 89:6348 (1988).
63. R. Kubo, J. Phys. Soc. Japan 12:570 (1959).
64. T. H. Dunning and V. McKoy, J. Chem. Phys. 47:1735 (1967).
65. P. Jørgensen and J. Linderberg, Int. J. Quantum Chem. 4:587 (1970).
66. D. L. Yeager and P. Jørgensen, Chem. Phys. Lett. 65:77 (1979).
67. P. Albertsen, P. Jørgensen, and D. L. Yeager, Mol. Phys. 41:409 (1980).
68. A. Banerjee, J. Kenney III, and J. Simons, Int. J. Quantum Chem. 16:1209 (1979).
69. R. McWeeny, Int. J. Quantum Chem. 23:405 (1983); M. Jaszunski and R. McWeeny, Mol. Phys. 46:863 (1982).
70. E. Dalgaard, J. Chem. Phys. 72:816 (1980); E. Dalgaard, Phys. Rev. A 26:42 (1982).
71. R. Graham, D. L. Yeager, J. Olsen, P. Jørgensen, R. Harrison, S. Zarrabian, and R. Bartlett, J. Chem. Phys. 85:6544 (1986).
72. J. E. Holmström and L. Johansson, Ark. Fys. 40:133 (1969); L. Johansson, *ibid.* 23:119 (1962).
73. S. Zarrabian and R. Bartlett, Chem. Phys. Lett. 153:133 (1988).
74. J. Geertsen, J. Oddershede, and G. E. Scuseria, Int. J. Quantum Chem. S21:475 (1987).
75. R. Graham and D. L. Yeager, Int. J. Quantum Chem. 31:99 (1987).
76. I. Martinson, A. Gaupp, and L. J. Curtis, J. Phys. B 7:L463 (1974).
77. P. Albertsen, P. Jørgensen and D. L. Yeager, Int. J. Quantum Chem. S14:249 (1980).
78. J. E. Gready, G. N. B. Bacskay and N. S. Hush, Chem. Phys. 23:9 (1977).
79. H. Werner and W. Meyer, Phys. Rev. A13:3 (1976).
80. T. Voegel, J. Hinze and J. Tobin, J. Chem. Phys. 790:107 (1979).
81. W. J. Stevens and J. P. Billingsley II, Phys. Rev. A6:855 (1972).
82. H. Kolker and H. H. Michels, J. Chem. Phys. 43:1027 (1965).
83. F. Maeder and W. Kutzelnigg, Chem. Phys. 42:95 (1979).
84. H. P. Kelly, Phys. Rev. 136:B896 (1964).
85. P. Jørgensen, J. Oddershede, P. Albertsen and N. H. F. Beebe, J. Chem. Phys. 68:2533 (1978).
86. G. H. F. Diercksen and A. J. Sadlej, Chem. Phys. 65:407 (1982).
87. J. D. S. Sims and J. R. Rumble Jr., Phys. Rev. 8A:2231 (1973).
88. M. J. Scott and E. Zaremba, Phys. Rev. A21:12 (1980).

89. L. J. Bartolotti, J. Chem. Phys. 80:5687 (1984).
90. P. Albertson, P. Jørgensen, and D. L. Yeager, Int. J. Quantum Chem. S14:249 (1980).
91. D. Yeager, J. Olsen, and P. Jørgensen, Int. J. Quantum Chem. S15:151 (1981).
92. J. Nichols and D. Yeager, Chem. Phys. Lett. 84:77 (1981).
93. D. Lynch, M. Herman, and D. Yeager, Chem. Phys. 64:69 (1982).
94. A. Rizzo, R. Graham, and D. Yeager, J. Chem. Phys. 89:1533 (1988).
95. M. Jaszunski, A. Rizzo, and D. Yeager, Chem. Phys. 136:385 (1989).
96. M. Jaszunski, A. Rizzo, and D. Yeager, J. Chem. Phys. 89:3063 (1988).
97. M. Jaszunski, A. Rizzo, and D. Yeager, Chem. Phys. Lett. 149:79 (1988).
98. M. Jaszunski and D. Yeager, Phys. Rev. A40:1651 (1989).
99. N. Rahman, A. Rizzo, and D. Yeager, Chem. Phys. Lett. 166:565 (1990).
100. P. Jørgensen and J. Simons, J. Chem. Phys. 79:334 (1983).
101. T. Helgaker, J. Almlöf, H. J. Aa. Jensen, and P. Jørgensen, J. Chem. Phys. 84:6266 (1986).
102. "Geometrical Derivatives of Energy Surfaces and Molecular Properties", P. Jørgensen and J. Simons, ed., Reidel, Dordrecht (1986).
103. R. L. Graham and D. Yeager, J. Chem. Phys. 94:2884 (1991).

MBPT AND COUPLED CLUSTER APPROACHES TO PARITY NONCONSERVATION IN ATOMS: A SURVEY OF RECENT DEVELOPMENTS

Steven A. Blundell

University of California
Lawrence Livermore National Laboratory
Livermore, CA 94550, USA

INTRODUCTION

Experiment and theory have recently advanced to the point where table-top, laser experiments on atoms can make quantitative statements about weak interaction coupling constants, complementing information otherwise obtained only from high-energy accelerator-based experiments. The atomic experiments measure *parity nonconservation* (PNC) in heavy atoms. A complex many-body calculation is required to relate the measurements to the underlying fundamental interactions. In this article, I discuss the recent improvements in atomic theory which have enabled atomic PNC to yield information with accuracy comparable to that obtained in high-energy experiments.

Until 1956, it was believed that the laws of physics possessed mirror symmetry, that is, that the mirror image of any physical process was also a possible process. For example, the existence of right-handed sugar should imply the possible existence of its mirror image, left-handed sugar. In quantum mechanics, this means that the interactions were assumed to be invariant under the parity operator P, which replaces the spatial co-ordinate (x, y, z) by $(-x, -y, -z)$, or equivalently that P commutes with the Hamiltonian, $[P, H] = 0$. Consequently, the expectation value of P in any time-dependent process would be constant, and parity is said to be conserved. Furthermore, a (non-degenerate) stationary state would be a parity eigenstate, with eigenvalue +1 (symmetric) or −1 (anti-symmetric). The electromagnetic interaction is known to conserve parity.

In 1956, Lee and Yang[1] questioned whether the weak interactions responsible for β-decay conserved parity, and in 1957 parity nonconservation was discovered in β-decay.[2] Thus parity conservation was known not to be a fundamental principle of physics. (Earlier, in 1950, Purcell and Ramsey[3] had also questioned parity conservation on a fundamental level by pointing out that the possible existence of an electric dipole moment on a neutron, although violating parity conservation, remained a purely experimental question.)

The weak interaction in an atom means that the atomic eigenstate no longer has definite parity, but contains small admixtures of states of opposite parity. The physics of the atom is dominated by the electromagnetic interaction, and the admixture coefficients are very small,

Applied Many-Body Methods in Spectroscopy and Electronic Structure
Edited by D. Mukherjee, Plenum Press, New York, 1992

typically of order 10^{-11}. Nevertheless, an atom has a very slight intrinsic handedness; the mirror image atom does not exist, and parity is no longer conserved.

Originally, measurements on atoms were proposed to detect PNC at the level predicted by the standard model[4] of electroweak interactions. Agreement at 20% or better has now been found[5] in a number of elements (Cs, Tl, Pb and Bi) in experiments performed in various laboratories throughout the world. Nowadays, the emphasis is slightly different: one hopes to be able to make a *precision* test of the standard model, and place quantitative constraints on new physics beyond standard model. The importance of PNC is then as a discriminant, enabling the effects of the weak neutral current to be separated from those of the dominant electromagnetic interaction. In this way, we can use the atom as a tool to probe interactions normally accessible only in high-energy accelerators. How accurately this can be done depends on our ability to calculate the structure of the heavy neutral atoms for which results have been reported.

A recent experiment[6] in Cs has reported an accuracy of about 2.5%, somewhat more accurate than the other measurements. This element is also the most favorable theoretically, being described fairly well in the single-particle approximation as a single valence 6*s* electron outside a relatively tightly bound Xe-like core. I will concentrate almost entirely on this element in this review, because at present it offers the best chance of a precision test of weak interaction physics.

An important set of experiments are under way to detect effects that not only violate parity conservation, but also time-reversal invariance.[5] This involves searching for a permanent atomic electric dipole moment, an effect not yet found. While the atomic theory of these effects parallels closely that to be discussed here, existing calculations have not yet reached the same level of sophistication as for PNC in Cs, and the interested reader is referred elsewhere for details.[9]

WEAK NEUTRAL CURRENT INTERACTION

β-decay is mediated by the exchange of a charged $W^{\pm}$ boson. This *charged current* interaction does not contribute to lowest order in a stable atom because the identity of the particles involved changes. However, the standard model of electroweak interactions predicted a weak *neutral current* interaction, first discovered in 1973,[10] mediated by Z^0 exchange. This gives the diagonal interactions in Figs. 1(a-b) which contribute to PNC in an atom in leading order. A third PNC mechanism exists, Fig. 1(c), in which PNC inside the nucleus is communicated electromagnetically to the atomic electrons; the nuclear PNC here is parametrized in terms of an odd-parity vector moment, the *anapole* moment.[11,12]

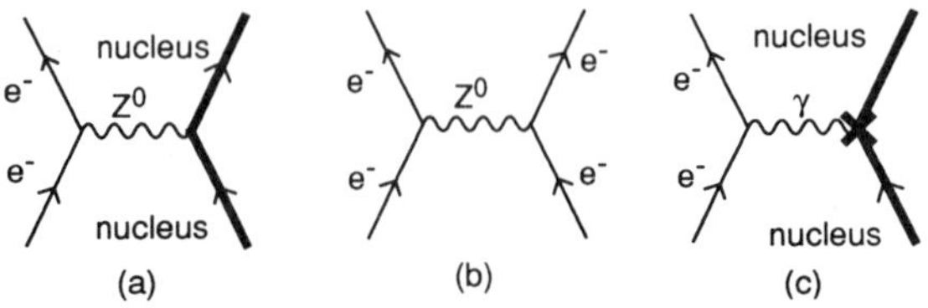

Fig. 1. Interactions responsible for PNC in atoms. Processes (a) and (b) are weak neutral current processes. In process (c), the current is not the usual one, but an "anapole current."

In a pioneering paper in 1974, Bouchiat and Bouchiat[13] showed that the small effect of the weak neutral current in atoms grows roughly as Z^3 and could in fact be observable in heavy atoms. Since then, the effect has been observed in a variety of ways in a range of heavy atoms. The atomic theory problem would of course be greatly simplified by a measurement on a hydrogenic system, but despite great efforts no effect has yet been measured.[14]

The first stage in any atomic structure calculation is to determine effective PNC Hamiltonians. From Fig. 1(a) one finds two terms,[13]

$$h^{\mathrm{PNC}} = \frac{G_{\mathrm{F}}}{2\sqrt{2}} Q_{\mathrm{W}} \rho(r) \gamma_5, \tag{1}$$

$$h_{\mathrm{A}}^{\mathrm{PNC}} = \frac{G_{\mathrm{F}}}{\sqrt{2}} K \rho_{\mathrm{A}}(r) \vec{\alpha} \cdot \vec{\mathbf{I}}, \tag{2}$$

where $G_{\mathrm{F}} \approx 2.22\ 10^{-14}$ a.u. is the Fermi coupling constant, $\rho(r)$ and $\rho_{\mathrm{A}}(r)$ are suitable nuclear densities normalized to unity, $\mathbf{I}$ is the nuclear spin, and K and Q_{W} are coupling constants, Q_{W} being the *weak charge* of the nucleus. These are contact interactions which vanish outside the nucleus, reflecting the fact that the Z^0 boson is very massive with Compton wavelength of order 10^{-18} m. Fig. 1(b) gives a two-body interaction,

$$H_{\text{e-e}}^{\mathrm{PNC}} = -\frac{G_{\mathrm{F}}}{4\sqrt{2}} (1 - 4\sin^2\theta_{\mathrm{W}}) \sum_{i \neq j} (\gamma_5^i + \gamma_5^j)(1 - \vec{\alpha}_i \cdot \vec{\alpha}_j) \delta^3(\vec{\mathbf{x}}_i - \vec{\mathbf{x}}_j). \tag{3}$$

The anapole term, Fig. 1(c), gives a Hamiltonian identical in form to (2),[12] and effectively modifies the value of K (with the approximation that the nuclear density is the same for each term).

Approximations have already been made in obtaining these Hamiltonians. In (1) and (2) we have taken the nonrelativistic limit for the nucleus and the spherical average of the nuclear density. Both these approximations are expected to be good at the 0.1% level or better, and are justified at present considering the ≥1% error from electron correlation.

As first pointed out by the Bouchiats,[13] term (1) is the dominant one in an atom. While in (1) the nucleons contribute coherently to Q_{W}, for the nuclear-spin-dependent term (2) their contributions cancel in pairs leaving only that of the outer unpaired nucleon. The term (2) is thus suppressed by a factor of about $1/Z$. Furthermore, the appropriate coupling constant in the standard model is about 10 times smaller for (2) than for (1), and overall (2) enters at the several tenths of a percent level (0.2% in Cs, neglecting the anapole moment contribution). The electron-electron interaction (3) is in fact very small, less than 0.05% for Cs.[8] Khriplovich *et al.*[12] have found that the anapole term makes the dominant contribution to (2), boosting the size of this term to about 1% to 1.5% of (1) in Cs.

The precision test of the standard model is made by extracting the weak charge Q_{W} as accurately as possible from experiment for comparison with the prediction of the standard model. The anapole term is also of interest, but mainly for nuclear physics. In this article, I will regard (2) and (3) as small corrections to be taken into account in extracting Q_{W} from experiment, and restrict the discussion solely to term (1). In fact, the calculation of (2) and (3) can be made to adequate precision at Dirac-Fock level,[8] leaving the electron correlation associated with (1) as the largest uncertainty in the entire calculation. Moreover, the nuclear-spin dependence of (2) provides a discriminant which enables K to be separated experimentally from Q_{W}, since term (2) contributes differently to different hyperfine transitions, while (1) does not. Appropriate linear combinations of results on various hyperfine transitions can therefore either isolate or eliminate (2).[8,15] Finally, Khriplovich[16] has pointed out that interference between (1) and the atomic hyperfine interaction can lead to a further small nuclear-spin-dependent term.

In the standard model in lowest order, the weak charge is given in terms of the Weinberg angle θ_W by[13]

$$Q_W = -N + (1 - 4\sin^2\theta_W)Z. \tag{4}$$

Higher-order (loop) electroweak corrections modify this value, in Cs by about 5%.[17] The aim, which is now just being realized, is to extract Q_W to at least the level of loop corrections, for then the test of the standard model becomes the most stringent. The point is that new physics enters in these loop corrections through the virtual excitation of pairs of new particles and their antiparticles. The loop corrections can in fact be quite sensitive to the assumed model. More details are given in the penultimate section.

ATOMIC MEASUREMENTS

The effect of the small, PNC terms (1)-(3) in the atomic Hamiltonian is that the atomic states Ψ are no longer eigenstates of parity but contain small opposite-parity admixtures Ψ^{PNC}

$$\Psi \to \tilde{\Psi} = \Psi + \Psi^{PNC}. \tag{5}$$

This parity mixing has the observable consequence that the usual parity selection rules for radiative transitions are violated. In particular, a magnetic dipole (M1) amplitude between two states of the same nominal parity now contains a small electric dipole component $E1^{PNC}$ arising through the weak interaction. Experiments have sought to measure $E1^{PNC}$ through its interference with a larger, parity conserving, electromagnetic amplitude. In both the original proposal by the Bouchiats,[13] and the recent accurate experiment by the Boulder group,[6] the interference is with a Stark amplitude $E1^{STARK}$ induced by an applied electric field. Specifically, it is the ratio $E1^{PNC}/\beta$ of $E1^{PNC}$ to the vector Stark polarizability β for the transition that is measured. Another class of experiments[18] measures the rotation of the plane of linearly polarized light due to the different absorption rates of left- and right-handed circularly polarized light. This effectively determines $E1^{PNC}/M1$.

To first order in h^{PNC}, the opposite-parity admixture is

$$\left|\Psi^{PNC}\right\rangle = \sum_X \frac{\left|X\right\rangle\left\langle X\right|H^{PNC}\left|\Psi\right\rangle}{E - E_X}, \tag{6}$$

where X denotes excited many-body states of opposite parity to Ψ, and H^{PNC} is a sum of the one-body operators h^{PNC}. The induced E1 amplitude is then

$$E1^{PNC} = \left\langle\Psi_2^{PNC}\middle|D\middle|\Psi_1\right\rangle + \left\langle\Psi_2\middle|D\middle|\Psi_1^{PNC}\right\rangle$$

$$= \left\{\sum_X \frac{\left\langle\Psi_2\middle|D\middle|X\right\rangle\left\langle X\middle|H^{PNC}\middle|\Psi_1\right\rangle}{E_1 - E_X} + \begin{pmatrix} D \leftrightarrow H^{PNC} \\ E_1 \to E_2 \end{pmatrix}\right\}, \tag{7}$$

where D is the dipole operator. β is also given by a double-perturbation expression (for the Cs $6S$-$7S$ transition)[13]

$$\beta = \frac{1}{6}\sum_n \left\{ \langle 7S \| D \| nP_{1/2} \rangle \langle nP_{1/2} \| D \| 6S \rangle \left(\frac{1}{E_{7S} - E_{nP_{1/2}}} - \frac{1}{E_{6S} - E_{nP_{1/2}}} \right) \right.$$

$$\left. + \frac{1}{2} \langle 7S \| D \| nP_{3/2} \rangle \langle nP_{3/2} \| D \| 6S \rangle \left(\frac{1}{E_{7S} - E_{nP_{3/2}}} - \frac{1}{E_{6S} - E_{nP_{3/2}}} \right) \right\}. \tag{8}$$

The task for atomic theory in Cs is to calculate $E1^{PNC}/\beta$ in terms of the weak charge Q_W. While the energies and some dipole matrix elements are known experimentally, the matrix elements of H^{PNC} have to be calculated. Early attempts using semi-empirical methods produced a 10% range of final values for $E1^{PNC}$. Considerable progress has been made recently by combining semi-empirical methods with *ab initio* MBPT techniques,[19,20] using the S-state hyperfine interaction to gauge the normalization of the wavefunction at the origin for use in the matrix element of H^{PNC}. The most accurate such calculation[20] claims an accuracy of 2% in Cs and agrees with the purely *ab initio* calculations to be discussed here.

In the next section I discuss two basic strategies that have been employed to evaluate the double perturbation expression (7).

TWO APPROACHES TO THE DOUBLE PERTURBATION PROBLEM

The most obvious approach is to calculate directly from (7), saturating the sum over intermediate many-body states X. This approach is especially straightforward for the 6S-7S transition in Cs, because the most important intermediate $X = P_{1/2}$ states are those which, in the single-particle approximation, are described by a single valence $p_{1/2}$ electron outside a closed-shell core. This simplicity facilitates greatly an *ab initio* approach to the matrix elements in (7). In fact, about 100.2% of the sum comes from such states,[8] the remainder coming from more complex open-shell (autoionizing) states in which the core is excited. Furthermore, the sum over principal quantum number n converges rather quickly, with about 98% of the sum coming from just the $X = 6P_{1/2}$, $7P_{1/2}$, $8P_{1/2}$ and $9P_{1/2}$ states. Blundell *et al.*[8] have used a pair correlation approach of Coupled-Cluster (CC) type for these four states, and less sophisticated methods for the *tail*, $n = 10$–∞ (including the continuum).

This approach would not work so well, however, for the other elements of experimental interest. For Bi and Pb, the intermediate states X are of a complex, open-shell nature for which sufficiently accurate computer codes have not yet been developed. Even for Tl, which is nominally of single-valence-electron type, having a single $6p_{1/2}$ electron outside a closed $6s^2$ core, the excitations from the $6s^2$ core shell make a substantial contribution (about 20%)[21] and are difficult to handle using existing procedures.

A second approach is to use perturbed single-particle states, as first suggested in connection with the atomic PNC problem by Sandars.[22] The weak Hamiltonian h^{PNC} is of one-body type and can formally be included in the definition of the single-particle states, which now become parity-mixed. Since h^{PNC} is a pseudo-scalar, the single-particle states remain eigenstates of angular momentum, however. One now proceeds with any conventional many-body approach for a dipole matrix element, using these parity-mixed states instead of the usual parity-pure ones. At each stage it is usually convenient to linearize the expressions in the weak interaction. This parity-mixed approach can in principle be applied to any method, such as multi-configuration DF[23] or Equations of Motion.[24] Here I will confine the discussion to an MBPT method which has recently been refined to the 1% level.[7,8]

While the parity-mixed technique successfully simplifies the diagrammatic book-keeping by effectively removing one perturbation from the diagrammatic expansion, it is nevertheless still rather complicated to implement. Consider the evaluation of a perturbation diagram, containing in general intermediate sums over parity-mixed single-particle states. After linearization in the weak interaction, one obtains a sum of terms formed by replacing the parity-mixed states by parity-pure ones, and then substituting one state at a time by its opposite-parity admixture. In this way, even a single second-order diagram expands into a sum of six terms.

It may be possible eventually to implement a full parity-mixed coupled cluster procedure. The development of practical procedures to solve these equations, however, is a rather daunting task, and remains to be done. A code along these lines is however presently being developed by Liu and Kelly.[25]

One of the most difficult aspects of the atomic PNC calculation is gauging the theoretical error involved. For this reason, it is important to have as many different approaches to the calculation as possible. Here I will describe the successful implementation of the above two methods for Cs; the two approaches agree with one another to within the quoted 1% theory error.

GENERAL FEATURES OF *AB INITIO* SCHEMES FOR ATOMIC PNC

Before describing in detail the many-body schemes that have so far been applied to PNC in Cs, I will discuss some of the general features that are required for this problem.

Relativity

The dominant PNC interactions (1)-(2) occur inside the nucleus, where the potential is many times the rest mass energy of the electron, for example, about 30 times for Cs. Consequently the details of the wavefunction are highly relativistic in the region where the interaction occurs. It is essential to use s and $p_{1/2}$ solutions to the Dirac equation in forming matrix elements of h^{PNC}. Most schemes that have been used recently are in fact fully relativistic, being based upon solutions to the Dirac equation, and it is therefore necessary at least in principle to understand how to formulate the many-body problem from the viewpoint of the relativistic quantum field theory for the atom, quantum electrodynamics (QED).

The QED Hamiltonian for the atom, acting in the Fock space of the electron-positron and photon fields, can be written[26]

$$H_{\mathrm{QED}} = \int d^3x\,\psi^\dagger(x)\left(c\vec{\alpha}\cdot\vec{\mathbf{p}} + \beta c^2 + V_{\mathrm{nuc}}(r)\right)\psi(x)$$
$$+\frac{1}{2}\int d^3x d^3x' \frac{\psi^\dagger(x)\psi^\dagger(x')\psi(x')\psi(x)}{|\vec{\mathbf{x}}-\vec{\mathbf{x}}'|}$$
$$+\int d^3x\,\psi^\dagger(x)\vec{\alpha}\cdot\vec{\mathbf{A}}(x)\psi(x)\,. \tag{9}$$

We have assumed that the nucleus is fixed at the origin and is the source of a classical external potential $V_{\mathrm{nuc}}(r)$ in which the atomic electrons move and interact. The Hamiltonian contains an instantaneous Coulomb interaction between electrons and a coupling to the (transverse) electromagnetic field. This corresponds to the use of the *Coulomb* or *radiation gauge*,[26] in which only the two physical degrees of freedom of the photon are quantized. The theory is gauge invariant, however, and in principle any gauge can be used. To this Hamiltonian must be added counter terms to remove the divergences that occur in higher-order expressions, and also kinetic terms for the radiation field.

In nonrelativistic MBPT, one chooses an arbitrary potential $U(r)$ to approximate the interaction of one atomic electron with the remainder, and adds and subtracts $U(r)$ to and from the atomic Hamiltonian. We adopt a similar procedure here, partitioning H_{QED} into[27]

$$H_{\mathrm{QED}} = H_0 + V \tag{10}$$

$$H_0 = \int d^3x\, \psi^\dagger(x)\left(c\vec{\alpha}\cdot\vec{\mathbf{p}} + \beta c^2 + V_{\mathrm{nuc}}(r) + U(r)\right)\psi(x) \tag{11a}$$

$$V = \frac{1}{2}\int d^3x d^3x' \frac{\psi^\dagger(x)\psi^\dagger(x')\psi(x')\psi(x)}{|\vec{\mathbf{x}} - \vec{\mathbf{x}}'|} - \int d^3x \psi^\dagger(x) U(r)\psi(x)$$

$$+\int d^3x \psi^\dagger(x)\vec{\alpha}\cdot\vec{\mathbf{A}}(x)\psi(x)\,. \tag{11b}$$

In these expressions, $\vec{\mathbf{A}}(x)$ is the photon field operator, given by

$$\vec{\mathbf{A}}(x) = \sum_{\lambda=1}^{2}\int \frac{d^3k}{(2\pi)^3}\frac{1}{2k_0}\left(a_\lambda(\vec{\mathbf{k}})\vec{\varepsilon}_\lambda(\vec{\mathbf{k}})e^{-ik\cdot x} + a_\lambda^\dagger(\vec{\mathbf{k}})\vec{\varepsilon}_\lambda^{\,*}(\vec{\mathbf{k}})e^{ik\cdot x}\right), \tag{12}$$

and $\psi(x)$ is the electron field operator,

$$\psi(x) = \sum_{n_+} b_{n_+} e^{-i\varepsilon_{n_+}t}\psi_{n_+}(\vec{\mathbf{x}}) + \sum_{n_-} d_{n_-}^\dagger e^{-i\varepsilon_{n_-}t}\psi_{n_-}(\vec{\mathbf{x}}). \tag{13}$$

The states ψ_{n+} and ψ_{n-} here are positive- and negative-energy Dirac eigenstates for the potential $V_{\mathrm{nuc}}(r) + U(r)$. (More properly the sum becomes an integral for the positive- and negative-energy continua.)

It is now possible to develop a systematic perturbation expansion using, for example, the prescription of Gell-Mann, Low and Sucher[28,29] for field theoretical bound-state level shifts. We will not go into the details here.[30] The result is a partially understood perturbation series, containing in a consistent fashion the Coulomb correlation energy, the Breit interaction, retardation terms, and radiative corrections. It is rigorous, apart from the omission of nuclear recoil terms, which require further study at this time.

For a heavy neutral atom like Cs, the dominant physics is associated with the Coulomb interaction, which leads to a complex many-body problem. It is necessary to treat the Coulomb interaction nonperturbatively, while the remaining terms may be added in low order perturbation theory. Below we will treat the Coulomb interaction using high-order MBPT and coupled-cluster (CC) techniques, while adding the Breit interaction in lowest order; the remaining terms will be neglected.

One feature of the QED atomic perturbation series is particularly important: there exists a subset of terms which is diagram by diagram equivalent to the usual nonrelativistic perturbation series, except that the single-particle states involved are Dirac states, and the intermediate-state sums are restricted to the positive-energy half of the spectrum only. In the nonrelativistic limit, this set of terms goes into the nonrelativistic perturbation series, and the remaining terms vanish. It is clear that such a set of terms exists from considering the substitution of the positive energy part of ψ, Eq. (13), into the Coulomb interaction part of V, Eq. (11b). We will call this set of terms the *no-virtual-pair Coulomb* (NPC) approximation. It is equivalent to solving the *configuration-space* equation

$$H_{\mathrm{NPC}}\Psi(\vec{\mathbf{x}}_1,\vec{\mathbf{x}}_2,\ldots,\vec{\mathbf{x}}_N) = E\Psi(\vec{\mathbf{x}}_1,\vec{\mathbf{x}}_2,\ldots,\vec{\mathbf{x}}_N), \tag{14}$$

where H_{NPC} is the NPC Hamiltonian of Sucher[31]

$$H_{\mathrm{NPC}} = H_0' + V_{\mathrm{NPC}}, \tag{15}$$

$$H_0' = \sum_{i=1}^{N} \Lambda_i^+\left(c\vec{\alpha}_i\cdot\vec{\mathbf{p}}_i + \beta_i c^2 + V_{\mathrm{nuc}}(r_i) + U(r_i)\right)\Lambda_i^+, \tag{16a}$$

$$V_{\mathrm{NPC}} = \sum_{i>j}^{N} \Lambda_i^+\Lambda_j^+\frac{1}{r_{ij}}\Lambda_i^+\Lambda_j^+ - \sum_{i=1}^{N}\Lambda_i^+U(r_i)\Lambda_i^+, \tag{16b}$$

and the $\Lambda_+(i)$ are projection operators on to the positive-energy Dirac states in the potential $V_{\mathrm{nuc}}(r) + U(r)$. Without these projection operators, the NPC equation would possess no normalizable, bound-state solutions, a phenomenon known as *continuum dissolution*.[31] This arises because the unprojected V_{NPC} would allow transitions into the negative energy sea that are in reality forbidden by the Pauli exclusion principle. Negative energy states do play a role in the QED perturbation series, but their inclusion in intermediate state sums are subject to special rules.[27] For level shifts in neutral atoms, negative energy effects enter in order α^3 Rydbergs and can be treated perturbatively.

The close analogy between the no-virtual-pair Coulomb approximation and nonrelativistic MBPT means that much of the machinery of nonrelativistic MBPT can be taken over in solving the relativistic correlation problem. Our relativistic CC method is formally equivalent to the usual nonrelativistic one, but we use relativistic states and restrict excited state sums to the positive-energy half of the spectrum.

In order to solve (14) it is necessary to find a representation for the $\Lambda_+(i)$. While some success has been achieved in finding a configuration space representation,[32] the most extensive calculations have been performed by generating an essentially complete relativistic basis set and implementing the restriction to positive-energy states directly. Two types of basis set can be distinguished: analytic,[33] based on sets of known analytic functions, and numerical.[34,35] The numerical basis set of the Notre Dame group[34] is obtained by placing the atom in a cavity of large radius, so that the whole spectrum is discretized. The small and large components of the Dirac wavefunction are then expanded in terms of N piecewise polynomials (or *basis splines*) $B_i(r)$,

$$\begin{pmatrix} g(r) \\ f(r) \end{pmatrix} = \sum_{i=1}^{M} \begin{pmatrix} c_i \\ d_i \end{pmatrix} B_i(r),$$

and a variational principle used to determine the expansion coefficients. This gives a set of N positive- and N negative-energy *pseudo-states*, of which the lowest few positive-energy members approximate closely the lowest-lying bound states and energies. Sum rules may be used to test the completeness of the basis set. One can also test for convergence of a given calculation as the number of basis functions is increased, and compare results calculated with different basis sets. Agreement for the relativistic second-order correlation energy in He has been found to 7 or 8 digits using two different types of numerical basis set.[34,35] Thus accurate numerical techniques exist for handling positive-energy projection operators.

Open-shell formalism

The calculation becomes more complex as the number of valence electrons increases. For complex open-shell systems, some sort of (nearly) degenerate perturbation theory formulation with a multi-configuration model space is generally required to handle the strongly interacting valence configurations. Even in Cs, the simplest type of open-shell system with a single valence electron, there are constraints on the method suitable for atomic PNC. One requires a method that can generate excited states with a given angular symmetry, for example the 7S, $7P_{1/2}$, $8P_{1/2}$, etc., and furthermore compute matrix elements between these states. It is highly advantageous to use the same set of single-particle states for each atomic state. The calculations described here employ a V^{N-1} DF potential, the DF potential for the closed-shell N-1 electron core. Such a potential has the advantage that whole classes of perturbation theory diagrams vanish,[36] considerably reducing calculational labour. In the CC method, we follow the general open-shell strategy outlined by Lindgren and Morrison,[36] and calculate each state independently in a single-state model-space approach.

Excitation of Deep Core States

The PNC interaction takes place inside the nucleus, and many-body effects involving deep core electrons are quite important. For example, in Cs, many-body effects involving excitations of the $1s$ electron contribute 0.4% to the $\langle 6S | H^{PNC} | 6P_{1/2} \rangle$ matrix element, while the $2s$ contributes 0.7%. Likewise, the $1s$ and $2s$ states each contribute about 0.8% to the $6s$ hyperfine constant, described by a contact interaction in the nonrelativistic limit. Ideally, therefore, one should excite the *whole* atom when calculating correlation corrections. A slight simplification is permitted by noting that the dominant effect of the core arises through a specific subset of many-body terms, the RPA, that can readily be separated off and handled in a complete fashion. Even this technique, however, allows one to drop from the main calculation only the n=1 and 2 shells, and possibly the $3s$ state, at the level of accuracy required.

To accommodate the large number of core electrons participating in the correlation, it is essential to take full advantage of the spherical symmetry of the atom, working explicitly only with radial states and summing over spin-angular quantum numbers analytically. Assuming this to be done, let us estimate the computer memory required for an all-order pair correlation calculation. Suppose that we work with a basis set with n states per angular momentum value, and include all angular momenta up to $L = 7$ ($k_{13/2}$ and $k_{15/2}$), which reasonably saturates the partial wave summations. Now, a pair excitation *channel* for core states a and b to angular momenta κ_n and κ_m with multipole L,* $ab \rightarrow \kappa_n \kappa_m\ (L)$, can be described by a set of n^2 coefficients (in either a configuration interaction (CI) or coupled cluster approach). Including all core states for Cs, selection rules permit about 17000 core-core channels, and about 1300 core-valence channels (i.e. where b is a valence state) for each ($j = 1/2$) valence state of interest. Using $n = 25$, we require about 12 MWords of memory for the calculation, which can then be performed in fast memory with the present generation of supercomputers. It should be clear, however, that a similarly complete treatment of triple excitations would require excessive storage, and triples will therefore require a modified treatment. Finally, it is worth noting that

* The multipole is defined similarly to the multipole of a Slater integral in a Coulomb matrix element.

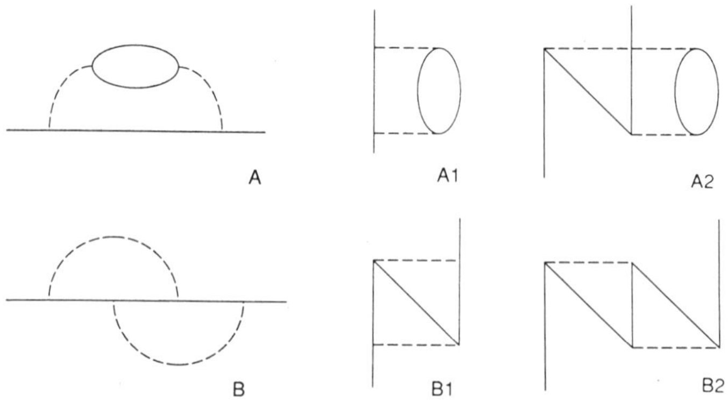

Fig. 2 Feynman and Goldstone diagrams for the second-order removal energy.

the number of channels in a fully relativistic calculation, as discussed here, is much greater than that in a nonrelativistic one because of the need to distinguish between, for example, $p_{1/2}$ and $p_{3/2}$ states.

Basis Set

An important feature of atomic PNC is that there are two length scales in the problem: the nuclear region, which is where h^{PNC} acts, and distances of order 1 a.u. which determine electric dipole matrix elements. It is important to choose a basis set which can represent both regions accurately. While the relative merits of the different types of relativistic basis set are still under investigation, the piecewise polynomial approach has one advantage in this respect: the basis splines are *localized* functions, non-zero only between two *knot* points which can be chosen arbitrarily. It is therefore straightforward to improve accuracy in, say, the nuclear region by putting more knot points there.

Blundell *et al.*,[8] in their CC-type calculation, have used a piecewise polynomial basis set consisting of $n = 25$ basis functions for s- and p-states, and $n = 20$ for higher angular momenta. The greater number of s- and p-states reflects the need for more accuracy inside the nucleus. Otherwise, the number of basis functions was chosen to reduce the basis set truncation error to at most 1% of the correlation.

STRUCTURE OF PERTURBATION THEORY IN CESIUM

It is instructive to consider a detailed breakdown of the second- and third-order MBPT contributions to the valence ionization energy in Cs in order to understand what types of effects are most important. The V^{N-1} DF diagrams for the second-order valence energy are shown in Fig. 2, both in the Feynman form (A and B) which follows from the *time-dependent* development of MBPT,[37] and in the Goldstone form (A1, A2, B1, B2) more usual in atomic and molecular physics.[36] The Goldstone diagrams follow by taking all possible *time-orderings* of the Feynman diagrams; thus A1 and A2 are the 2! orderings of the 2 Coulomb interactions in A, and A = A1 + A2. The Feynman form corresponds then to a grouping of Goldstone diagrams. Another grouping is that of Brandow,[38] in which the *exchange variant* pairs (A1 + B1) and (A2 + B2) are each summed and represented by single diagrams (A1 and A2 respectively, say). Numerical values for Cs and Tl are given in Table I. Physically, A is the

TABLE I. Values of the Goldstone diagrams of Fig. 2 for the second-order removal energy of the 6s electron in Cs. (From Ref. 39)

	Energy (a.u.)	Percentage of total energy
A1	−0.02168	15.1%
B1	0.00236	−1.6%
A2	0.00219	−1.5%
B2	−0.00069	0.5%
Sum	−0.01782(2)	12.5%

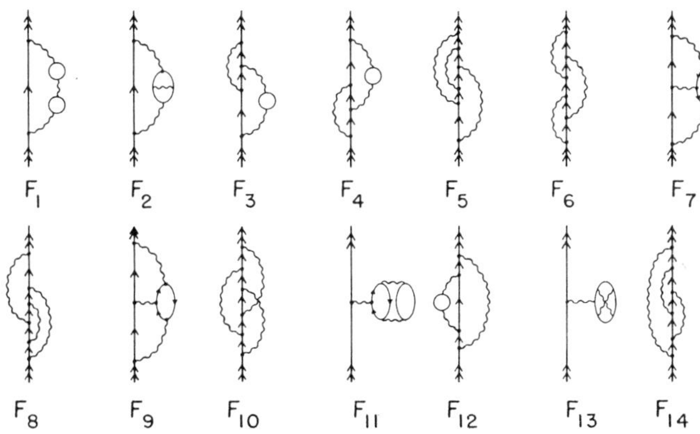

Fig. 3 Feynman diagrams for third-order removal energy.

energy shift due to the electrostatic polarization of the core by the valence electron, while B is the exchange interaction between the valence electron and the core.

Fig. 3 shows the 14 Feynman diagrams for the third-order valence removal energy when a V^{N-1} DF potential is used, and numerical values are presented in Table II.[39] Diagrams F1 and F2 are very large, comparable to the second order diagram A, indicating that perturbation theory will not converge quickly at all, and that at least some infinite sequences have to be summed. In fact, as shown in Table III, the first few orders of perturbation theory form an irregularly convergent alternating series with the result after third order in error by about twice as much as that after second order. At the level of precision aimed at here, there are collective effects that must be summed. Using the time-dependent formulation, the Novosibirsk group[40] have summed to all orders the dominant effects in F1 and F2 to obtain quite good agreement with experiment at the 1% level. However, to proceed systematically to the 0.1% level, it is clear that all effects in third order (with the possible exception of F6 and F14) will have to be incorporated.

The coupled-cluster (CC) formalism[40] provides a suitable framework for summing classes of perturbation theory diagrams to all orders in a complete and systematic way. It is a

TABLE II. Values of the Feynman diagrams of Fig. 3 for the third-order removal energy of the 6s electron in Cs, expressed as a percentage of the total ionization energy. (From Ref. 39)

Diagram	Value	Diagram	Value
F1	-8.0%	F8	-0.2%
F2	4.9%	F9	-3.4%
F3	0.2%	F10	0.3%
F4	0.2%	F11	-0.7%
F5	-0.2%	F12	-0.1%
F6	0.0%	F13	0.1%
F7	2.9%	F14	0.0%

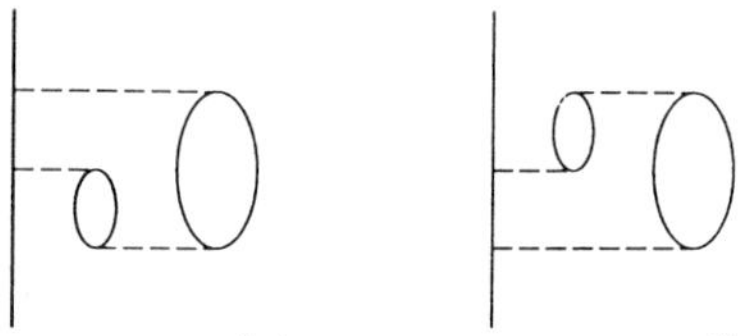

Fig. 4 Third-order removal energy diagrams. (b) contains a triple excitation.

method for summing to all orders one-, two-, three-, ..., N-body effects by means of a hierarchy of coupled equations for excitation amplitudes. The hierarchy is in practice usually truncated after two- or three-body effects. It is possible to identify precisely which set of MBPT diagrams are included after any truncation. A good introductory treatment of the CC method in the atomic context is given in the textbook by Lindgren and Morrison.[36]

We now wish to understand the third-order energy in terms of the CC method. One can find the degree of excitation contained in any given energy diagram by removing the uppermost interaction to give the underlying wavefunction diagram, and then inspecting its energy denominators. In this way one sees that the third-order energy diagram of Fig. 4(a) contains only double excitations, or two-body effects, while that of Fig. 4(b) contains a triple excitation. Correspondingly, the standard CC formalism including singles and doubles (CCSD) will miss Fig. 4(b) but pick up Fig. 4(a). Similarly one can show that the entire second-order energy is a two-body effect, while the third-order ionization energy consists of both two- and three-body effects. Note that both Figs. 4(a) and 4(b) are time-orderings of the dominant diagram F1 (Fig. 3) and are correspondingly both quite large. In fact, the three-body part accounts for roughly half of $E^{(3)}$, and is a –2% effect overall.[39] Therefore, a conventional CCSD approach is significantly in error for Cs, and it is essential for better than 1% accuracy to build in the set of third-order diagrams associated with triple excitations.*

As will be shown in the next section, there always exist simplifications which reduce the triple excitation energy denominator in Fig. 4(b) to a double excitation one, suggesting that

* Note that for the core correlation energy, or in general for the total correlation energy of a closed-shell system, where the diagrams have no external legs, both the second- and third-order energies are contained in CCSD.

really Fig. 4(b) is a two-body effect, and that there exists an extension of the standard CC formulation which incorporates this effect at the pair correlation level. One may also observe that Fig. 4(b) is just the Hermitian conjugate of Fig. 4(a). In so-called Hermitian formulations[42] of the CC method, these effects are treated symmetrically, and the entire third-order energy is picked up in the pair approximation. We shall discuss a variation on these methods in the next section. In Hermitian formulations, the *genuine* three-body effects enter for the first time in the fourth-order valence removal energy.

TABLE III. Behaviour of low orders of MBPT for Cs in units of a.u. Numbers in parentheses indicate numerical errors. (From Ref. 39)

Order	Energy	Accumulated energy
0th	−0.12737	−0.12737
2nd	−0.01782(2)	−0.14519(2)
3rd	0.00570(17)	−0.13949(17)
4th (partial)	−0.00370(20)	−0.14319(26)
Experiment		−0.14310

COUPLED-CLUSTER CALCULATIONS ON CESIUM

In the CC method a *cluster operator* S is introduced such that the exact wavefunction Ψ is given in terms of the zeroth-order one $\Psi^{(0)}$ by a normal-ordered exponential,

$$\Psi = \{\exp(S)\}\Psi^{(0)} \equiv \Omega\Psi^{(0)}. \tag{17}$$

This *ansatz* embodies the theorem that, in *intermediate normalization*,† the disconnected parts of the wave operator Ω are merely powers $S^n/n!$ of the connected part S. We consider here the case of one valence electron outside closed shells. The cluster operator satisfies a generalized Bloch equation[36,42]

$$[S, H_0]P = (V\Omega - \Omega PV\Omega P)_{\text{conn}}, \tag{18}$$

where H_0 is the model Hamiltonian of which $\Psi^{(0)}$ is an eigenstate, V is the perturbation, P is a projection operator on to $\Psi^{(0)}$, and $Q = 1 - P$. In the standard CCSD approach, one solves (18) in the approximation that S is restricted to single and double excitations $S = S_1 + S_2$,

$$S_1 = \sum_{ma} \rho_{ma} a_m^\dagger a_a + \sum_{m} \rho_{mv} a_m^\dagger a_v, \tag{19a}$$

$$S_2 = \frac{1}{2} \sum_{mnab} \rho_{mnab} a_m^\dagger a_n^\dagger a_b a_a + \sum_{mna} \rho_{mnav} a_m^\dagger a_n^\dagger a_v a_a. \tag{19b}$$

Here and later (m, n, r, s) refer to excited orbitals (including valence), (a, b, c) to core orbitals, and (v, w) to valence orbitals. Further details and a derivation of the equations satisfied by the excitation coefficients in S_1 and S_2 can be found in e.g. Ref. 36.

† $\langle\Psi^{(0)}|\Psi^{(0)}\rangle = \langle\Psi|\Psi^{(0)}\rangle = 1$, but $\langle\Psi|\Psi\rangle \neq 1$.

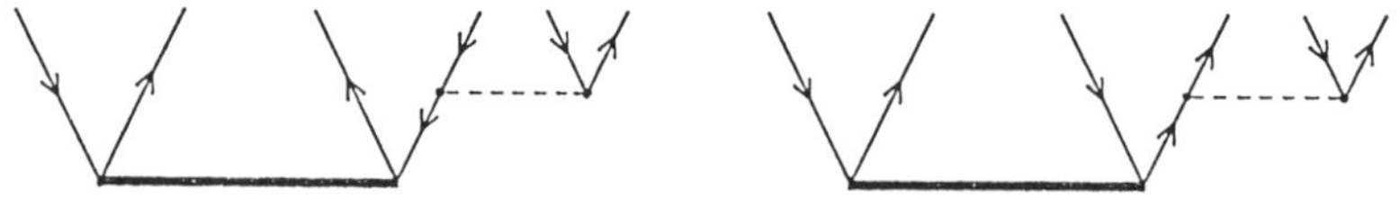

Fig. 5 Leading contributions to S_3 (for core excitations).

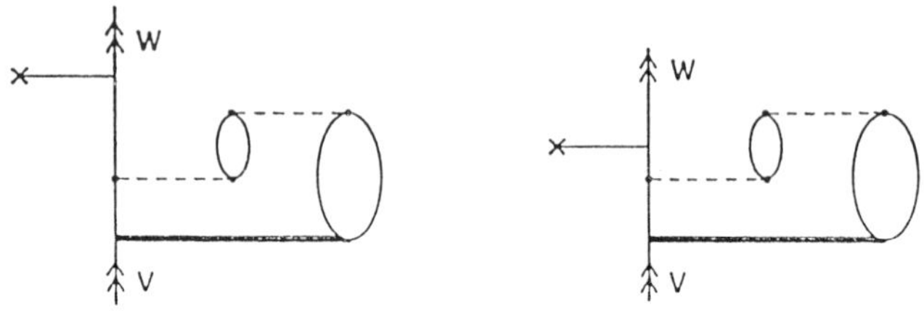

Fig. 6 Two matrix-element diagrams assoicated with a wave-operator diagram from S_3.

We now consider how to extend CCSD to incorporate all missing terms in the third- and fourth-order valence ionization energy for a one-valence-electron system. The key is to consider the *lowest-order* triple excitation operator S_3 which satisfies

$$[S_3, H_0]P = \{VS_2\}_{3,\mathrm{conn}}, \tag{20}$$

as shown in Fig. 5. S_3 in turn effectively makes a contribution to the RHS of (18) which becomes modified to

$$[S_1 + S_2, H_0]P = \left(V\Omega - \Omega PV\Omega P + V\frac{1}{D_3}\{VS_2\}_{3,\mathrm{conn}} \right)_{1,2,\mathrm{conn}}, \tag{21}$$

for singles and doubles, where D_3 is the energy denominator associated with S_3. Because we have now included all terms (single through quadruple excitations) in the second-order wave operator, and because for the fourth-order removal energy only the single and double excitation parts of the third-order wave operator contribute, this equation is complete through fourth-order for removal energies. Owing to the impossibility of storing triple excitation coefficients discussed earlier, (21) seems the most practical approach to including genuine triple excitations in a reasonably complete way.

A nonrelativistic atomic CCSD code has been developed by the Gothenburg group and applied to e.g. Na[44] and K.[45] In the Na calculation, the missed third-order terms[46] and a selection of plausibly dominant fourth-order triples were added to removal energies. For Cs, however, the only CC-type of calculation so far has been by the author, who has developed a fully relativistic implementation of a subset of (21). Two approximations to (21) were made in this work. The first was to drop the non-linear terms in the exponential expansion of Ω, which are expensive to evaluate, enter first in fourth order, and constitute about 5% of correlation. Second, the extra term involving S_3 was applied only to single excitations (and modified along the lines of the Hermitian CC formulations). This means that we pick up the missed terms in the third-order ionization energy, but we omit fourth-order genuine triple excitations, which probably also enter at the level of 5% of correlation.

To understand the modification to the S_3 term, consider the associated matrix element diagrams in Fig. 6. There are two time-orderings in which the external one-body operator is either above or below the final Coulomb interaction. Upon adding the two diagrams, the triple excitation energy denominator in each simplifies to a double excitation one,

$$\sum_{bcmnrs} \frac{z_{wm}\tilde{g}_{bcrs}\tilde{g}_{mrnb}\tilde{\rho}_{nsvc}}{\varepsilon_v+\varepsilon_b+\varepsilon_c-\varepsilon_r-\varepsilon_m-\varepsilon_s}\left(\frac{1}{\varepsilon_v-\varepsilon_m}+\frac{1}{\varepsilon_b+\varepsilon_c-\varepsilon_r-\varepsilon_s}\right)$$

$$=\sum_{bcmnrs} \frac{z_{wm}\tilde{g}_{bcrs}\tilde{g}_{mrnb}\tilde{\rho}_{nsvc}}{(\varepsilon_v-\varepsilon_m)(\varepsilon_b+\varepsilon_c-\varepsilon_r-\varepsilon_s)}, \tag{22}$$

and the diagram now contains the combination $\tilde{g}_{bcrs}/(\varepsilon_b+\varepsilon_c-\varepsilon_r-\varepsilon_s)$ which is the lowest order approximation to a coefficient in S_2, $(\tilde{\rho}_{rsbc})^*$. The tilde notation indicates the inclusion of exchange, $\tilde{g}_{bcrs}=g_{bcrs}-g_{cbrs}$. By use of a *factorization theorem*,[36] one may show that such energy denominator simplifications occur whenever the basic wave-operator diagram in Fig. 5 is attached to any other unit, and that furthermore the final Coulomb interaction may be replaced exactly by an all-order pair coefficient from S_2 (with removal of the denominator D_3).

Putting all approximations and modifications together, the Bloch equations effectively solved are thus,

$$[S_1,H_0]P=\left(V(1+S)-SPVSP+S_2^\dagger(VS_2)_{3,\text{conn}}\right)_{1,\text{conn}}, \tag{23a}$$

$$[S_2,H_0]P=(V(1+S)-SPVSP)_{2,\text{conn}}. \tag{23b}$$

The modified S_3 term derived here arises naturally in Hermitian CC formulations.[42] However, such formulations stress the use of a normalization convention different from the intermediate normalization used here. Correspondingly, we find a residual normalization denominator that must be included when evaluating matrix elements, as discussed next.

Here we describe the method used for calculating matrix elements in the Cs calculation by the author.[8,47] We start from the basic expression

$$M_{wv}=\frac{\langle\Psi_w|Z|\Psi_v\rangle}{\sqrt{\langle\Psi_w|\Psi_w\rangle\langle\Psi_v|\Psi_v\rangle}}, \tag{24}$$

and consider the case in which $Z=\sum_{i=1}^N z_i$ is a sum of one-body operators, and where Ψ_w and Ψ_v are (in lowest order) states with one valence electron outside closed shells. The denominator may then be shown[48] to cancel disconnected terms from the numerator leaving however a residual normalization term,

$$M_{wv}=M_{\text{core}}+\frac{M_{\text{val}}}{\sqrt{(1+N_w)(1+N_v)}}, \tag{25}$$

where

$$M_{\text{core}}=\delta_{wv}\left\langle 0_C\left|\Omega_w^\dagger Z\Omega_v\right|0_C\right\rangle_{\text{conn}}, \tag{26a}$$

$$M_{\text{val}} = \left\langle 0_C \left| \overline{a_w (\Omega_w^\dagger Z \Omega_v) a_v^\dagger} \right| 0_C \right\rangle_{\text{conn}} , \tag{26b}$$

$$N_v = \left\langle 0_C \left| \overline{a_v (\Omega_v^\dagger \Omega_v) a_v^\dagger} \right| 0_C \right\rangle_{\text{conn}} . \tag{26c}$$

$|0_C\rangle$ is the core determinant, $|\Psi_v\rangle = \Omega_v a_v^\dagger |0_C\rangle$, and the bar notation indicates that a_w and $a_v^\dagger$ contract into the intervening parentheses. Since we are using the linearized approximation $\Omega = 1 + S_1 + S_2$ we can substitute directly into (25) without obtaining an infinite series. (Mårtensson-Pendrill and Ynnerman[49] have shown how some infinite sequences in the full non-linear case can be summed by simple techniques.) This substitution gives the set of terms employed on Li-like systems by Blundell *et al.*[48] However, we can also build in the most important effect of the omitted non-linear terms by modifying the matrix element formalism to incorporate the RPA directly.[47] This can be achieved by everywhere replacing the bare vertex for the one-body operator z by the RPA vertex,

$$z_{ij}^{\text{RPA}}(\omega) = z_{ij} + \sum_{am} \left(\frac{\tilde{g}_{imja} z_{am}^{\text{RPA}}(\omega)}{\varepsilon_a - \omega - \varepsilon_m} + \frac{\tilde{g}_{iajm} z_{ma}^{\text{RPA}}(\omega)}{\varepsilon_a + \omega - \varepsilon_m} \right), \tag{27}$$

as shown diagrammatically in Fig. 7. The evaluation frequency here is fixed at the experimental value $\omega = (E_w - E_v)^{\text{expt}}$ (or zero for diagonal matrix elements). Care must be taken to avoid double counting, as discussed in more detail in Ref. 47.

The final set of terms can be conveniently grouped into six categories: (i) first-order (V^{N-1} DF) matrix element, (ii) RPA correction, (iii) Brueckner Orbital (BO) terms, in which many-body corrections are applied to the external valence state, (iv) Structural Radiation (SR) terms, which in lowest order give the terms remaining in the third-order matrix element after removing the third-order BO and RPA terms,[50] (v) other numerator terms, not included above, and (vi) normalization terms. The diagrams are shown in Fig. 8. We also give a seventh term, the Breit correction, obtained by adding the frequency-independent Breit interaction to the Coulomb interaction in the DF procedure. The resulting 'Dirac-Fock-Breit' (DFB) equations are solved using conventional point-by-point integration techniques. The change in the first-order matrix element when re-evaluated with DFB valence orbitals is then the tabulated Breit correction; it is only significant for the hyperfine constants of some states.

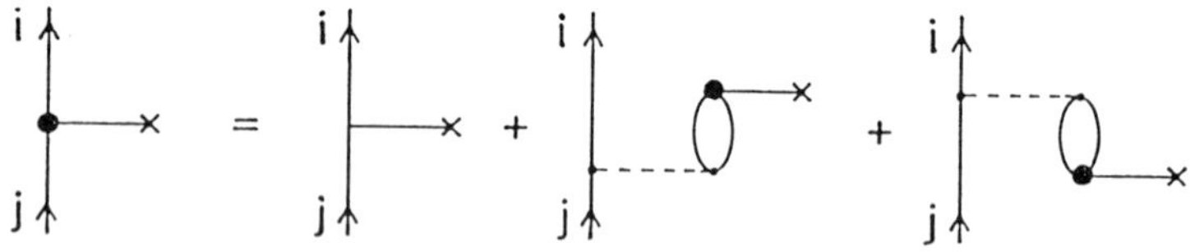

Fig. 7 Definition of RPA vertex for a one-body operator.

Results[47] for valence ionization energies are given in Table IV, for hyperfine constants in Table V, and for allowed electric dipole amplitudes in Table VI. The basis set parameters for this calculation were as described in a previous section; the entire core was excited, with L ≤ 7. The theoretical predictions for allowed electric dipole amplitudes are in many cases more accurate than the measured values, with the exception possibly of the small matrix elements of the type D(6S-7P). The energy results correspond to roughly 5% of correlation. Because of the large basis set, and the complete treatment of the core, most of this remaining 5% is given

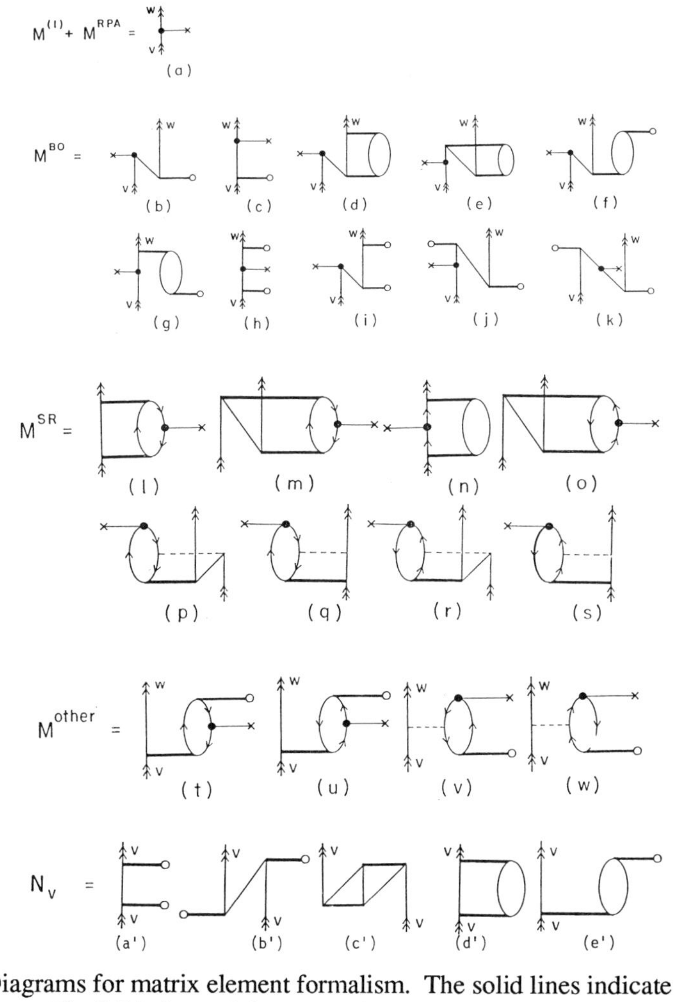

Fig. 8 Diagrams for matrix element formalism. The solid lines indicate all-order coefficients. The RPA form of the external one-body operator is used throughout (see Fig. 7).

TABLE IV Valence removal energies (a.u.).

State	ε_v^{DHF}	$\delta\varepsilon_v$	Sum	Experiment[a]	$\Delta(\%)$
$6s$	0.127 37	0.015 21	0.142 57	0.143 10	-0.37
$7s$	0.055 19	0.003 26	0.058 45	0.058 65	-0.33
$6p_{1/2}$	0.085 62	0.006 36	0.091 98	0.092 17	-0.21
$7p_{1/2}$	0.042 02	0.001 83	0.043 85	0.043 93	-0.17
$8p_{1/2}$	0.025 12	0.000 80	0.025 92	0.025 96	-0.14
$6p_{3/2}$	0.083 79	0.005 72	0.089 51	0.089 64	-0.15
$7p_{3/2}$	0.041 37	0.001 66	0.043 03	0.043 10	-0.16
$8p_{3/2}$	0.024 81	0.000 74	0.025 55	0.025 58	-0.13

[a]C. E. Moore, *Atomic Energy Levels*, Natl. Bur. Stand. Ref. Data Ser., Natl. Bur. Stand. (U.S.) Circ. No. 35 (U.S. GPO, Washington, D.C., 1971), Vol. I.

TABLE V Magnetic Dipole Hyperfine Constants (MHz) for ^{133}Cs, $I = \frac{7}{2}$, g_I = 0.7377208. Conversion factor: 1 a.u. = 6.579684 × 10^9 MHz.

State	$M^{(1)}$	M^{RPA}	M^{BO}	M^{SR}	M^{other}	M^{norm}	M^{Breit}	Sum	Exp[a]
$6s$	1426.81	292.83	596.85	4.66	26.23	-56.39	-0.00	2291.00	2298.16
$7s$	392.05	79.64	76.58	0.76	6.54	-11.49	-0.05	544.04	545.90(9)[b]
$6p_{1/2}$	161.09	39.93	84.67	9.67	3.06	-4.49	-1.25	292.67	291.90(13)
$7p_{1/2}$	57.68	13.68	20.54	3.15	1.01	-1.46	-0.39	94.21	94.35(4)
$8p_{1/2}$	27.11	6.32	8.34	1.39	0.46	-0.66	-0.17	42.79	42.97(10)
$6p_{3/2}$	23.944	18.853	16.226	-9.321	0.890	-0.675	-0.131	49.785	50.275(3)[c]
$7p_{3/2}$	8.650	6.700	3.948	-3.083	0.303	-0.220	-0.043	16.255	16.605(6)
$8p_{3/2}$	4.087	3.146	1.610	-1.415	0.140	-0.100	-0.020	7.447	7.58(1)

[a] E. Arimondo, M. Inguscio and P. Violino, Rev. Mod. Phys. 49, 31 (1977).
[b] S.L. Gilbert, R.N. Watts and C.E. Wieman, Phys. Rev. A 27, 581 (1983).
[c] C. Tanner and C.E. Wieman, Phys. Rev. A 38, 1616 (1988).

TABLE VI Reduced E1 transition matrix elements for length (L) and velocity (V) forms of dipole operator (a.u.).

Transition		$M^{(1)}$	M^{RPA}	M^{BO}	M^{SR}	M^{other}	M^{norm}	Sum	Exp[a]
$6s-6p_{1/2}$	L	-5.278	0.303	0.404	-0.040	-0.005	0.090	-4.525	-4.52(1)
	V	-4.129	0.106	-0.505	-0.051	-0.002	0.090	-4.492	
$6s-7p_{1/2}$	L	0.372	-0.134	0.028	0.017	0.002	-0.006	0.279	0.284(2)
	V	0.263	-0.077	0.084	0.010	0.001	-0.006	0.275	
$6s-8p_{1/2}$	L	0.1326	-0.0838	0.0194	0.0109	0.0010	-0.0016	0.0787	
	V	0.0835	-0.0515	0.0407	0.0057	0.0004	-0.0015	0.0772	
$6s-6p_{3/2}$	L	-7.426	0.413	0.582	-0.053	-0.006	0.121	-6.370	-6.36(1)
	V	-5.761	0.119	-0.722	-0.078	-0.007	0.121	-6.328	
$6s-7p_{3/2}$	L	-0.695	0.187	-0.052	-0.024	-0.002	0.011	-0.575	-0.583(10)
	V	-0.508	0.098	-0.151	-0.018	-0.002	0.011	-0.571	
$6s-8p_{3/2}$	L	-0.283	0.118	-0.036	-0.015	-0.001	0.004	-0.214	
	V	-0.196	0.067	-0.076	-0.009	-0.001	0.004	-0.212	
$7s-6p_{1/2}$	L	-4.413	-0.037	0.140	0.006	0.000	0.077	-4.228	
	V	-3.854	-0.189	-0.247	-0.039	0.000	0.077	-4.251	
$7s-7p_{1/2}$	L	-11.009	0.088	0.410	-0.008	-0.002	0.189	-10.332	
	V	-9.653	-0.108	-0.650	-0.041	0.000	0.188	-10.264	
$7s-8p_{1/2}$	L	-0.921	0.052	-0.059	-0.005	-0.001	0.017	-0.917	
	V	-0.788	-0.007	-0.125	-0.012	0.000	0.017	-0.916	
$7s-6p_{3/2}$	L	-6.671	-0.042	0.144	0.008	-0.001	0.112	-6.451	
	V	-5.899	-0.298	-0.351	-0.064	-0.004	0.113	-6.503	
$7s-7p_{3/2}$	L	15.345	-0.117	-0.634	0.010	0.002	-0.248	14.357	
	V	13.352	0.174	0.917	0.063	0.005	-0.247	14.265	
$7s-8p_{3/2}$	L	1.605	-0.071	0.109	0.007	0.001	-0.028	1.624	
	V	1.382	0.022	0.224	0.019	0.002	-0.028	1.620	

[a] L. Shabanova, Yu. Monakov and A. Khlyustalov, Opt. Spectrosc. (USSR) 47, 1 (1979).

by the omitted correlation terms, namely the fourth-order genuine triples, and the fourth-order non-linear singles and doubles terms.

$E1^{PNC}$ BY SUM OVER INTERMEDIATE STATES

We evaluate now off-diagonal matrix elements of H^{PNC} using the all-order procedure described above and attempt to saturate the sum over X in Eq. (7). The contributions from the states X = 6-9$P_{1/2}$ are summarized in Table VII. Note the rather rapid convergence, with most of the sum coming from the $6P_{1/2}$ and $7P_{1/2}$ states, and only about 1.7% coming from the $9P_{1/2}$ state. The error estimation for these terms is critical, and is discussed further below.

The small 'tail' ($10P_{1/2}$–$\infty P_{1/2}$ including the continuum) contribution is estimated using less sophisticated techniques. We evaluate

$$E1^{PNC}(\text{tail}) = \sum_{n=10}^{\infty} \frac{\langle 7s|h^{PNC}_{RPA}|np_{1/2}\rangle\langle np_{1/2}|D_{RPA}|6s\rangle}{\varepsilon_{7s}-\varepsilon_{np_{1/2}}} + \begin{pmatrix} h^{PNC}_{RPA} \leftrightarrow D_{RPA} \\ \varepsilon_{7s} \rightarrow \varepsilon_{6s} \end{pmatrix}, \tag{28}$$

TABLE VII. Contributions to $E1^{PNC}$ in Cs from intermediate states X = 6–9$P_{1/2}$. Units: energies E in a.u., dipole matrix elements $\langle Xm = 1/2\,|D_z|\,Ym = 1/2\rangle$ in a.u., matrix elements of h^{PNC} in units of $i(-Q_W/N)\times 10^{-11}$ a.u.; final contributions to $E1^{PNC}$ in $i|e|a_0(-Q_W/N)\times 10^{-11}$.

n	$\langle 7S\,|D|\,nP_{1/2}\rangle$	6S perturbed $\langle nP_{1/2}\,|h^{PNC}|\,6S\rangle$	$E_{6S} - E_{nP_{1/2}}$	Contribution
6	1.723	−0.0549	−0.05060	1.868
7	4.221	0.0312	−0.09872	−1.333
8	0.376	0.0210	−0.11665	−0.068
9	0.143	0.0159	−0.12539	−0.018
n	$\langle 7S\,|h^{PNC}|\,nP_{1/2}\rangle$	7S perturbed $\langle nP_{1/2}\,|D|\,6S\rangle$	$E_{7S} - E_{nP_{1/2}}$	Contribution
6	−0.0267	1.8564	0.03353	−1.480
7	0.0151	−0.1189	−0.01460	0.123
8	0.0102	−0.0356	−0.03253	0.011
9	0.0077	−0.0168	−0.04127	0.003
TOTAL				−0.893

where the states are now *single-particle* DF states, and operators are the RPA forms (27) of the weak operator (1) and the dipole operator. The evaluation frequency for the RPA parts should in principle be set to the energy difference of the two states involved in each matrix element, but in practice there is very little sensitivity to this frequency. The tail term is, however, rather sensitive to the RPA correction to the dipole matrix element, and we conservatively choose this correction as the error on the tail, quoting $E1^{PNC}$(tail) = −0.019[6] in the usual units. Further work may show how to refine this value with more confidence.

In addition to the X = $P_{1/2}$ sequence discussed above, there are also autoionizing $P_{1/2}$ states in which the core is excited. One such class of states in fact contributes already to the lowest-order perturbation-theory expression, those in which a core $p_{1/2}$ state is excited to the $7s$ valence state. The lowest-order contribution of such states is

$$E1^{PNC}(\text{core}) = \sum_{n=2}^{5} \frac{\langle 7s|h^{PNC}|np_{1/2}\rangle\langle np_{1/2}|D|6s\rangle}{\varepsilon_{7s}-\varepsilon_{np_{1/2}}} + \begin{pmatrix} h^{PNC} \leftrightarrow D \\ \varepsilon_{7s} \rightarrow \varepsilon_{6s} \end{pmatrix}, \tag{29}$$

which evaluates to only a −0.2% effect, negligible at the present level of theory error. These terms are suppressed partly because of the large energy required for excitation of the tightly bound Xe-like core, and partly by reduced dipole matrix elements between valence and core states. Their small size is convenient, because many-body corrections to such terms may be substantial and are complicated to evaluate. We assign an error equal to the value of the term, $E1^{PNC}$(core) = +0.002[2].

Finally, we have considered the effect of the Breit interaction at DF level as described in more detail in the next section. At present the Breit correction, $E1^{PNC}$(Breit) = +0.002[2], is negligible in comparison with the error in the main term (X = 6-9$P_{1/2}$), which we now discuss.

TABLE VIII. Error estimation on main term in sum-over-states approach (Table VII). Modifications to energy denominator (E), dipole matrix elements (D), the particular dipole matrix element $D_{6S-7P_{1/2}}$, and matrix elements of $h^{\rm PNC}$ as described in the text. Units: $i|e|a_0(-Q_{\rm W}/N) \times 10^{-11}$

E	D	$D_{6S-7P_{1/2}}$	$h^{\rm PNC}$	$\rm E1^{PNC}$
theory	raw	theory	raw	−0.893
exp	raw	theory	raw	−0.900
exp	e-fit	theory	e-fit	−0.894
exp	e-fit	theory	hfs-fit	−0.890
exp	e-fit	exp	e-fit	−0.893
exp	e-fit	exp	hfs-fit	−0.888
exp	e-fit	theory	sr-fit	−0.893
exp	e-fit	exp	sr-fit	−0.891
exp	raw	exp	raw	−0.898
theory	raw	exp	raw	−0.891
			Final choice:	−0.894(6)

Since there are no rigorous bounds available for estimating error in a correlation calculation of this type, we necessarily proceed by making plausible assumptions. The general idea is to repeat the calculation for X = 6-9$P_{1/2}$ many times making small adjustments to input energies and matrix elements, and to consider the resulting scatter. One important error in the matrix element involves the BO term (Fig. 8), which is closely related to the ionization energy in that both are determined primarily by the valence (v) single-excitation coefficients ρ_{vr} for $v \rightarrow \kappa_v$. One observes an approximate scaling behaviour in which, as correlation effects are added, the valence correlation energy $\delta\varepsilon$ and the BO term change by roughly the same fractions. Thus one can 'fit' experimental removal energies by rescaling the valence single excitation coefficients in the ratio $\delta\varepsilon^{\rm expt}/\delta\varepsilon^{\rm calc}$, and then re-evaluate the matrix elements. This 'energy fit' is one type of modification to matrix elements. In a similar way, we can adjust the BO contribution of s-electron hyperfine constants to fit the experimental values, in this way effectively adjusting the normalization at the origin. Since the PNC also acts at the nucleus, we rescale the BO contribution to PNC in the same ratio. This gives 'hfs fitted' PNC matrix elements. Furthermore, other error terms are associated with non-BO parts of the matrix element. Here we can, somewhat arbitrarily, rescale SR parts of matrix element to fit hyperfine constants, and rescale the SR part of a PNC matrix element in the same ratio. We also note that some disagreement was found between calculated and experimental dipole matrix elements of the type D(6S-6P$_{1/2}$) (see Table VI), and consider the effect of substituting the experimental value for the calculated one. Finally, we can use either experimental or theoretical energies in combination with either 'raw' matrix elements or matrix elements fitted or modified in the various ways as described above. This gives a scatter of values with a width ±0.6% for the main term as in Table VIII.

Combining errors in quadrature, we quote as our final value from this method,

$$\rm E1^{PNC}(\text{direct sum}) = -0.909[9]\,(Q_W/-N)\,i|e|a_0 10^{-11}, \tag{30}$$

TABLE IX. Summary of various contributions to $E1^{PNC}$ in Cs from the sum-over-states method. Units: $i|e|a_0(-Q_W/N) \times 10^{-11}$.

Term	Contribution
$6-9P_{1/2}$	−0.894(6)
$E1^{PNC}$(tail)	−0.019(6)
$E1^{PNC}$(core)	0.002(2)
Breit	0.002(2)
Total	−0.909(9)

as summarized in Table IX. Since we actually require $E1^{PNC}/\beta$ to interpret the Cs experiment, we have repeated the above procedure for β, Eq. (8), finding

$$\beta(\text{direct sum}) = 27.00[20]a_0^3. \tag{31}$$

We shall not present the details for β here.[51] Note that, in performing an analogous scatter experiment on β, we found no noticeable correlation between β and $E1^{PNC}$, so that the error in $E1^{PNC}/\beta$ is to be found by combining the separate errors.

Another check on the value of $E1^{PNC}$ comes from calculating it in an entirely different way, as we now describe.

PARITY-MIXED MANY-BODY APPROACH

A number of different groups have worked on this approach and for an historical perspective the reader is referred to Blundell *et al.*.[52] The first complete calculations along the lines to be described were by members of the Novosibirsk group.[7,53] The first step in this procedure is the solution of the PNC-DF equations to give parity-mixed core and excited single-particle orbitals,

$$\left(h_D + V_{nuc} + \tilde{V}_{DF} + h^{PNC}\right)\tilde{\psi} = \varepsilon\tilde{\psi}, \tag{32}$$

$$\tilde{\psi} = \psi + \psi^{PNC}.$$

Here $\tilde{V}_{DF}$ is the usual V^{N-1} DF potential evaluated however with parity-mixed core orbitals arising from the self-consistent solution of (32),

$$\langle i|\tilde{V}_{DF}|j\rangle = \sum_{core}\left(\langle i\tilde{c}|r_{12}^{-1}|j\tilde{c}\rangle - \langle i\tilde{c}|r_{12}^{-1}|\tilde{c}j\rangle\right), \tag{33}$$

(for arbitrary states i and j). Eqn. (32) is linearized in h^{PNC} and solved in two stages, the first being just the normal DF equations,

$$(h_D + V_{nuc} + V_{DF})\psi = \varepsilon\psi, \tag{34a}$$

$$(\varepsilon - h_D - V_{nuc} - V_{DF})\psi^{PNC} = \left(h^{PNC} + V_{DF}^{PNC}\right)\psi. \tag{34b}$$

The PNC-DF potential in (34b) is the part of $\tilde{V}_{DF}$ linear in h^{PNC}. One evaluates $E1^{PNC}$ at this level of approximation from

$$E1^{PNC}(\text{PNC - DF}) = \left\langle 7s \middle| D \middle| 6s^{PNC} \right\rangle + \left\langle 7s^{PNC} \middle| D \middle| 6s \right\rangle. \tag{35}$$

The well-established result[7,54,55] $E1^{PNC}$(PNC-DF) = $-0.927.10^{-11}$ $i|e|a_0$ $(Q_W/-N)$ is only about 3% different from the final total, and all subsequent correlation corrections are quite small, making this approach appear attractive. Note, however, that an important subset of correlation terms has already been included at this level. The PNC-DF potential is equivalent to including RPA corrections to h^{PNC}; dropping this potential from (34b) modifies $E1^{PNC}$ to -0.740 in the usual units, showing that a ~25% correlation correction has already been included at this stage.

The PNC-DF procedure defines a complete, orthonormal parity-mixed set of single-particle states. We follow the procedure introduced by the Notre Dame group[56] of generating explicitly a "complete" parity-mixed basis set. First we generate a conventional parity-pure DF basis set by the B-spline technique,[34] as described earlier. Then, for each state $|i\rangle$ in this set, we generate its opposite-parity admixture by direct summation over the formal solution to (34b)

$$\left| i^{PNC} \right\rangle = \sum_j \frac{\left| j \right\rangle \left\langle j \middle| h^{PNC} + V_{DF}^{PNC} \middle| i \right\rangle}{\varepsilon_j - \varepsilon_i}. \tag{36}$$

Negative-energy states, representing virtual pair creation, may be included in the sum over j here. Their overall effect is small, << 0.1%, when using the length form of the dipole operator, but amounts to about 1% with the velocity form.[57] (In the earlier approach, using solutions to the no-virtual-pair CC equations, we used only the length form of the dipole operator.)

We are now in a position to apply parity-mixed MBPT for an E1 matrix element, regarding the PNC-DF as a lowest-order approximation. We shall consider in turn the following set of many-body or relativistic corrections: (i) RPA corrections to the dipole operator, (ii) BO corrections to the valence orbitals, (iii) residual terms from third-order perturbation theory, including normalization terms, and (iv) the Breit correction. Numerical values are summarized in Table X.

(i) There are two parts to the RPA dipole operator in parity-mixed perturbation theory, the usual one (27), and a part linear in h^{PNC} formed by replacing one state at a time in (27) by its opposite-parity admixture. Writing $D_{RPA} = D + \Delta D + \Delta D^{PNC}$, we can express the RPA correction as

$$E1^{PNC}(\text{RPA}) = \left\langle 7s \middle| \Delta D \middle| 6s^{PNC} \right\rangle + \left\langle 7s^{PNC} \middle| \Delta D \middle| 6s \right\rangle + \left\langle 7s \middle| \Delta D^{PNC} \middle| 6s \right\rangle. \tag{37}$$

Diagrammatically, ΔD^{PNC} arises from opposite-parity substitutions inside the RPA bubbles. These 'internal' substitutions are suppressed by energy-denominator arguments, since they implicitly involve excitations from the core, while the 'external' substitutions (the other two terms above) bring in instead an energy denominator involving the small differences $\varepsilon_{6,7s} - \varepsilon_{np1/2}$.

(ii) Formally, the BO contribution follows from adding the nonlocal (proper) self-energy operator Σ to the DF Hamiltonian and resolving for normalized *quasi-particle* valence orbitals ϕ_v and corresponding energy eigenvalues,[58]

$$\left(h_{\mathrm{D}} + V_{\mathrm{nuc}} + V_{\mathrm{DF}} + \Sigma(\varepsilon'_v)\right)\phi_v = \varepsilon'_v \phi_v. \tag{38}$$

If Σ is the exact self energy, ε'_v is the exact valence removal energy. The self energy is given by the sum of all possible one-particle-irreducible diagrams,[37] that is, diagrams that cannot be split into two by cutting a single line. Solution of (38) is then equivalent to summing chains of self-energy diagrams. This formalism follows most naturally from the time-dependent development of MBPT. The treatment of the self energy in this way was first advocated in PNC calculations by the Novosibirsk group.[53] Note that $\Sigma(\varepsilon)$ is a function of an energy parameter ε, and (38) must be solved in a self-consistent manner such that ε is the final eigenvalue. In a perturbative treatment this brings in derivatives of the self energy, which in time-independent MBPT correspond to *folded* diagrams.

We consider the solution of (38) in a number of stages. First, we approximate Σ by the leading term $\Sigma^{(2)}$,

$$\left\langle i\left|\Sigma^{(2)}(\varepsilon)\right|j\right\rangle = \sum_{amn} \frac{g_{aimn}\tilde{g}_{mnaj}}{\varepsilon_a + \varepsilon - \varepsilon_m - \varepsilon_n} - \sum_{abm} \frac{g_{abjm}\tilde{g}_{imab}}{\varepsilon_a + \varepsilon_b - \varepsilon_m - \varepsilon}, \tag{39}$$

and solve (38) to first order in $\Sigma^{(2)}$. The result can be expressed in terms of the lowest-order Brueckner modification $|\delta v\rangle$ to the DF state, $|\phi_v\rangle \approx |v\rangle + |\delta v\rangle$,

$$|\delta v\rangle = \sum_{j \neq v} \frac{|j\rangle\left\langle j\left|\Sigma^{(2)}(\varepsilon_v)\right|v\right\rangle}{\varepsilon_v - \varepsilon_j}, \tag{40}$$

where the sum over states is over the DF basis set. The corresponding matrix-element correction is the third-order BO term,

$$Z^{(3)}_{wv}(BO) = \langle \delta w|z|v\rangle + \langle w|z|\delta v\rangle. \tag{41}$$

Calculations of this correction for standard properties (hyperfine constants, allowed E1 matrix elements) have now been performed by a number of groups using relativistic basis sets,[59,60,61] and numerical agreement is very good.

In the parity-mixed treatment of interest here, Σ separates into a normal and a PNC part, $\Sigma \rightarrow \Sigma + \Sigma^{\mathrm{PNC}}$, the latter corresponding to opposite-parity substitutions *inside* the self-energy diagram. The Brueckner corrections $|\delta v\rangle$ then acquire opposite-parity admixtures $|\delta v^{\mathrm{PNC}}\rangle$. It is useful to distinguish between 'internal' and 'external' PNC-BO contributions, given to lowest order by

$$\left|\delta v^{\mathrm{PNC}}\right\rangle = \left|\delta v^{\mathrm{PNC}}\right\rangle_{\mathrm{ext}} + \left|\delta v^{\mathrm{PNC}}\right\rangle_{\mathrm{int}}, \tag{42a}$$

$$\left|\delta v^{\mathrm{PNC}}\right\rangle_{\mathrm{ext}} = \sum_{j \neq v} \frac{\left(|j\rangle\left\langle j^{\mathrm{PNC}}\right| + \left|j^{\mathrm{PNC}}\right\rangle\langle j|\right)\Sigma^{(2)}(\varepsilon_v)|v\rangle}{\varepsilon_v - \varepsilon_j} + \sum_{j \neq v} \frac{|j\rangle\left\langle j\left|\Sigma^{(2)}(\varepsilon_v)\right|v^{\mathrm{PNC}}\right\rangle}{\varepsilon_v - \varepsilon_j}, \tag{42b}$$

$$\left|\delta v^{\rm PNC}\right\rangle_{\rm int} = \sum_{j\neq v} \frac{|j\rangle\langle j|\Sigma^{(2){\rm PNC}}(\varepsilon_v)|v\rangle}{\varepsilon_v - \varepsilon_j}. \tag{42c}$$

For each type of correction, internal or external, one can define a third-order BO contribution to $\mathrm{E1^{PNC}}$,

$$\mathrm{E1^{PNC}(BO\text{-}3)} = \left\langle \delta w^{\rm PNC}\middle|z\middle|v\right\rangle + \left\langle w^{\rm PNC}\middle|z\middle|\delta v\right\rangle + \left\langle \delta w\middle|z\middle|v^{\rm PNC}\right\rangle + \left\langle w\middle|z\middle|\delta v^{\rm PNC}\right\rangle. \tag{43}$$

There is now agreement on the numerical value of these terms for Cs,[7,56,62] $\mathrm{E1^{PNC}}$(BO-3,ext) = –0.0058, $\mathrm{E1^{PNC}}$(BO-3,int) = –0.003, in the usual units. While the external contribution is a large 6% effect, the 'internal' contribution is very small, 0.3%, once more suppressed by energy denominator considerations. We shall see below that the non-linear effects of the self-energy are rather important, but this agreement is an important demonstration that the numerical issues surrounding parity-mixed basis sets are now well understood.

Thus far we have considered only linear effects of the self energy. We now wish to consider a parity-mixed treatment of the full quasi-particle equation in which Σ is chained. Since the term from $\Sigma^{\rm PNC}$ is negligible at the present level of error, we shall separate it out and treat it only to lowest order as discussed above. Linearizing in $h^{\rm PNC}$ then gives,

$$\left(\varepsilon_v' - h_{\rm D} - V_{\rm nuc} - V_{\rm DF} - \Sigma(\varepsilon_v')\right)\phi_v^{\rm PNC} = \left(h^{\rm PNC} + V_{\rm DF}^{\rm PNC}\right)\phi_v. \tag{44}$$

We solve Eqs. (38) and (44) numerically by an iterative procedure which starts from the DF level solutions. Having thus found ϕ_v and $\phi_v^{\rm PNC}$ we build in cross-terms with the RPA correction to the dipole operator by finally evaluating,

$$\mathrm{E1^{PNC}(BO)} = \left\langle \phi_{7s}^{\rm PNC}\middle|D+\Delta D\middle|\phi_{6s}\right\rangle + \left\langle \phi_{7s}\middle|D+\Delta D\middle|\phi_{6s}^{\rm PNC}\right\rangle + \left\langle \phi_{7s}\middle|\Delta D^{\rm PNC}\middle|\phi_{6s}\right\rangle$$
$$- \mathrm{E1^{PNC}(PNC\text{-}DF)} - \mathrm{E1^{PNC}(RPA)}. \tag{45}$$

We have evaluated[8] (38), (44) and (45) numerically for the case $\Sigma = \Sigma^{(2)}$, giving $\mathrm{E1^{PNC}}$(BO) = –0.014. The reduction compared to $\mathrm{E1^{PNC}}$(BO-3) is due partly to the RPA cross terms (a –1.5% effect) but mostly to the chaining (a –3% effect).

Finally, we need to consider contributions to the self energy operator beyond second order. The effect of the higher-order self energy can be estimated crudely by putting $\Sigma = \lambda\,\Sigma^{(2)}$, where $\lambda = 0.80$ (s-wave) and $\lambda = 0.84$ ($p_{1/2}$-wave) are chosen to fit energy levels, and resolving the parity-mixed quasi-particle equation. As can be seen from the value of λ, $\Sigma^{(2)}$ is a substantial overestimate of Σ. Nevertheless, the final result for $\mathrm{E1^{PNC}}$(BO) is changed by only a few tenths of a percent (of the total $\mathrm{E1^{PNC}}$). This suggests that while it is essential to chain the self energy, higher-order treatment of the self-energy is less important. This is certainly a big simplification for future more precise work.

It is rather easy to see why the chaining of the self energy is so important. If one considers the implicit sum over states (7) in the parity-mixed procedure, it may be seen that chaining has the effect of correcting the energy denominator from its DF value to the physical value. This is important, because the DF energies are substantially (~10%) in error. Less obvious is the reason for the insensitivity to Σ, which appears to be due to an accidental, though happy, cancellation of numerator corrections with corresponding denominator modifications as Σ is varied. A similar cancellation is observed in the scatter analysis described in the previous section.

The Novosibirsk group, who originally chained the self energy in this way, have added[7] an explicit set of higher-order terms to $\Sigma^{(2)}$ given by iterating the effects of the dominant diagrams F1 and F2 in Fig. 3. Their final result agrees quite well with ours, as do individual terms in a breakdown (see Table X).

(iii) Next we add residual third-order diagrams, and normalization terms. The residual third-order terms are of SR type, as in Fig. 8 but with the all-order coefficients replaced by Coulomb interactions. Once more 'internal' substitutions are suppressed, so we have evaluated only the terms in which the external valence lines are replaced by their opposite-parity admixture. The resulting term $E1^{PNC}(SR)$ is quite small, ~0.5%.

In origin, the normalization term $E1^{PNC}(\text{norm})$ is related to the normalization denominator in (26). There are some differences between that formalism and this, however. Clearly, normalization terms such as Fig. 8(a', b') are not required as they are part of the normalization of the quasi-particle orbital ϕ_v. More subtly, the sign of the normalization term in Fig. 8(c') must be changed.[51] The final result, using the values of Figs. 8(c', d') with all-order coefficients, is a –0.8% effect.

(iv) Finally, we consider the effect of the Breit interaction at DF level by adding a 'Dirac-Fock-Breit' potential to the DF equations, as discussed earlier for hyperfine constants. The PNC-DF equation (34b) is similarly modified to

$$\left(\varepsilon - h_D - V_{nuc} - V_{DF} + V_{DFB}\right)\psi^{PNC} = \left(h^{PNC} + V_{DF}^{PNC}\right)\psi. \tag{46}$$

In principle, there is also a Breit modification to the PNC-DF potential on the RHS of (46), but since the Breit correction overall is so small, consideration of this term does not seem justified at present.

TABLE X. Various contributions to $E1^{PNC}$ in Cs from parity-mixed MBPT. See text for more details. Units: $i|e|a_0(-Q_W/N) \times 10^{-11}$

Term	Present (Ref. 8)	Ref. 7	Ref. 62
PNC-DF	−0.927		−0.927
RPA (internal)	0.035		0.035
RPA (external)	0.002	−0.886[a]	0.002
BO (internal)	−0.014(8)	−0.022	
BO-3 (external)	−0.003	−0.003	
SR	−0.006(4)	−0.003	
norm	0.008	0.006	
Breit	0.002(2)		
TOTAL	−0.903(9)	−0.908	

[a]This value represents the sum of PNC-DF, RPA (internal) and RPA (external).

The final total is shown in Table X along with a comparison with other calculations. The agreement is in general very good. Further, the total also agrees quite closely with the sum-over-states value given in Table IX. We take the average of the values from each method, and quote as our final recommended value,

$$\mathrm{E1^{PNC}(theory)} = -0.906[9](Q_W/-N)i|e|a_0 10^{-11}. \tag{47}$$

COMPARISON WITH THE STANDARD ELECTROWEAK MODEL

To give some idea how the atomic measurements agree with the standard model, we now infer the value of the weak charge Q_W of the ^{133}Cs nucleus from the most recent and accurate measurement[6] of $\mathrm{E1^{PNC}}$,

$$\left.\frac{\mathrm{E1^{PNC}}}{\beta}\right|_{\mathrm{expt}} = \begin{cases} -1.513(50) \ \ \mathrm{mV/cm} \ \ (\mathrm{F}=3 \rightarrow \mathrm{F}'=4), \\ -1.639(48) \ \ \mathrm{mV/cm} \ \ (\mathrm{F}=4 \rightarrow \mathrm{F}'=3). \end{cases} \tag{48}$$

The barely significant discrepancy on the two hyperfine transitions may be due to nuclear-spin-dependent effects from (2); these effects can be eliminated to good approximation by averaging the two values.[8,15] Taking β(theory) from Eq. (31) and $\mathrm{E1^{PNC}}$(theory) from Eq. (47), we then find

$$Q_W = -71.04(1.58)[0.88]. \tag{49}$$

The error given is a combination of atomic experiment (parentheses) and atomic theory (square brackets). There have been a number of calculations of one-loop electroweak radiative corrections to Q_W.[63] A recent calculation by Marciano and Rosner[64] including one-loop corrections in the standard model gives $Q_W = -73.20 \pm 0.13$, in agreement with the value from atomic PNC to within one standard deviation.

It is likely that the experimental error will be reduced to about 0.5% within a year or so,[65] at which point a critical test of the standard model may become possible without further improvement in atomic theory. The hope is that one may be sensitive to new physics beyond the standard model entering through radiative corrections. Kennedy and Lynn[66] have shown that the effect of new physics on loop corrections can be described in a model-independent way in terms of a small number of parameters. Using the notation of Ref. 67, Marciano and Rosner[64] have given for Cs,

$$Q_W = -73.20 \pm 0.13 - 0.8S - 0.005T, \tag{50}$$

where S here describes new physics entering in a weak isospin conserving manner, and T that entering in a weak isospin breaking manner. Physics associated with a heavy top quark mass enters via T as

$$T \approx 0.257\left(\frac{m_{\mathrm{t}}^2 - 140\mathrm{GeV}^2}{m_{\mathrm{W}}^2}\right). \tag{51}$$

The small coefficient of T in (50), which results partly by accidental cancellations, renders Q_W very insensitive to variations in the unknown mass of the top quark.[8,68] This puts atomic PNC in a unique position, for other tests of electroweak radiative corrections, such as a precise measurement of the mass of the $W^{\pm}$ boson, or the measurement of asymmetries in the scattering of positrons and polarized electrons, are sensitive to S and T with roughly equal coefficients,[64] so that it is not possible to discuss new physics independently of the mass of the top quark. A clear deviation of theory and experiment for atomic PNC, however, indicates immediately a deviation from the standard model entering through S. Nevertheless, the goal should be to fit as many accurate experiments as possible to the parametrization of radiative corrections, and with a little more accuracy, which may soon be forthcoming on the experimental side, atomic PNC may be able to make a significant contribution to this process.

Atomic PNC also provides the most stringent constraints on the masses of additional hypothetical Z bosons, which enter independently of the parametrization of radiative corrections discussed above. A discussion is given in Refs. 69 and 70.

FUTURE PROSPECTS

There is considerable hope of improving the atomic theory, at least to the 0.5% level, probably less. The hope is that such an improvement could occur from the inclusion of omitted fourth-order terms in the CC procedure, notably genuine triple excitations (and here also the nonlinear terms in the CCSD approximation). Studies on small atomic and molecular systems have shown a rapid improvement in accuracy upon including triples, e.g. for the molecule BH_3, CCSD gives about 99.2% of the correlation energy, and the inclusion of triples accounts for most of the remainder, with no more than about 0.01% coming from quadruples.[71] If a similar situation persists for a large system like Cs, then calculations to the 0.5% level should indeed be possible. It remains, however, a rather hard technical challenge. As suggested in Eq. (21), the most practical procedure for including genuine triples in a reasonably complete way is probably to add extra terms to the RHS of the doubles equation. This circumvents the computer memory problem, although one is left with an expensive calculation which will probably force additional approximations, such as the truncation of core states or the basis set, or the selection of certain dominant triple excitation terms. If this can be done, however, there is real hope of a substantial improvement in the accuracy obtained.

Another interesting possibility is the use of the isotopic effect to avoid the need for atomic theory altogether.[72] To a first approximation, the atomic theory cancels upon dividing results obtained on identical transitions for different isotopes. As pointed out recently by Fortson, Pang and Wilets,[73] however, the atomic theory does not quite cancel, because there is a significant contribution due to the change in neutron distribution from isotope to isotope (the nuclear density $\rho(r)$ in (1) is dominated by the neutron density). It remains an open question whether enough is known about changes in neutron density along sequences of isotopes that this approach can lead to greater accuracy than is already provided by a 1% atomic theory in the direct method.

REFERENCES

1. T. D. Lee and C. N. Yang, Phys. Rev. **105**, 1671 (1956).
2. E. M. Purcell and N. F. Ramsey, Phys. Rev. **78**, 807 (1950).
3. C. S. Wu, E. Ambler, R. W. Hayward, D. D. Hoppes and R. P. Hudson, Phys. Rev. **105**, 1413 (1957).
4. S. Weinberg, Phys. Rev. Lett. **19**, 1264 (1967); Rev. Mod. Phys. **52**, 515 (1980)
   A. Salam, in: *Elementary Particle Theory, Proc. 8th Nobel Symposium,* p.367, N. Svartholm, ed., Almquist and Wiksell Förlag, Stockholm (1968).

5. See, for example, the review of E. D. Commins, Physica Scripta **36**, 468 (1987).
6. M. C. Noecker, B. P. Masterson, and C. E. Wieman, Phys. Rev. Lett. **61**, 310 (1988).
7. V. A. Dzuba, V. V. Flambaum, and O. P. Sushkov, Phys. Lett. A **141**, 147 (1989).
8. S. A. Blundell, W. R. Johnson, and J. Sapirstein, Phys. Rev. Lett. **65**, 1411:1414 (1990).
9. See, for example, A.-M. Mårtensson-Pendrill and P. Öster, Physica Scripta **36**, 444:52 (1987) an references therein.
10. F. J. Hasert et al., Phys. Lett. **46B**, 138 (1973).
11. Ya. B. Zel'dovich, Zh. E.T.F. **33**, 1531 (1957)
L. M. Barkov, I. B. Khriplovich and M. S. Zolotorev, Comments At. Mol. Phys. **8**, 79 (1979)
I. B. Khriplovich, Comments At. Mol. Phys. **23**, 189:99 (1989).
12. V. V. Flambaum, I. B. Khriplovich and O. P. Sushkov, Phys. Lett. B **146**, 367 (1984).
13. M. A. Bouchiat and C. Bouchiat, Phys. Lett. **48**, 111:4 (1974).
14. R. W. Dunford, R. R. Lewis and W. C. Williams, Phys. Rev. A **18**, 2421 (1978).
15. P. A. Frantsuzov and I. B. Khriplovich, Z. Physik D **7**, 297 (1988).
16. I. B. Khriplovich, private communication at ICAP, Ann-Arbor, Michigan (1990).
17. W. Marciano and A. Sirlin, Phys. Rev. D **29**, 75 (1984); B. W. Lynn, PhD Thesis, Columbia University, unpublished (1982).
18. See, for example, the review by E. N. Fortson and L. L. Lewis, Phys. Rep. **113**, 289 (1984).
19. C. Bouchiat and C.-A. Piketty, Europhys. Lett. **2**, 511:8 (1986).
20. A. C. Hartley and P. G. H. Sandars, J. Phys. B **23**, 1961:74 (1990); 2649:61 (1990).
21. D. V. Neuffer and E. D. Commins, Phys. Rev. A **16**, 844 (1977).
22. P. G. H. Sandars, J. Phys. B **10**, 2983:95 (1977).
23. E. P. Plummer and I. P. Grant, J. Phys. B **18**, L315:20 (1985).
24. C. P. Botham, S. A. Blundell, A.-M. Mårtensson-Pendrill, J. Phys. B **23**, 3417:36 (1990).
25. Z. W. Liu and H. P. Kelly, work in progress.
26. See, for example, S. S. Schweber, *An Introduction to Relativistic Quantum Field Theory*, Harper and Row, New York (1961).
27. J. Sapirstein, Physica Scripta **36**, 801 (1987).
28. M. Gell-Mann and F. Low, Phys. Rev. **84**, 350 (1951).
29. J. Sucher, Phys. Rev. **107**, 1448 (1957).
30. P. J. Mohr, in: *Relativistic, Quantum Electrodynamic, and Weak Interaction Effects in Atoms*, AIP Conference Proceedings **189**, New York (1989).
31. J. Sucher, Int. J. Quant. Chem. **24**, 3 (1984).
32. E. Lindroth, Physica Scripta **36**, 485:92 (1987); E. Lindroth, J.-L. Heully, I. Lindgren and A.-M. Mårtensson-Pendrill, J. Phys. B **20**, 1679:96 (1987).
33. I. P. Grant, Phys. Rev. A **25**, 1230:2 (1982).
S. P. Goldman and G. F. W. Drake, J. Phys. B **16**, L183 (1983).
H. M. Quiney, I. P. Grant and S. Wilson, J. Phys. B **18**, 577:87, 2805:15 (1985); J. Phys. B **22**, L15:19 (1989); J. Phys. B **23**, L271:8 (1990).
34. S. A. Blundell, W. R. Johnson and J. Sapirstein, Phys. Rev. A **37**, 307:15 (1988); W. R. Johnson and J. Sapirstein, Phys. Rev. Lett. **57**, 1126:9 (1986).
35. S. Salomonson and P. Öster, Phys. Rev. A **40**, 5559:67, 5548:58 (1989).
36. I. Lindgren and J. Morrison, *Atomic Many-Body Theory*, 2nd ed., Springer-Verlag, Berlin (1986).
37. A. Fetter and J. D. Walecka, *Quantum Theory of Many-Particle Systems*, McGraw-Hill, New York (1971).
38. B. H. Brandow, Rev. Mod. Phys. **39**, 771 (1967).
39. S. A. Blundell, W. R. Johnson, and J. Sapirstein, Phys. Rev. A **42**, 3751:62 (1990).

40. V. A. Dzuba, V. V. Flambaum and O. P. Sushkov, Phys. Lett. A **140**, 493 (1989)
V. A. Dzuba, V. V. Flambaum, A. Ya. Kraftmakher, and O. P. Sushkov, Phys. Rev. Lett. A **142**, 373 (1989).
41. F. Coester, Nucl. Phys. **1**, 421 (1958)
F. Coester and H. Kümmel, Nucl. Phys. **17**, 477 (1960).
42. I. Lindgren, J. Phys. B (in print); I. Lindgren and D. Mukherjee, Phys. Rep. **151**, 93 (1987).
43. I. Lindgren, Int. J. Quant. Chem. S **12**, 33 (1978).
44. S. Salomonson and A. Ynnerman, Phys. Rev. A (accepted for publication).
45. A.-M. Mårtensson-Pendrill, L. R. Pendrill, S. Salomonson, A. Ynnerman and H. Warston, J. Phys. B **23**, 1749 (1990).
46. S. A. Blundell, private communication.
47. S. A. Blundell, W. R. Johnson and J. Sapirstein, Phys. Rev. A (in press).
48. S. A. Blundell, W. R. Johnson, Z. W. Liu, and J. Sapirstein, Phys. Rev. A **40**, 2233 (1989).
49. A.-M. Mårtensson-Pendrill and A. Ynnerman, Physica Scripta **41**, 329:47 (1990).
50. S. A. Blundell, D. S. Guo, W. R. Johnson, and J. Sapirstein, At. Data Nucl. Data Tables **37**, 1103 (1987).
51. S. A. Blundell, W. R. Johnson and J. Sapirstein, in preparation.
52. S. A. Blundell, A. C. Hartley, Z.W. Liu, A.-M. Mårtensson-Pendrill, and J. Sapirstein, in *Proceedings from the Workshop on Coupled Cluster Theory at the Interface of Atomic Physics and Quantum Chemistry, Harvard, 6-11 August 1990*, Theoretica Chimica Acta, R. J. Bartlett, ed., Springer (1991).
53. V. A. Dzuba, V. V. Flambaum, P. G. Silvestrov and O. P. Sushkov, Physica Scripta **35**, 69-70 (1987); J. Phys. B **20**, 3297:311 (1987).
54. A.-M. Mårtensson-Pendrill, J. de Physique (Paris) **46**, 1949:59 (1985).
55. W. R. Johnson, D. S. Guo, M. Idrees, and J. Sapirstein, Phys. Rev. A **32**, 2093 (1985).
56. S. A. Blundell, W. R. Johnson, Z. W. Liu, and J. Sapirstein, Phys. Rev. A **37**, 1395:400 (1988).
57. S. A. Blundell and B. W. Lynn, in preparation.
58. See, for example, A. B. Migdal, *Theory of Finite Fermi Systems and Applications to Atomic Nuclei*, Interscience, New York (1967).
59. W. R. Johnson, M. Idrees, and J. Sapirstein, Phys. Rev. A **35**, 3218:26 (1987).
60. V. A. Dzuba, V. V. Flambaum and O. P. Sushkov, Phys. Lett. A **140**, 493 (1989).
V. A. Dzuba, V. V. Flambaum, A. Ya. Kraftmakher, and O. P. Sushkov, Phys. Rev. Lett. A **142**, 373 (1989).
61. A. C. Hartley and A.-M. Mårtensson-Pendrill, Z. Physik D **15**, 309:19 (1990).
62. A. C. Hartley, E. Lindroth, and A.-M. Mårtensson-Pendrill, J. Phys. B **23**, 3417:36 (1990).
63. W. J. Marciano and A. Sirlin, Phys. Rev. D **29**, 75 (1984); B. W. Lynn, PhD Thesis, Columbia University, unpublished (1982).
64. W. J. Marciano and J. L. Rosner, University of Chicago preprint EFI-90-55.
65. C. E. Wieman, private communication.
66. D. C. Kennedy and B. W. Lynn, Nucl. Phys. B **322**, 1:54 (1989).
67. M. E. Peskin and T. Takeuchi, Phys. Rev. Lett. **65**, 964 (1990).
68. P. G. H. Sandars, J. Phys. B **23**, L655:8 (1990).
69. D. London and J. L. Rosner, Phys. Rev. D **34**, 1530 (1986).
70. K. T. Mahanthappa and P. K. Mohapatra, Preprint Colo-Hep-228 (1990).
71. J. Paldus, J. Cizek, and I. Shavitt, Phys. Rev. A **5**, 50 (1972).
72. V. A. Dzuba, V. V. Flambaum and I. B. Khriplovich, Z. Physik D **1**, 243:5 (1986).
73. E. N. Fortson, Y. Pang, and L. Wilets, Phys. Rev. Lett. **65**, 2857 (1990).

THE COMPLEX-SCALING COUPLED-CHANNEL METHODS FOR ATOMIC AND MOLECULAR RESONANCES IN INTENSE EXTERNAL FIELDS

Shih-I Chu

Department of Chemistry
University of Kansas
Lawrence, KS 66045

INTRODUCTION

Resonance states are characterized by *complex* energies corresponding to poles of the resolvent operator $(E-\hat{H})^{-1}$ in the complex-energy plane of a non-physical higher Riemann sheet. Numerous techniques have been developed for computing these poles. One of the most powerful techniques popularized in the last decade is the method known as the complex scaling (coordinate-rotation, complex-coordinate, or dilatation) transformation.[1,2] As a result of the complex scaling transformation, $r \rightarrow re^{i\alpha}$, the eigenvalues corresponding to the bound states of $\hat{H}$ stay invariant, while the branch cuts associated with the continuous spectrum of $\hat{H}$ are rotated about their respective thresholds by an angle -2α (assuming $0<\alpha<\pi/2$), exposing the complex resonance states in appropriate strips of the complex energy plane. A crucial point from the computational point of view is that the eigenfunctions associated with the complex-scaling resonance wave functions are localized, i.e. square integrable. The square integrability led to the extension of well-established bound-state techniques to the determination of resonance energies (E_R) and widths (Γ) of metastable states.

In this article, we shall present some of our recent studies on atomic and molecular resonance states induced by strong external fields. In addition, a new complex-scaling technique -- the complex-scaling Fourier-grid Hamiltonian method: *without* the use of L^2 basis set expansion -- will be introduced. Because of its extreme simplicity and efficiency, the method is particularly useful for problems where no appropriate basis set can be found or highly excited-state resonances are involved.

INTENSITY-DEPENDENT COMPLEX QUASI-ENERGIES (SHIFTS AND WIDTHS) OF LOW-LYING EXCITED STATES OF ATOMIC HYDROGEN

Recent experiments on multiphoton ionization (MPI) and above-threshold ionization (ATI) of atoms in strong fields show significant energy shifts and broadenings of ATI peaks.[3,4] Further, for laser pulse widths of less than 1 psec, the electron energy spectrum can exhibit fine structure in the individual ATI peaks. These observations suggest that the structure of excited states plays a significant role in determining the properties of ATI. This creates the necessity of analyzing the excited-state atomic energy-level structure in the presence of strong fields.

Applied Many-Body Methods in Spectroscopy and Electronic Structure
Edited by D. Mukherjee, Plenum Press, New York, 1992

In this section, we review briefly the L^2 non-Hermitian Floquet method[5,6] and present the nonperturbative results of complex quasi-energies for several low-lying states of atomic H.[7] The method permits nonperturbative and self-consistent treatment of intense-field effects (in that all atomic levels are simultaneously shifted and broadened by the external fields) and straightforward inclusion of free-free transitions and the effects of coupling among electronic continua.

Corresponding to the periodically time-dependent Hamiltonian

$$\hat{H}(\vec{r},t) = -(\hbar^2/2m)\nabla^2 - e^2/r + eFz\cos\omega t, \qquad (1)$$

describing the interaction of atomic H with a monochromatic, linearly polarized, coherent field of frequency ω and peak field strength F, an equivalent time-independent Hamiltonian $\hat{H}_F(\vec{r})$ may be obtained by an extension of the semiclassical Floquet Hamiltonian method.[6,8,9] The structure of $\hat{H}_F$ has been documented elsewhere[5] and is reproduced in Fig. 1 for convenience of discussion. The Floquet Hamiltonian $\hat{H}_F$ shows a tridiagonal block structure, consisting of the diagonal $A \pm n\omega I$ (n=0, ±2, ±4,...) blocks and the off-diagonal B blocks. Each diagonal block is composed of angular momentum blocks S, P, D,..., representing the projection of the atomic electronic Hamiltonian onto states of $\ell = 0, 1, 2, \ldots$, and $V_{\ell,\ell'}$'s are electric dipole coupling matrix elements. Thus, in the case of atomic H, the S block consists of the 1s, 2s, 3s,...ns,... bound states and the entire ks Coulomb continuum. The Hamiltonian of Fig. 1 has no discrete spectrum, and the time evolution is dominated by poles of the resolvent $(E-\hat{H}_F)^{-1}$ near the real axis but on higher Riemann sheets.

$$H_F = \begin{pmatrix} \ddots & & & & & \\ & A+4\omega I & B & 0 & 0 & 0 \\ & B^T & A+2\omega I & B & 0 & 0 \\ & 0 & B^T & A & B & 0 \\ & 0 & 0 & B^T & A-2\omega I & B \\ & 0 & 0 & 0 & B^T & A-4\omega I \\ & & & & & & \ddots \end{pmatrix}$$

where

$$A = \begin{pmatrix} S & V_{SP} & 0 & 0 & 0 \\ V_{PS} & P-\omega I & V_{PD} & 0 & 0 \\ 0 & V_{DP} & D & V_{DF} & 0 \\ 0 & 0 & V_{FD} & F-\omega I & V_{FG} \\ 0 & 0 & 0 & V_{GF} & G & \\ & & & & & \ddots \end{pmatrix}$$

and

$$B = \begin{pmatrix} 0 & 0 & 0 & 0 & 0 \\ V_{PS} & 0 & V_{PD} & 0 & 0 \\ 0 & 0 & 0 & 0 & 0 \\ 0 & 0 & V_{FD} & 0 & V_{FG} \\ 0 & 0 & 0 & 0 & 0 & \\ & & & & & \ddots \end{pmatrix}$$

Fig. 1. Structure of the time-independent Floquet Hamiltonian for atomic MPI/ATI (Adapted from ref. 7).

These complex poles, which correspond to decaying complex quasi-energy states (QES's), may be found directly from the analytically continued Floquet Hamiltonian, $\hat{H}_F(\alpha)$, obtained by the complex scaling transformation $r \rightarrow re^{i\alpha}$. This transformation effects an analytical continuation of $(E-\hat{H}_F)^{-1}$ into the lower half-plane on an appropriate higher Riemann sheet, allowing the complex QES to be determined by solution of a non-Hermitian eigenproblem. The real parts of the complex eigenvalues of $\hat{H}_F(\alpha)$ provide the ac Stark shifts, whereas the imaginary parts determine directly the total MPI widths (rates). In practice, the atomic blocks are made discrete by use of a finite subset of the complete Laguerre basis $r^{\ell+1}e^{\lambda r}L_n^{2\ell+2}(\lambda r)$, where λ is an adjustable parameter and n = 0, 1, 2,.... This yields a Pollaczeck quadrature representation of the bound and continuum contributions to the spectral resolution of the hydrogenic Hamiltonian. In practice, the convergence of MPI calculations may achieve arbitrary precision by systematically increasing the basis size and the number of angular momentum blocks.

Table 1. Intensity-dependent complex Quasi-energies[a].

States	F_{rms}	E_R	$-\Gamma/2$
2s	1.0 (−4)[b]	−0.12499970	−0.1383 (−7)
	5.0 (−4)	−0.1249926	−0.1853 (−7)
	1.0 (−3)	−0.124970	−0.8769 (−7)
	2.0 (−3)	−0.124876	−0.1124 (−5)
	3.0 (−3)	−0.124709	−0.5125 (−5)
	4.0 (−3)	−0.124455	−0.1432 (−4)
	5.0 (−3)	−0.124102	−0.3041 (−4)
	7.5 (−3)	−0.12277	−0.1092 (−3)
	1.0 (−2)	−0.12125	−0.5898 (−3)
2p	1.0 (−4)	−0.12499905	−0.7705 (−9)
	5.0 (−4)	−0.1249762	−0.5371 (−8)
	1.0 (−3)	−0.124905	−0.7432 (−7)
	2.0 (−3)	−0.124626	−0.1159 (−5)
	3.0 (−3)	−0.124177	−0.5693 (−5)
	4.0 (−3)	−0.123581	−0.1726 (−4)
	5.0 (−3)	−0.122876	−0.4047 (−4)
	7.5 (−3)	−0.12136	−0.6558 (−3)
	1.0 (−2)	−0.11898	−0.4256 (−2)
3s	1.0 (−4)	−0.5555488	−0.1846 (−6)
	5.0 (−4)	−0.0555383	−0.5219 (−5)
	1.0 (−3)	−0.055486	−0.2096 (−4)
	2.0 (−3)	−0.055279	−0.8397 (−4)
	3.0 (−3)	−0.054933	−0.1892 (−3)
	4.0 (−3)	−0.054448	−0.3370 (−3)
	5.0 (−3)	−0.053823	−0.5278 (−3)
	6.0 (−3)	−0.05306	−0.7624 (−3)
	7.5 (−3)	−0.05165	−0.1197 (−2)
3p	1.0 (−4)	−0.05555525	−0.2443 (−6)
	5.0 (−4)	−0.0555478	−0.6329 (−5)
	1.0 (−3)	−0.055525	−0.2540 (−4)
	2.0 (−3)	−0.055432	−0.1025 (−3)
	3.0 (−3)	−0.055280	−0.2341 (−3)
	4.0 (−3)	−0.055062	−0.4248 (−3)
	5.0 (−3)	−0.05479	−0.6807 (−3)
	6.0 (−3)	−0.05445	−0.1010 (−2)
3d	1.0 (−4)	−0.05555580	−0.1336 (−6)
	5.0 (−4)	−0.0555615	−0.3369 (−5)
	1.0 (−3)	−0.055579	−0.1355 (−4)
	2.0 (−3)	−0.055642	−0.5541 (−4)
	3.0 (−3)	−0.055728	−0.1288 (−3)
	4.0 (−3)	−0.055812	−0.2390 (−3)
	5.0 (−3)	−0.05587	−0.3923 (−3)

a) Adopted from ref. 7.

Table 1 shows the intensity-dependent complex quasienergies (E_R, $-\Gamma/2$) of the perturbed low-lying excited states (2s, 2p, 3s, 3p, 3d) of atomic H at $\lambda = 530$ nm. The complex quasi-energies for the ground state in intense laser fields for a range of frequencies can be found in ref. 10. The E_R's are the ac Stark-shifted energies, whereas Γ's are the total MPI widths (rates). Up to five Floquet blocks (A, $A\pm2\omega$, $A\pm4\omega$) and (25s, 25p, 25d, 25f, 25g) basis functions for each block are used in these calculations to achieve convergence. Strong mixings with some other states have already occurred for each atomic state at the largest F_{rms} listed. Beyond these field intensities, the quasi-energy eigenvector components are spread among many Floquet states, and the identities of atomic states can no longer be discerned. Figure 2 depicts the intensity-dependent ac Stark behavior of these low-lying excited states. We note that all the low-lying states (except 3d) are shifted upward as the field intensity increases.

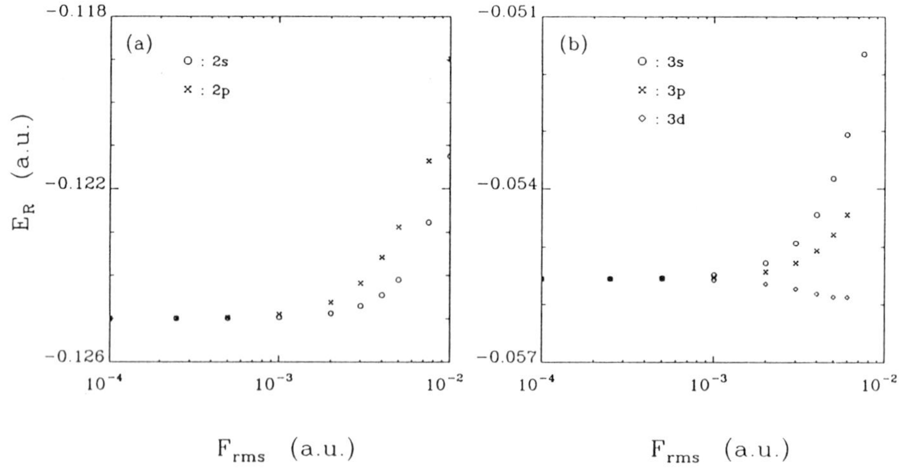

Fig. 2. Intensity-dependent energy-level-shift behavior for low-lying excited states (2s, 2p) and (3s, 3p, 3d) of atomic H. (Adapted from ref. 7).

Shifting and splitting of atomic states can be accounted for by perturbation theory for $F_{rms}<5\times10^{-4}$ a.u. For $F_{rms} \geq 10^{-3}$ a.u., higher-order effects become significant, and nonperturbative treatment, such as the L^2 non-Hermitian Floquet calculations presented here, is required to achieve convergence. For low lying states, we found that both $\vec{A}\cdot\vec{P}$ and $\vec{A}^2$ terms (in the velocity gauge) are important for the behavior of a.c. Stark shifts (in both weak and strong fields.)

A.C. STARK SHIFTS OF HIGHLY EXCITED STATES

To explore the a.c. Stark shift behavior for highly excited states in strong fields, we have recently developed a generalized Floquet method,[7] using the Sturmian basis. Figure 3 shows the ac Stark shifts of $n = 11$ atomic states for $\ell = 0, 1, 2, 3$. Forty Sturmian basis functions are used for each angular momentum block ($\ell = 0, 1, 2, 3, 4$) and five Floquet blocks ($A,, A \pm 2\omega I, A \pm 4\omega\ell$) are used in the Floquet eigenvalue analysis. Figure 3 reveals several essential energy-shift behaviors of excited states: (a) All the excited levels shown are shifted upward and closely follow the shift caused by the pondermotive potential $\bar{V}$ (shown by dotted curves)

$$\bar{V} = e^2F^2/4m\omega^2$$

in the weaker-field region. This effective potential $\bar{V}$ has its origin in the A^2 term (where A is the vector potential) and can be shown to be equal to the average quiver kinetic energy picked up by an electron of mass m and charge e driven sinusoidally by the fields. Our results lend further support to the view[11] that all Rydberg states and the continuum are upshifted by the same amount, described by $\bar{V}$. However, this description appears valid only in the weak-field regime where no strong mixings exist among atomic states. (b) Above some critical field strengths (F_c), the atomic energy levels (for a given n but different ℓ) split, and significant deviation from the A^2 curve occurs. The critical field strength F_c depends on n and decreases rather rapidly as n increases, as can be seen from these figures. One should therefore use the A^2 shift law with caution in the interpretation of energy-level shifts in high-intensity MPI/ATI experiments. (c) For $F > F_c$, strong mixings exist among nearby atomic states, and the level identities usually cannot be discerned. Similar behavior was born out in our study for highly excited states ($n > 50$).[7] Such behavior is thus expected to prevail for all Rydberg states.

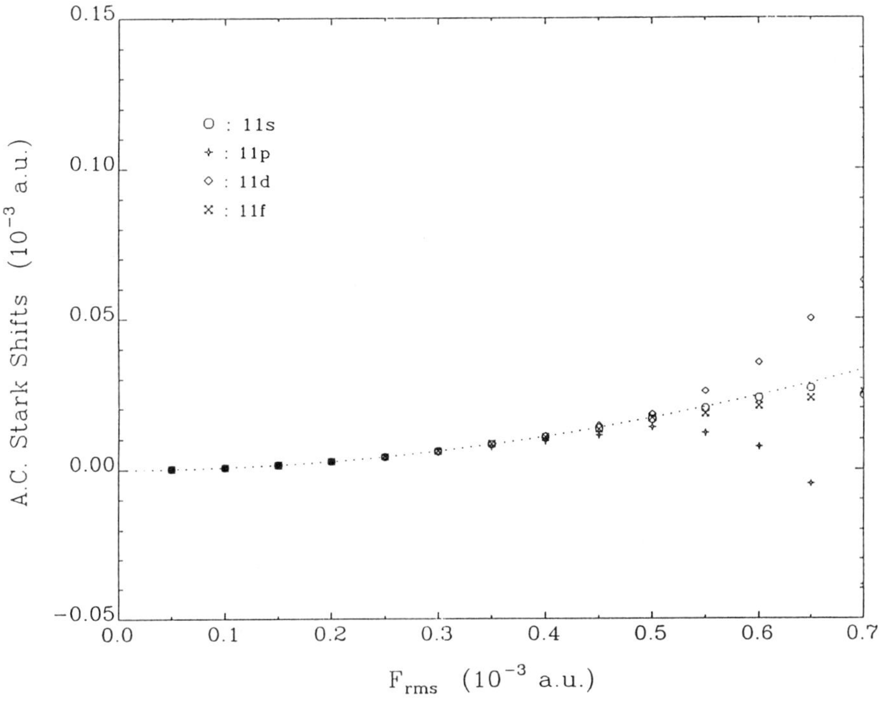

Fig. 3. A.C. Stark shifts of n=11 atomic states for ℓ=0,1,2,3. (Adapted from ref. 7).

AUTOIONIZING RESONANCES OF ATOMIC HYDROGEN IN INTENSE MAGNETIC FIELDS

The problem of the effects of the complex scaling transformation on the Zeeman Hamiltonian was first studied by Chu[12] and Avron et al.[13] Consider a hydrogen-like atom in a strong uniform magnetic field B. For the magnetic field strength $\gamma(\equiv B/B_0,\ B_0 = m_e^2 e^3 c/\hbar^3 = 2.35 \times 10^9$ Gauss) in the range 10^{-4} to 10^4, the system can be accurately described by the nonrelativistic single-particle Hamiltonian,

$$H_{Zeeman} \equiv H_Z = H_0 + V_c, \tag{2}$$

where H_0 is the free-electron Zeeman Hamiltonian, and V_c is the Coulomb potential. In the cylindrical coordinates and for a magnetic field in the z direction we have

$$H_0 = H_{\perp}(\rho,\phi) + P_z^2/2\mu, \tag{3}$$

and

$$V_c = -Ze^2/(\rho^2 + z^2)^{1/2}, \tag{4}$$

where

$$H_{\perp}(\rho,\phi) = P_{\perp}^2/2\mu + \omega_c \ell_z/2 + \mu\omega_c^2\rho^2/8, \tag{5}$$

μ is the reduced mass of electron and proton, Z is the nuclear charge, $\omega_c = eB/\mu c$ is the cyclotron frequency, and ℓ_z is the z component of the orbital angular momentum.

Physically, for large B, the electron will move in tightly bound Landau orbit in the x-y plane (described by the Hamiltonian $H_{\perp}$), although it can be autoionized or photodetached only in the loosely bound z direction. The problem is thus essentially one-dimensional. The operator $H_{\perp}(\alpha) \equiv H_{\perp}(\rho e^{i\alpha}, \phi)$ is dilatation analytic in α for $\alpha < \pi/4$ and its spectrum $\sigma[H_{\perp}(\alpha)]$ is independent of α and consists of discrete Landau states with energies $(N + 1/2)\hbar\omega_c$, $N = 0, 1, 2, \ldots$. When the longitudinal kinetic term $P_z^2/2\mu$ is added to $H_{\perp}$, we obtain the dilatated free-electron Zeeman spectrum,

$$\sigma[H_0(\alpha)] = \sigma[H_{\perp}(\alpha)] + \sigma[P_z^2 e^{-2i\alpha}/2\mu].$$

The spectrum of $H_0(\alpha)$ has an analytical continuation from $\alpha = 0$ onto nonphysical sheets with $\alpha < \pi/4$. As α "tuned up" from 0, the continuous spectrum swings out into the lower half plane by angle 2α, whereas the Landau thresholds stay invariant.[12]

Now we add on the Coulomb potential V_c and consider the spectrum of the total Zeeman Hamiltonian H_Z, Eq. (2). As the Coulomb potential is relatively compact with respect to H_0, the spectrum $\sigma[H_Z]$ differs from $\sigma[H_0]$ only in the point or discrete spectrum. The Landau states in H_0 are now perturbed and shifted by the Coulomb potential and become "quasi-Landau" states. In addition, below each quasi-Landau level there exists a series of Coulomb levels. All the Coulomb series except the first (N=0) are autoionizing resonances as they are bound states imbedded in the continuum (or continua) of all the preceding Landau levels. The quasi-Landau levels are in fact branching points of the continuous spectra. The effect of the dilatation transformation on the Zeeman Spectrum $H_Z(re^{i\alpha})$ in spherical coordinates or $H_Z(\rho e^{i\alpha}, ze^{i\alpha}, \phi)$ in cylindrical coordinates is illustrated in Figure 4. Thus, the point spectrum, including bound states

(N=0 Coulomb series), quasi-Landau thresholds, and autoionizing resonances ($N \geq 1$ Columb series) are invariant, whereas the continua rotate about their respective quasi-Landau thresholds by an angle 2α onto the lower half complex plane, exposing autoionizing resonances for a range of values of α. These autoionizing resonances can be determined by means of the complex-coordinate coupled-Landau-channel (CCCLC) method[14,15] briefly described below.

The Complex-Coordinate Coupled-Landau-Channel Formalism for Autoionizing Resonances

For the Zeeman Hamiltonian H_Z, Eq. (2), the azimuthal quantum number m and the "z-parity" $\hat{\pi}$ (i.e., the parity w.r.t. the reflection through the x-y plane) are good quantum numbers. For given values of m and π and for $\gamma \geq 1$, it is convenient to expand the solutions of the Schrodinger equation $H\Psi = E\Psi$ in Landau channel states $\Phi_N^m(\rho,\phi)$ that are eigenfunctions of the two-dimensional harmonic oscillator Hamiltonian $H_\perp$:

$$\Psi(\rho,\phi,z) = \sum_N \Phi_N^m(\rho,\phi)\psi_N(z). \qquad (6)$$

Defining the "zero" of energy as $(m + |m| + 1)\hbar\omega_c/2$ for a given m, and substituting Eq.(6) into $H\Psi = E\Psi$, yields a set of coupled equations for the wave functions $\psi_N(z)$:

$$[P_z^2/2\mu + V_{N,N}(z) + N\hbar\omega_c]\ \psi_N(z) + \sum_{N'\neq N} V_{N',N}(z)\psi_{N'}(z) = E\psi_N(z). \qquad (7)$$

where

$$V_{N',N}(z) = -Ze^2\int_0^\infty \rho d\rho \int_0^{2\pi} d\phi\ \Phi_{N'}^{*m}(\rho,\phi)\ (\rho^2+z^2)^{-1/2}\Phi_N^m(\rho,\phi). \qquad (8)$$

Neglect of $V_{N',N}(N\neq N')$ leads to a set of uncoupled equations, giving rise to the adiabatic approximation solutions for $\psi_N(z)$. For $N \geq 1$, all the states can autoionize via decay into the continuum (continua) of the lower Landau channel(s) due to interchannel Coulomb coupling $V_{N',N}(z)$, $N \neq N'$.

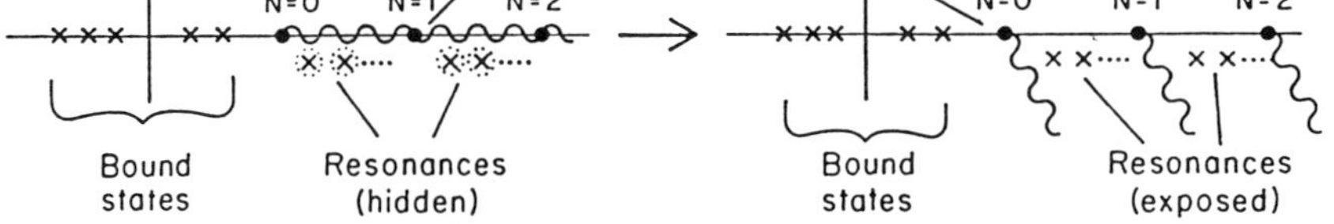

Fig. 4. Effects of a dilatation transformation on the spectrum, $\sigma(H_z)$, of an atomic Zeeman Hamiltonian H_z. The bound states (the Coulomb series below the first (N=0) quasi-Landau threshold) are invariant to the transformation, whereas the continua rotate about their respective quasi-Landau thresholds, exposing complex "resonance" Coulomb series (above the N=0 Landau threshold) in appropriate strips of the complex energy plane. (Adapted from Ref. 14).

To determine these autoionizing resonances, we apply the complex scaling transformation to the dissociative coordinate z, $z \to ze^{i\alpha}$, in the one-dimensional Hamiltonian operator in Eq. (7), and expand the longitudinal wave function $\psi_N(z)$ in terms of some appropriate L^2-basis functions $|Nn\rangle \equiv g_n^{(N)}(z)$ such as displaced Gaussians. This reduces the complex scaling coupled equaitons to a symmetric non-hermitian matrix eigenvalue problem:

$$\mathrm{Det}|(H_\alpha)_{N'n',Nn} - E(S)_{N'n',Nn} I| = 0. \qquad (9)$$

where S is the overlap matrix, and the rotated ($z \to ze^{i\alpha}$) Hamiltonian matrix elements are given by

$$\langle N'n'|H_\alpha|Nn\rangle = -e^{-2i\alpha}(\hbar^2/2\mu)\langle N'n'|d^2/dz^2|Nn\rangle\, \delta_{NN'} \qquad (10)$$

$$+ N\hbar\omega_c \delta_{nn'}\delta_{NN'} + \langle N'n'|V_{N',N}(ze^{i\alpha})|Nn\rangle.$$

The desirable resonance states can then be identified by the stationary points of the α trajectories. In the present case, the resonance eigenvalues are found to be stable over a wide range of rotational α angles (0.30 - 0.50 radians).

For magnetic fields in the range of 10^9 to 10^{12} Gauss (in the context of astrophysical interest),[16] we have previously calculated the positions and widths of the autoionizing states below the first two excited (N=1,2) Landau thresholds. The number of Landau channels used in the calculation to achieve convergencce varied from 12 for lower field strengths ($\gamma \leq 1$) to 2 for the higher field strengths ($\gamma \geq 500$), while the number of basis functions $g_n^{(N)}(z)$ used in each Landau channel varied from 10 for lower field strengths to 50 for higher field strengths. Table 2 displays the converged resonance energies (E_R) and widths (Γ) of several autoionizing states for different magnetic fields γ corresponding to the $m^\pi = 0^+$ subspace. It is

Table 2. Converged resonance energies E_R (relative to the Landau channel threshold $N\hbar\omega_c$) and (half) widths $\Gamma/2$ for several autoionizing states of atomic hydrogen ($m^\pi = 0^+$) in intense magnetic fields. (Adapted from ref. 14)

N	n_z	γ	E_R(Ryd)	$\Gamma/2$(Ryd)
1	0	2	-1.184	0.54 (-1)*
		10	-2.294	4.35 (-2)
		100	-5.370	2.58 (-2)
		500	-9.174	1.71 (-2)
1	1	2	-0.279	0.45 (-2)
		10	-0.364	2.10 (-3)
		100	-0.472	0.57 (-3)
		500	-0.535	0.22 (-3)
2	0	2	-0.856	0.25 (-1)
		10	-1.683	2.17 (-2)
		100	-4.069	1.42 (-2)
2	1	2	-0.242	0.25 (-2)
		10	-0.327	1.40 (-3)
		100	-0.438	0.42 (-3)

*0.54 (-1) = 0.54×10^{-1}.

found that the autoionizing width decreases with increasing Landau excitation quantum number N, "hydrogenic" quantum number n_z, and field strength γ. The widths of the lowest autoionizing state (i.e., N=1, n_z=0) follow approximately a simple power law ($\Gamma \propto \gamma^{-1/4}$) over the field range considered. In contrast, the electromagnetic decay of the Landau excited states in very high fields dominantely proceeds via $\Delta N = 1$, $\Delta m = 1$ cyclotron transitions with decay widths[17] proportional to γ^2. Thus, for the lowest Landau excited states, the autoionization dominates the electromagnetic decay for $\gamma < 200$, whereas the electromagnetic decay takes over autoionization at very high field strengths $\gamma > 200$.

More recently, a significant advance has been made in the laboratory observation of quasi-Landau spectrum of diamagnetic Rydberg atoms in the classically chaotic regime[18] as well as in the positive energy regime.[19] Extension of the CCCLC method to the study of quasi-Landau resonance states in the laboratory fields is in progress and will be reported elsewhere.

THE COMPLEX-SCALING FOURIER-GRID HAMILTONIAN METHOD

In this section we discuss the basic element of a complex scaling method recently proposed *without* the use of basis set expansion.[20] Because of its extreme simplicity in numerical implementation, the method has decisive advantages for problems where no appropriate basis set can be found or a large number of basis sets are required. The method makes use of the recent advancement of the calculation of bound state eigenvalues and eigenfunctions using the Fourier transform method.[21]

Consider for simplicity a one-dimensional system described by the Hamiltonian

$$\hat{H}(x) = \hat{T} + V(\hat{x}), \tag{11}$$

where $\hat{T} = \hat{p}^2/2m$ is the kinetic energy operator. Under the complex scaling transformation, $x \rightarrow x\, e^{i\alpha}$,

$$\hat{H}(x) \rightarrow \hat{H}(\alpha) \equiv \hat{H}(\hat{x}e^{i\alpha}) = e^{-2i\alpha}\hat{T} + V(\hat{x}e^{i\alpha}). \tag{12}$$

It is known that the kinetic energy operator is best represented in the momentum representation. Since

$$\langle k'|\hat{T}|k\rangle = (\hbar^2 k^2/2m)\, \delta(k-k'), \tag{13}$$

$\hat{T}$ is diagonal in the $|k\rangle$ representation. Here $|k\rangle$ are the eigenfunctions of $\hat{p}$, namely,

$$\hat{p}|k\rangle = \hbar k|k\rangle,$$

and satisfy the orthonormal and completeness relationships, respectively,

$$\langle k|k'\rangle = \delta(k-k') \tag{14}$$

and

$$\hat{I}_k = \int_{-\infty}^{\infty} |k\rangle\langle k|\, dk. \tag{15}$$

On the other hand, the potential energy term $V(\hat{x})$ is diagonal in the coordinate representation,

$$<x'|V(\hat{x})|x> = V(x)\ \delta(x-x'). \quad (16)$$

Here $|x>$ are the eigenfunctions of the coordinate operator,

$$\hat{x}|x> = x|x>, \quad (17)$$

and satisfy the relationships

$$<x|x'> = \delta(x-x') \quad (18)$$

and

$$\hat{I}_x = \int_{-\infty}^{\infty} |x><x|dx. \quad (19)$$

In the coordinate or Schrodinger representation, the Hamiltonian operator in eq. (12) becomes

$$<x|\hat{H}(\alpha)|x'> = e^{-2i\alpha}<x|\hat{T}|x'> + V(xe^{i\alpha})\ \delta(x-x'). \quad (20)$$

Using the identity operator, eq. (15), eq. (20) can be rewritten as

$$<x|\hat{H}(\alpha)|x'> = e^{-2i\alpha}\int_{-\infty}^{\infty}<x|k>\ T_k<k|x'>\ dk + V(xe^{i\alpha})\ \delta(x-x')$$

$$= (e^{-2i\alpha}/2\pi)\int_{-\infty}^{\infty} e^{ik(x-x')}T_k dk + V(xe^{i\alpha})\ \delta(x-x'), \quad (21)$$

where $T_k = \hbar^2k^2/2m$.

To discretize the continuous range of coordinate values x, we adopt the Fourier grid method.[21] Here a uniform discrete spatial grid

$$x_i = i\ \Delta x\ (i=1,2,\ldots,N) \quad (22)$$

will be used where N is an *odd* integer number. The orthogonality condition (18) and the identity operator (19) may now be written as

$$\Delta x\ <x_i|x_j> = \delta_{ij} \quad (23)$$

and

$$\hat{I}_x = \sum_{i=1}^{N} |x_i>\ \Delta x\ <x_i|. \quad (24)$$

The grid size and spacing in coordinate space determines the reciprocal grid size in momentum space. Thus $\Delta k = 2\pi/N\Delta x$. The central point in the momentum space grid is chosen to be k=0, and the grid's points are evenly distributed about zero.

The discretized version of the complex scaling operator (21) now has the following form (after some simplification and renormalization)

$$H_{ij}(\alpha) = <x_i|\hat{H}(\alpha)|x_j> = (2e^{-2i\alpha}/N)\sum_{\ell=1}^{n}\cos[2\pi\ell(i-j)/N]T_\ell + V(x_ie^{i\alpha})\delta_{ij}, \quad (25)$$

where $n = (N-1)/2$, and $T_\ell = (2/m)(\hbar\pi\ell/N\Delta x)^2$. This leads to the following *complex* secular equations for the resonance eigenvalues W_ν.

$$\sum_j [H_{ij}(\alpha) - W_\nu \delta_{ij}] \Psi_j^\nu = 0, \tag{26}$$

where the eigenvectors Ψ_j^ν give *directly* the amplitude of the normalized solutions of the resonance wave functions Ψ^ν evaluated at the grid points x_j, namely,

$$\Psi_j^\nu = \langle x_j | \Psi^\nu \rangle = \Psi^\nu(x_j e^{i\alpha}). \tag{27}$$

In a recent communication,[20] we have applied this method to the calculation of the resonance energies and widths (E_R, $-\Gamma/2$) for the tunneling in the anharmonic oscillator

$$V(x) = x^2/4 - \lambda x^3 \tag{28}$$

previously studied by several workers using the complex-scaling basis-set-expansion method and the complex-scaling finite difference method. Table 3 shows the comparison of the present calculation (CSFGH) with former studies. It is seen that the present method reproduces exactly the results obtained by the complex-scaling basis-set-expansion (CSB) method[22] and converges to at least 9 decimal digits. Further, the CSFGH method appears more accurate and far more efficient than the finite difference method[23] (CSFD). (In this calculation, we used N=61 space grid points, while in the finite difference method quoted in ref. 23, N = 4000.)

In the next section, we discuss the extension of the CSFGH to multi-channel problem -- with particular application to the study of multiphoton and above-threshold dissociaiton of molecules in intense laser fields.

Table 3*

Resonance energies (E_R-i $\Gamma/2$) (in a.u.) for the anharmonic oscillator $V(X) = \frac{1}{4}x^2 - 0.034$ computed using various theoretical methods.

N$^{a)}$	CSB$^{b)}$	CSFD$^{c)}$	CSFGH$^{d)}$
0	$0.48567937-0.286698 \times 10^{-5}i$	$0.4856786-0.30 \times 10^{-5}i$	$0.4856793718 -0.28669781 \times 10^{-5}i$
1	$1.3915748 -0.134193 \times 10^{-2}i$	$1.391572 -0.13425 \times 10^{-2}i$	$1.391574841 -0.13419340 \times 10^{-2}i$
2	$2.1321356 -0.687626 \times 10^{-1}i$	$2.13213 -0.68762 \times 10^{-1}i$	$2.132135564 -0.68762610 \times 10^{-1}i$
3	$2.817874 -0.363974\ i$	$2.81786 -0.363967\ i$	$2.817874151 -0.36397367\ i$
4	$3.586675 -0.777257\ i$	$3.58665 -0.77724\ i$	$3.58667470 -0.77725688\ i$

a) vibrational quantum number

b) complex scaling-basis set expansion method

c) complex-scaling finite-difference method

d) complex-scaling Fourier-grid Hamiltonian method

* Adapted from ref. 20.

MULTIPHOTON AND ABOVE-THRESHOLD DISSOCIATION IN INTENSE LASER FIELDS -- COMPLEX QUASI-VIBRATIONAL ENERGY FORMALISM AND COMPLEX-SCALING FOURIER-GRID HAMILTONIAN METHOD

As a second major application of the complex-coordinate coupled-channel formalism, we shall consider the problem of the determination of multiphoton induced molecular resonance states in the presence of intense laser fields. It has long been known that multiphoton dissociation (MPD) of polyatomic molecules is an efficient process and can occur in relatively weak infrared laser fields. In contrast, MPD of small molecules, particularly diatomic molecules, is a very slow and inefficient process, due to the low density and anharmonicity of vibrational states. Indeed, MPD of diatomic molecules from the *ground* vibrational states of diatomic molecules has never been observed experimentally until 1986. The only exception, as far as diatomic molecules are concerned, is the experimental observation of two-photon dissociation from highly *excited* vibrational states of HD^+.[24] Theoretical studies have shown that MPD from the weaker-bound *high* vibrational levels exhibiting large amplitude vibration is usually far more efficient than from those (tighter-bound) low-lying levels.[25,26]

Due to the recent advent of high power lasers, with intensities ranging from 10^{12} to 10^{15} W/cm^2(peak electric fields up to 10 V/A readily achievable, there is a rapidly growing new interest in the study of nonlinear multiphoton dynamics in small molecules, notably molecular hydrogen.[27-29] The latter process involves multiphoton ionization (MPI) of H_2 followed by multiphoton dissociation of H_2^+. A particularly interesting and novel nonlinear optical phenomenon is the observation of the so-called above-threshold dissociation (ATD), a process where molecules can absorb more photons than necessary to dissociate the chemical bond. The appearance of equally-spaced multiple peaks in the fragment dissociation spectrum is analogous to the observation of additional peaks in the electron spectra (a phenomenon called above-threshold ionization (ATI))[3,4] widely studied in the multiphoton ionization of atoms. In addition, chemical bonds can be "softened" in the presence of intense fields, leading to efficient dissociation of nearly all of the vibrational levels. The presence of additional interatomic degrees of freedom in molecules thus enrich the problem of the nonlinear interaction of molecules with intense laser fields.

Considerable theoretical works have been advanced in the description of molecular photoabsorption and laser-induced resonances in the presence of strong laser fields.[30] For the study of MPD of molecules, we have previously developed two theoretical methods. The first method, called the *complex quasi-vibrational energy* (QVE) formalism,[31] is a nonperturbative approach based on the generalization of the Floquet theory to include the complete set of continuum as well as bound vibronic states. This has the effect of giving each of the dressed vibronic levels an intensity-dependent part (width) in addition to the usual field-induced A.C. Stark shifts. Proper examination of the frequency and intensity dependence of these complex QVE's gives rise to rates for MPD processes and is equivalent to infinite-order perturbation theory, self-consistent in that shifts and widths of all levels are simultaneously determined. The second approach[25,26] is an extension of the inhomogeneous differential equation (IDE) of Dalgarno and Lewis for numerical evaluation of the infinite sum over intermediate states.[32] The IDE method was found to be powerful for weak-field non-resonant MPD calculation.

Prompted by the recent high-intensity experiments on MPD/ATD of diatomic molecules,[27-29] we have recently further developed efficient non-perturbative procedures for the determination of the resonance energies and (MPD) total widths from ground and arbitrary *excited* vibrational levels -- based on the extension of complex quasi-vibrational energy formalism and the complex-scaling Fourier-grid Hamiltonian method.[33] The element of the procedure proposed is described below.

Fig. 5 shows the potential energy curves of the ground ($1s\sigma_g$) and first excited ($2p\sigma_u$) states of H_2^+ as a function of internuclear separation R. Also displayed is a schematic diagram showing the MPD/ATD process at 266 nm from the ground vibrational level of the $1s\sigma_g$ states. The corresponding dressed-state (electronic-field potential energy curves) picture (solid lines: diabatic curves; dotted lines: adiabatic curves) is shown in Fig. 6. Each curve corresponds to $U_i(R) + n\,\hbar\omega$, where $U_i(R)$ are the electronic potential energy for $1s\sigma_g$ or $2p\sigma_u$ states, and n=0, -1, -2, -3 are the Fourier photon indices. Formally, the photodissociation or multiphoton dissociation between a bound and a repulsive electronic states is a half-collision process and can be regarded as a (diabatic) curve-crossing or an (adiabatic) avoided-crossing predissociation problem. The potential energy curves and the transition dipole moment used are the same as those given in ref. 25.

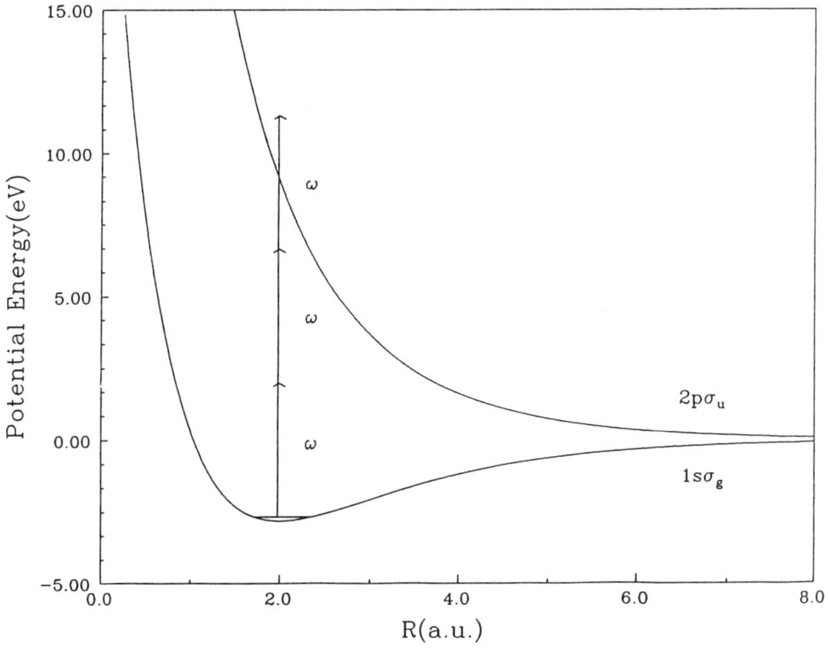

Fig. 5. Potential-energy curves for the ground ($1s\sigma_g$) and first excited ($2p\sigma_u$) states of H_2^+ as a function of internuclear separation R. Also displayed is a schematic diagram showing the absorption of one, two and three photons of wavelength 2660 A from the ground vibrational level of the ($1s\sigma_g$) state.

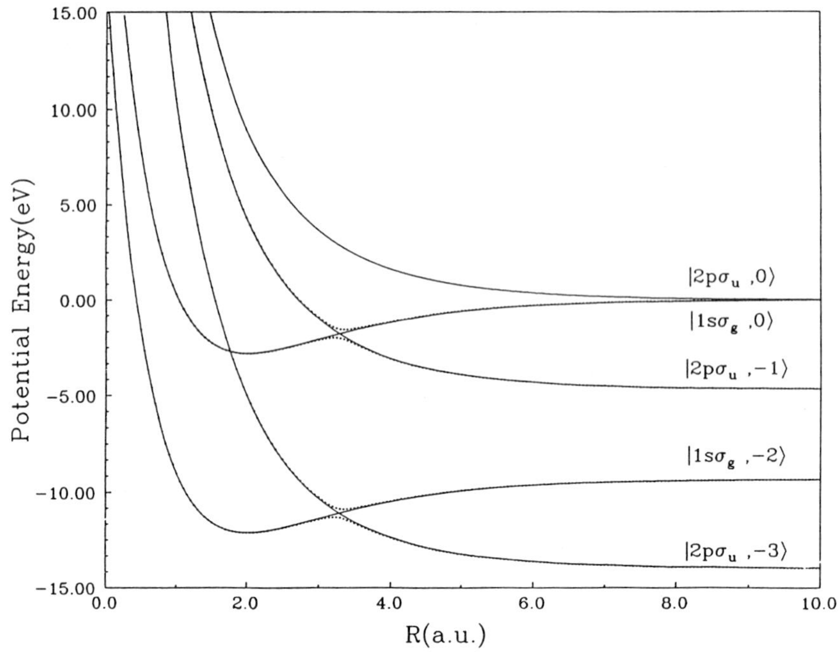

Fig. 6. Electronic-field potential-energy curves of the two electronic states of H_2^+ dressed by n=0,-1,-2,-3 photons of wavelength 2660 Å. Solid lines: diabatic curves. Dotted lines: adiabatic curves.

Fig. 7 shows the (time-independent) Floquet Hamiltonian appropriate for this problem in the length gauge. Here $\hat{T}_R$ is the kinetic energy operator, and $U_1(R)$ and $U_2(R)$ are, respectively, the $1s\sigma_g$ and $2p\sigma_u$ electronic potential energy. In the coordinate or Schrödinger representation, the complex-scaling ($R \rightarrow Re^{i\alpha}$) Hamiltonian operators in the diagonal Floquet blocks become

$$\langle R|\hat{H}_i(Re^{i\alpha})|R'\rangle \equiv \langle R|e^{-2i\alpha}\hat{T}_R + \hat{U}_i(Re^{i\alpha})\ |R'\rangle$$

$$= e^{-2i\alpha}\langle R|\hat{T}_R|R'\rangle + U_i(Re^{i\alpha})\ \delta(R-R'), \qquad (i=1,\ 2). \qquad (29)$$

which can be efficiently evaluated by the Fourier grid Hamiltonian method *without* the use of basis set expansion.[20] The electric-dipole coupling operator $\vec{\mu}_{12}(Re^{i\alpha})\cdot\vec{\varepsilon}_0$ in the off-diagonal blocks can be similarly readily computed, as it is also diagonal in the coordinate representation. The complex quasi-energy eigenvalues and eigenvectors can now be determined by the solution of the complex secular equations:

$$\sum_{j'} \{[\hat{H}_F(Re^{i\alpha})]_{jj'} - W_\nu\delta_{jj'}\}\ \psi_{j'}^\nu(\alpha) = 0, \qquad (30)$$

where the eigenvectors ψ_j^ν give *directly* the amplitude of the complex quasi-energy resonance wave functions ψ^ν evaluated at the grid points R_j,

$$\psi_j^\nu(\alpha) = \langle R_j|\psi^\nu(\alpha)\rangle = \psi^\nu(R_je^{i\alpha}). \qquad (31)$$

As a test of the reliability and efficiency of the complex-scaling Fourier grid Hamiltonian (CSFGH) method, Fig. 8 shows the results of weak-field ($\varepsilon_0 = 10^{-3}$a.u. or 7 x 10^{10} W/cm^2) one-photon dominant photodissociation from $H_2^+(1s\sigma_g$, v=0, 1, 2, ..., 15) and j=0 states at 266 nm. By judicious choice of R_{min} and R_{max}, and using only 51 space grid points (R_i), we are able to capture all (except the last two highest vibrational levels close to the dissociation limit) the complex quasi-vibrational energies (converged to at least 4 decimal digits). This CSFGH procedure is thus found to be simpler and far more efficient than the previous L^2-complex quasi-energy calculations for the same process, where at least 100 harmonic oscillator basis functions are required to achieve similar accuracy for high lying levels.[25]

$A+4\omega I$	B	o	o	o
B^T	$A+2\omega I$	B	o	o
o	B^T	A	B	o
o	o	B^T	$A-2\omega I$	B
o	o	o	B^T	$A-4\omega I$

WHERE

A=

$T_R + U_1(R)$	$\frac{1}{2}\vec{\mu}_{12}(R)\cdot\vec{\epsilon}_0$
$\frac{1}{2}\vec{\mu}_{21}(R)\cdot\vec{\epsilon}_0$	$T_R + U_2(R) - \omega I$

AND

B=

o	o
$\frac{1}{2}\vec{\mu}_{12}(R)\cdot\vec{\epsilon}_0$	o

Fig. 7. Structure of the Floquet Hamiltonian $\hat{H}_F$ for MPD/ATD.

It is instructive to also examine the resonance wave function behavior. Fig. 9 shows the resonance wave functions (without the use of complex scaling transformation, α=0) for v=5, 10, and 15 at 266 nm and field intensity 7 x 10^{10}W/cm^2. Under weak-field perturbation, all the wavefunctions still preserve the unperturbed nodal structures except they carry a continuous long tail -- indicating they are bound states embedded in the continuum. When the complex scaling transformation ($R \rightarrow Re^{i\alpha}$) is

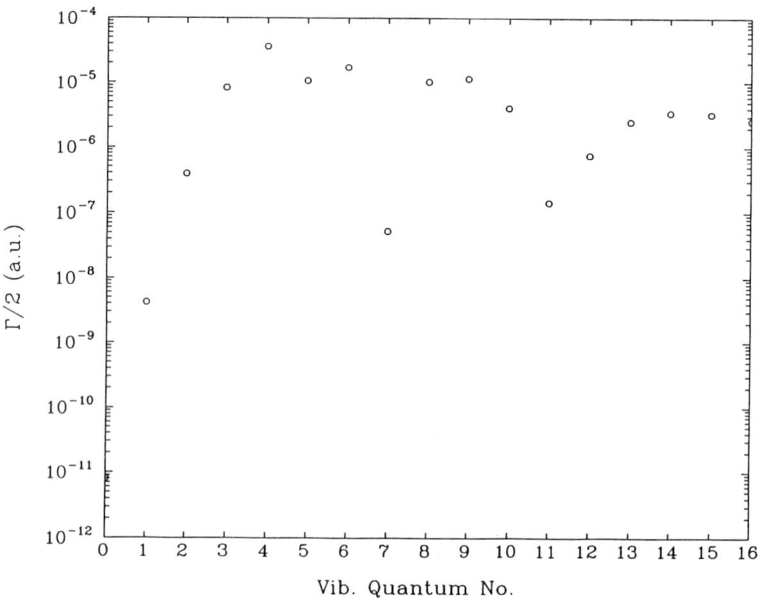

Fig. 8. The photodissociation (half-) widths $\Gamma/2$ ($\equiv -\mathrm{Im}(W)$) of $H_2^+(1s\sigma_g)$ as a function of the initial vibrational quantum number v. The field parameters are $\lambda = 2660$ Å, and $F_{rms} = 0.001$ a.u. At this field strength, the photodissociation is dominantly a one-photon process.

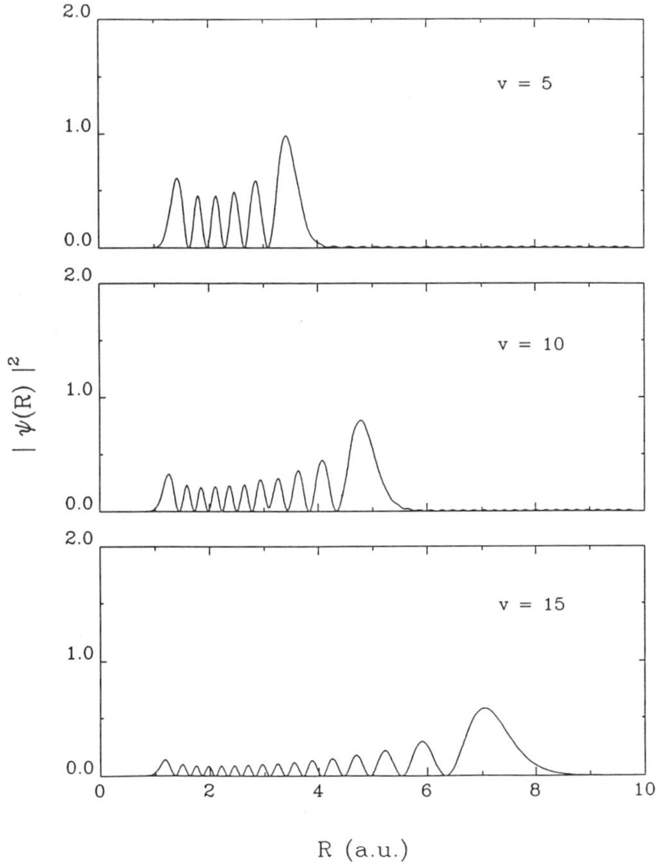

Fig. 9. $|\psi(R)|^2$ versus R for quasi-vibrational energy wave functions $\psi(R)$ correlated with v = 5, 10, and 15 vibrational levels of $H_2^+(1s\sigma_g)$ state. $\psi(R)$'s are eigenfunctions of the (real) Floquet Hamiltonian $\hat{H}_F$(Fig. 7) discretized in the coordinate representation. The field parameters are $I = 7 \times 10^{10}$ W/cm^2 and $\lambda = 2660$ Å.

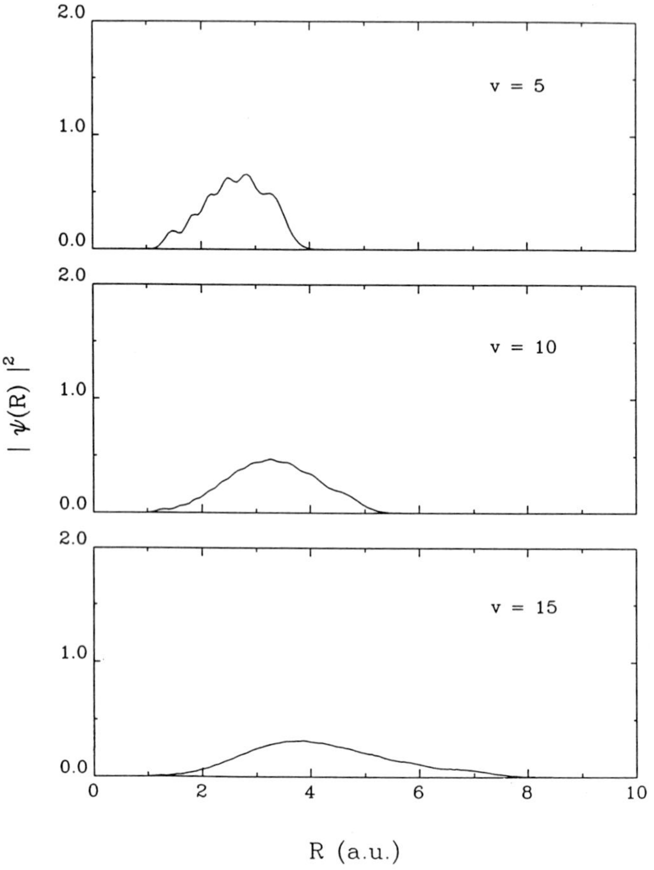

Fig. 10. $|\psi(R)|^2$ versus R for *complex* quasi-vibrational energy wave functions $\psi(R)$ correlated with v = 5, 10, and 15 vibrational levels of $H_2^+(1s\sigma_g)$ state. $\psi(R)$'s are the eigenfunctions of the complex-scaling Floquet Hamiltonian $\hat{H}_F(\alpha)$. Field parameters are the same as Fig. 9. The rotation angle α used is 0.08 radian. These resonance states are seen to be localized and broadened wave packets.

turned on, the resonance wave functions become localized (Fig. 10) in the coordinate space. This explains why the equal-spaced-grid version of the CSFGH method is capable of treating the excited-state resonances accurately with a relatively small number of mesh points.

Figure 11 shows the intensity-dependent MPD half-widths ($\Gamma/2$) as a function of the laser intensity I for the ground vibrational level (v=0) of the $H_2^+(1s\sigma_g)$ electronic state. At weaker fields, Γ/F_{rms}^2 (proportional to Γ/I) is seen to be independent of the laser intensity I, and the photodissociation is dominantly a one-photon process. Above some critical field intensity ($I \simeq 10^{12}$ W/cm^2), ATD sets in, and the process becomes highly nonlinear. More work about MPD/ATD dynamics of H_2^+ will be reported elsewhere.

ACKNOWLEDGMENT

The work described in this article was partially supported by the Guggenheim Fellowship and by the U.S. Department of Energy, Division of Chemical Sciences.

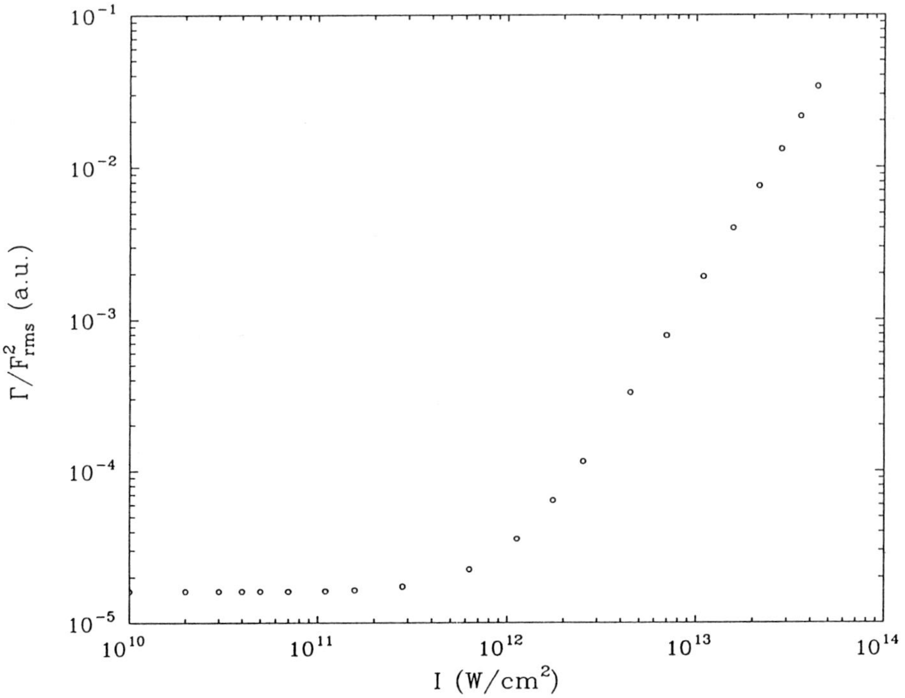

Fig. 11. Reduced widths (Γ/F_{rms}^2) versus intensity I, for the ground vibrational level (v=0) of the $H_2^+(1s\sigma_g)$ state at λ = 2660 Å.

REFERENCES

1. E. Balslev and J.M. Combes, Commun. Math. Phys. 22:280 (1971); A. Aguilar and J.M. Combes, Commun. Math. Phys. 22:265 (1971); B. Simon, Ann. Math. 97:247 (1973).
2. Proceedings of the 1978 Sanibel Workshop on Complex Scaling, Intern. J. Quantum Chem. 14:343-542 (1978).
3. P. Agostini, A. Antonetti, P. Breger, M. Crance, A. Migus, H.G. Muller, and G. Petite, J. Phys. B22:1971 (1989).
4. R.R. Freeman, P.H. Bucksbaum, H. Milchberg, S. Darrack, D. Schumacher, and M.E. Geusic, Phys. Rev. Lett. 59:1092 (1987).
5. S.I. Chu and W.P. Reinhardt, Phys. Rev. Lett. 39:1195 (1977); A. Maquet, S.I. Chu, and W.P. Reinhardt, Phys. Rev. A27:2946 (1983).
6. S.I. Chu, Adv. At. Mol. Phys. 21:197 (1985).
7. S.I. Chu, K. Wang, and E. Layton, J. Opt. Soc. Am. B7:425 (1990).
8. J.H. Shirley, Phys. Rev. 138:B979 (1965).
9. For a recent review on various generalizations of Floquet theories and techniques for the treatment of intense-field multiphoton and nonlinear optical processes, see, S.I. Chu, Adv. Chem. Phys. 73:739 (1989).
10. S.I. Chu and J. Cooper, Phys. Rev. A32:2769 (1985).
11. P. Avan, C. Cohen-Tannoudji, J. Dupont-Roc, and C. Fabre, J. Phys. (Paris) 37:993 (1976); L. Hollberg and J.L. Hall, Phys. Rev. Lett. 53:230 (1984).
12. S.I. Chu, Chem. Phys. Lett. 58:462 (1978).
13. J. Avron, I. Herbst, and B. Simon, Duke Math. J. 45:847 (1978).
14. S.K. Bhattacharya and S.I. Chu, J. Phys. B16:L471 (1983).
15. S.K. Bhattacharya and S.I. Chu, J. Phys. B18:L275 (1985).
16. R.H. Garstang, Rep. Prog. Phys. 40:105 (1977).
17. G. Wunner, H. Herold, and H. Ruder, J. Phys. B16:2973 (1983).
18. A. Holle, J. Main, G. Wiebusch, H. Rottke, and K.H. Welge, Phys. Rev. Lett. 61:161 (1988).
19. C.H. Iu, G.R. Welch, M.M. Kash, L. Hsu, and D. Kleppner, Phys. Rev. Lett. 63:1133 (1989).
20. S.I. Chu, Chem. Phys. Lett. 167:155 (1990).
21. C.C. Marston and G.G. Balint-Kurti, J. Chem. Phys. 91:3571 (1989).
22. K.K. Datta and S.I. Chu, Chem. Phys. Lett. 87:357 (1982).
23. O. Atabek and R. Lefebvre, Chem. Phys. Lett. 84:233 (1981).
24. A. Carrington and J. Buttenshaw, Mol. Phys. 44:267 (1981).
25. S.I. Chu, C. Laughlin, and K.K. Datta, Chem. Phys.. Lett. 98:476 (1983).
26. C. Laughlin, K.K. Datta, and S.I. Chu, J. Chem. Phys. 85:1403 (1986).
27. C. Cornaggia, D. Normand, J. Morellec, G. Mainfray, and C. Manus, Phys. Rev. A34:207 (1986).
28. T.S. Luk and C.K. Rhodes, Phys. Rev. A38:6180 (1988).
29. P.H. Bucksbaum, A. Zavriyev, H.G. Muller, and D.W. Schumacher, Phys. Rev. Lett. 64:1883 (1990).
30. J.O. Hirschfelder, R.E. Wyatt, and R.D. Coalson, ed., "Lasers, Molecules, and Methods," Wiley, New York, Adv. Chem. Phys. 73:1-978 (1989).
31. S.I. Chu, J. Chem. Phys. 75:2215 (1981).
32. A. Dalgarno and J.T. Lewis, Proc. Roy. soc. A233:70 (1955).
33. S.I. Chu, J. Chem. Phys. (in press).

MULTIREFERENCE COUPLED-CLUSTER APPROACH TO SPECTROSCOPIC CONSTANTS: MOLECULAR GEOMETRIES AND HARMONIC FREQUENCIES*

Uzi Kaldor

School of Chemistry, Tel Aviv University, 69 978 Tel Aviv, Israel

INTRODUCTION

The coupled-cluster (CC) method,[1-4] originally designed for closed-shell systems, has been extended to include open-shell systems, which cannot be described adequately by a single determinant.[5-22] The basic approach of the multireference method is to define an effective Hamiltonian in a low-dimensional model (or P) space, with eigenvalues approximating some desirable eigenvalues of the physical Hamiltonian. The effect of the complementary Q space is taken into account in the calculation of the effective Hamiltonian matrix elements, using an appropriate truncation of the wave operator. Two different approaches are commonly used. Most applications to date follow the state-universal or Fock space approach, with simultaneous calculation of many states having different numbers of valence electrons. The state-specific or Hilbert space method, on the other hand, treats a manifold of states with a constant number of valence electrons. The selection of the model space plays a crucial role in both methods. Intruder states, which spoil the convergence of the calculation, occur frequently. Careful construction of the model space may alleviate the problem. In particular, so-called incomplete model spaces are useful in many cases.

A detailed formal discussion of open-shell coupled cluster theories may be found in a recent review by Mukherjee and Pal.[23] A brief description of the Fock space method is presented below. The bulk of this chapter describes one type of applications carried out in our laboratory over the past few years, namely the calculation of geometries and vibrational frequencies of molecular systems. The systems to be discussed include the diatomics Li_2 and Na_2, where ground and excited states where calculated, as well as the N_3, NO_2, NO_3, and NS_2 molecules, where our main interest was in the geometry and harmonic frequencies of both radical and ionic species.

*Supported in part by the U.S.-Israel Binational Science Foundation.

THE FOCK SPACE COUPLED CLUSTER METHOD

Formalism

The Hamiltonian H of the system is separated in the conventional way into H_0, with known eigenfunctions, and a perturbation V,

$$H = H_0 + V \tag{1}$$

$$H_0|\alpha\rangle = E_0^{\alpha}|\alpha\rangle \ . \tag{2}$$

A d-dimensional model space P and its complement Q are defined by projection operators,

$$P = \sum_{\alpha\in P} |\alpha\rangle\langle\alpha| \ , \quad Q = 1 - P \ . \tag{3}$$

There will usually be d eigenfunctions of H with major components in the model space,

$$H\Psi^a = E^a\Psi^a \ , \tag{4}$$

$$P\Psi^a = \Psi_0^a \ , \quad a=1,2,...,d \tag{5}$$

where Ψ_0^a are linear combinations of $|\alpha\rangle$, $\alpha\in P$. The wave operator Ω transforms the model functions into exact ones,

$$\Omega\Psi_0^a = \Psi^a \ , \ a=1,2,...,d \ . \tag{6}$$

Intermediate normalization is assumed,

$$\langle\Psi^a|\Psi_0^a\rangle = \langle\Psi_0^a|\Psi_0^a\rangle = 1 \ . \tag{7}$$

The method used by us follows Lindgren's derivation.[10] The key equation is the generalized Bloch equation

$$[\Omega,H_0]P = V\Omega P - \Omega PWP, \tag{8}$$

where W is the effective interaction

$$W = V\Omega \ . \tag{9}$$

Alternatively, one may write (8) as

$$[\chi,H_0]P = QWP - \chi PWP \ , \tag{10}$$

where the correlation operator χ is defined by

$$\Omega = 1 + \chi \ . \tag{11}$$

The energies of interest are obtained by diagonalizing the effective Hamiltonian in the model space,

$$H_{eff}\Psi_0^a = E^a\Psi_0^a \ , \tag{12}$$

where

$$H_{eff} = PH\Omega P = P(H_0+W)P \ . \tag{13}$$

The correlation operator χ includes single, double, . . . , virtual excitations and may be written as

$$\chi = C_1 + C_2 + \ldots = \sum_{ij}\{a_i^\dagger a_j\}s_j^i + \frac{1}{2}\sum_{ijkl}\{a_i^\dagger a_j^\dagger a_\ell a_k\}s_{k\ell}^{ij} + \ldots \tag{14}$$

s_j^i , $s_{k\ell}^{ij}$, . . . , are excitation amplitudes, and the curly brackets denote normal order with respect to a reference (core) determinant. *All* terms, connected as well as disconnected, are included in (14). The operator used in CCM is the excitation operator S, related to Ω by

$$\Omega = \{\exp(S)\} = 1 + S + \frac{1}{2}\{S^2\} + \ldots \quad . \tag{15}$$

S is obtained by summing the rhs of (14) over *connected* terms only. Perturbative or non-perturbative schemes for calculating the excitation operator and correlation energies may be derived from either of the following two equations, which include connected terms only[10]

$$[S,H_0] = (QV\Omega - \chi PV\Omega)_{conn} , \tag{16}$$

or

$$[S,H_0] = W_{op,conn} - (\chi W_{cl})_{conn} . \tag{17}$$

$W_{op,conn}$ describes all connected diagrams which have some open (non-valence) lines, corresponding to P→Q transitions. W_{cl} diagrams, with no external non-valence lines, describe P→P transitions. The latter also appear in the effective Hamiltonian, which may be written as

$$H_{eff} = PH_0P + W_{cl} . \tag{18}$$

The second term in equation (16) or (17) gives rise to the so-called folded diagrams.

The Fock space approach is valence universal, meaning that one S operator is used for a manifold of states which may have different numbers of valence electrons. The operator may be partitioned according to the number of valence electrons,

$$S = S^{(0)} + S^{(1)} + S^{(2)} + \ldots \quad . \tag{19}$$

Haque and Mukherjee[17] have shown that partial decoupling of the equations occurs if normal order is assumed, as the equations for $S^{(n)}$ involve only $S^{(m)}$ elements with $m \leq n$. This decoupling is helpful in reducing the computational effort, and is used routinely in our calculations.

The H_{eff} or W_{cl} diagrams (see Eq. (18)) may be separated into core and valence parts,

$$H_{eff} = H_{eff}^{core} + H_{eff}^{val} , \tag{20}$$

where the first term on the rhs consists of diagrams without any external lines. The eigenvalues of H_{eff}^{val} will then give directly the transition energies from the core, with correlation effects included for both core and valence electrons. The physical significance of these energies depends on the nature of the model space. Thus, electron affinities may be calculated by constructing a model space with valence particles only,[24,25] ionization potentials are given using valence holes,[26,27] and both types are included for the purpose of getting excitations out of a closed-shell system.[28-31]

Structure of the Model Space

Most derivations of the coupled-cluster equations, as well as the original open-shell many-body perturbation theory (MBPT) of Brandow,[32] depend on a particular choice of the model space. The orbitals are classified as core, valence, or particles (unoccupied).

The core orbitals are always occupied, and all possible distributions of the remaining electrons in the valence orbitals give rise to determinants included in the P space (there may also be valence holes, i.e. unoccupied core orbitals, but the situation is not fundamentally different). Such model spaces have been called "complete".[33] This recipe is appropriate when the open-shell orbitals are close in energy, which is not the case for most atomic and molecular excited states. It is often impossible to select valence orbitals so that no Q space determinants (with one or more non-valence orbitals) lie close to or even within the energy range spanned by the P space. Thus, if we were to describe the 1s2s ^1S state of He by designating the 1s and 2s orbitals as valence, the P space would include the $1s^2$, 1s2s and $2s^2$ determinants, but exclude the various 1sns terms lying within its energy span. This situation leads to the so-called "intruder states",[34] which destroy the convergence of the expansion.

A general, incomplete model space MBPT has been proposed by Hose and Kaldor[33] and used in extensive calculations.[35] A similar CC method has been described by Jeziorski and Monkhorst.[14] Significant theoretical progress has been made in recent years in understanding incomplete model spaces.[23,36,37] While our main interest here is in the Fock space approach, it should be mentioned in passing that a general Hilbert space formalism has recently been presented by Meissner, Kucharski and Bartlett[38] and by Mukhopadhyay and Mukherjee.[39] The number of applications has also increased considerably.[28-31] A substantial number of CC applications with incomplete model spaces have been aimed at calculating one-electron excitation energies, where a natural choice of P determinants involves one hole and one particle with respect to the closed-shell ground state. This is a special case of "quasicomplete" model spaces.[11,37] Sinha et al.[40] demonstrated that the energy calculation for the particular case of a 1h-1p space is operationally equivalent to the complete-space procedure.

Mukherjee[36] has derived a linked-diagram expansion for Fock space coupled cluster in a general model space. The model space $P^{(m)}$ with m valence electrons is chosen on physical grounds, and may be incomplete. Model spaces $P^{(k)}$, $k<m$, are then constructed by deleting m-k orbitals in all possible ways from the $P^{(m)}$ determinants. An operator is designated k-open if it corresponds to a $P^{(k)} \rightarrow Q^{(k)}$ transition, where $Q^{(k)}$ is the complement of $P^{(k)}$; otherwise it is k-closed. The construction of the $P^{(k)}$ spaces causes all m-closed operators to be k-closed for all $k<m$; m-open operators may however be k-closed (in other words, an orbital change transforming every $P^{(k)}$ determinant to another $P^{(k)}$ function may take some $P^{(m)}$ determinant to a $Q^{(m)}$ term). The basic equations for the k-valence sector are then[36]

$$[S,H_0]^{(k)}_{m\text{-op}} = \{V\Omega - \Omega W\}^{(k)}_{m\text{-op,conn}} \tag{21}$$

$$\{\Omega W\}^{(k)}_{m\text{-cl}} = \{V\Omega\}^{(k)}_{m\text{-cl}} \quad . \tag{22}$$

Two differences between Eqs. (21)-(22) and (8)-(9) should be noted. The classification of the transitions at the k-valence level into $P \rightarrow P$ and $P \rightarrow Q$ has to be done according to their effect on m-valence states; and the equations for W are implicit [Eq. (22)] rather than explicit [Eq. (9)]. The former requires some additional, not very difficult, bookkeeping. The latter involves a few diagrams not encountered in complete model spaces, and the solution of a set of equations for W matrix elements. As the new diagrams are relatively simple and the equation system is of low dimension, incomplete model spaces are not more difficult to handle than complete ones.

The necessity of using incomplete model spaces may arise under different circumstances. A glance at the zero-order energies of the states to be investigated (the sums of orbital energies) frequently tells us that a complete model space approach is doomed to diverge, since it generates states in Q space which have energies in the range spanned by P space and significant coupling to P states. These intruder states, owing their existence to the one electron spectrum, may be expected in most molecular excited state calculations, and are discussed in connection to Li_2 states below. In other cases, the zero order spectrum may look harmless enough, but the CC equations fail to converge because of strong two-electron interactions giving rise to intruder states, as in the ^1S states of the beryllium atom.[41]

APPLICATIONS

A considerable number of applications implementing the methods described above have been reported. Many of them were summarized in recent reports[42–47] and will not be discussed here. Applications from our group include the ionization potentials and excitation energies (about ten per system) of Be,[28] Ne,[28] Mg,[29] Ar,[29] H_2O,[30] N_2[48] and O_2,[42] as well as ionization potentials and electron affinities of the alkali atoms Li, Na, K, Rb, and Cs.[25] Highly satisfactory results were obtained in all cases. The subject of the present chapter is spectroscopic properties, including rotational constants (determined by molecular geometry) and vibrational frequencies. These will be illustrated for two types of molecules: the alkali dimers Li_2[49] and Na_2,[50] and the radical and ionic species of the nitrogen-containing molecules NO_2,[51] NO_3,[52] N_3,[53] and NS_2.[54] Results of the calculations are given below and compared with experiment and with other theoretical methods.

Nine States of Li_2

Potential functions for the nine lowest states of the molecule were calculated. The basis started from the 5s5p1d set developed for the ionization potential and electron affinity of the Li atom,[25] and was augmented to better cover the δ sector and allow more flexibility in the large r domain (low exponents). To this end, the exponent of the most diffuse s and p functions was replaced by 0.0093504, a set of d functions on each atom with the exponent 0.07 was added (s component excluded), and functions taken from Schmidt-Mink, Müller and Meyer[55] were placed on the bond center. The latter include s and p_σ with exponent 0.004, and two sets of p_π and $d_{\pi,\delta}$ with exponents 0.12 and 0.04, for a total basis of 74 contracted Gaussian-type orbitals.

Table 1. Molecular constants of $^7Li_2$.

	method[a]	ref	R_e(A)	T_e(eV)	D_e(eV)	ω_e(cm^{-1})
$X^1\Sigma_g^+$						
	exp	56	2.673	--	1.056	351.4
	CCSD		2.67	--	1.061	351
	CI	59	2.671	--	1.034	350.5
	MCSCF	57	2.692	--	1.029	347.1
	ECP-CI	55	2.675	--	1.050	351.0
	ECP-CI	60	2.692	--	1.080	356.9
$a^3\Sigma_u^+$						
	CCSD		4.05	1.023	0.0380	75
	MCSCF	57	4.234		0.0362	61
	ECP-CI	55	4.182	1.010	0.0399	63.7
	ECP-CI	60	4.068		0.0439	66.4
$b^3\Pi_u$						
	exp	56	2.591	1.394	1.510	345.7
	CCSD		2.58	1.375	1.533	349
	MCSCF	57			1.424	339
	ECP-CI	55	2.595	1.396	1.506	345.9
	ECP-CI	60	2.555		1.543	354.1
$A^1\Sigma_u^+$						
	exp	56	3.108	1.744	1.160	255.5
	CCSD		3.10	1.750	1.158	257
	MCSCF	57	3.13		1.153	254
	ECP-CI	55	3.108	1.742	1.160	256.1
	ECP-CI	60	3.072		1.197	261.3
$^3\Sigma_g^+$						
	CCSD		3.06	2.032	0.876	252
	MCSCF	57	3.096		0.845	245
	ECP-CI	55	3.067	2.025	0.877	252.2
	ECP-CI	60	3.032		0.907	256.2
$^1\Sigma_g^+$						
	CCSD		3.56	2.530	0.378	137
	ECP-CI	55	3.655	2.496	0.406	129
$B^1\Pi_u$						
	exp	56	2.936	2.534	0.370	270.7
	CCSD		2.94	2.557	0.350	269
	MCSCF	58	2.97		0.303	260
	MCSCF	57	3.050		0.197	288
	ECP-CI	55	2.942	2.542	0.360	270.3
	ECP-CI	60	2.921		0.322	273.1
$^1\Pi_g$						
	CCSD		4.02	2.726	0.181	97
	MCSCF	57	4.161		0.161	83
	ECP-CI	55	4.073	2.728	0.174	91.8
	ECP-CI	60	4.046		0.176	92.4

[a] CCSD: coupled cluster, single and double excitations (present work).

CI: configuration interaction.

MCSCF: multiconfiguration self consistent field.

ECP: effective core potential, with one or more adjustable parameters.

The reference state for the calculation must have closed shells. Li_2^{++} was selected because it dissociates correctly, and all the states of interest are obtained from it by adding two electrons in the $2\sigma_g$, $3\sigma_g$, $2\sigma_u$, $1\pi_u$, and $1\pi_g$ orbitals. Some of the determinants thus formed are very high in energy (doubly excited relative to Li_2 ground state), and a complete model space calculation diverges. The model space is therefore constructed by selecting determinants according to their zero-order energy. It includes functions with at least one $2\sigma_g$ electron, as well as $2\sigma_u 1\pi_u$, 17 determinants in all. The $2\sigma_u{}^2$ function must be included in the model space for large internuclear separations.

The nine potential functions are shown in Figure 1. Table 1 presents a comparison of the CCSD molecular properties of $^7Li_2$ with experiment[56] and with the best previous calculations.[56-60] Included are the dissociation energy D_e, the equilibrium separation R_e, the adiabatic excitation energy T_e, and the vibrational constant ω_e. A cubic spline fit of the potential served to extract these constants, and ω_e was fitted from the lowest two or three vibrational levels.

Excellent agreement is obtained for the four experimentally known states, with error limits of 0.01 A for R_e, 0.02 eV for D_e and T_e, and 3 cm^{-1} for ω_e. This is better than any previous *ab initio* results, and almost as good as the best effective core potential work,[55] which uses empirical parameters to take care of the important core-core and core-valence correlations.

Five States of Na_2

Few theoretical studies of the excited states of Na_2 appear in the literature. The most extensive of the ground state calculations is that of Partridge et al.,[61] and Jeung[62] investigated a large number of excited states. Both calculations start by regarding Na_2 as a two electron problem, correlating the valence electrons by the configuration interaction (CI) method. Results obtained this way are not good enough, since core polarization of the valence electrons plays an important role in alkali atoms.[63] Jeung includes this effect by an approximate perturbation procedure, since all core electrons are represented by an effective core potential. Partridge et al.[61] include core correlation explicitly by allowing appropriate excitations in the CI function. They have to limit the CI state functions to those representing core-valence correlation; the inclusion of core-core correlation gives poor results. This is due to the size inextensivity of the CI method. A similar effect appears in the calculation of the alkali atoms electron affinities. The coupled cluster method is size extensive, and the electron affinities it predicts are improved (slightly) upon inclusion of core correlation.[25] The effect on spectroscopic constants of Na_2 was examined by us.

The five lowest states of the molecule are calculated. The 7s6p3d basis set developed for the ionization potential and electron affinity of the Na atom[25] is used, except that the d exponents are 1.3, 0.4 and 0.13. Na_2^{++} serves as the closed-shell reference, because it remains closed shell at all internuclear separations and dissociates

Table 2. Molecular constants of Na_2 compared with experiment.[64]

		R_e(A)	T_e(eV)	D_e(eV)[a]	ω_e(cm^{-1})
$X\,^1\Sigma_g^+$					
	3s correlation	3.21	--	0.651	147
	+2s2p corr.	3.05	--	0.702	164
	exp	3.08	--	0.747	159
$x\,^3\Sigma_u^+$	repulsive				
$a\,^3\Pi_u$					
	3s correlation	3.21	1.568	1.051	146
	+2s2p corr.	3.09	1.661	1.119	155
	exp	3.11	1.692	1.159	152
$A\,^1\Sigma_u^+$					
	3s correlation	3.77	1.661	0.958	112
	+2s2p corr.	3.65	1.787	0.993	117
	exp	3.64	1.820	1.031	117
$B\,^1\Pi_u$					
	3s correlation	3.58	2.458	0.161	112
	+2s2p corr.	3.47	2.622	0.158	118
	exp	3.42	2.519	0.322	124

[a] The dissociation energies of the excited states were not measured directly, and are calculated here from the atomic and molecular excitation energies.

correctly. The $4\sigma_g$, $4\sigma_u$, $2\pi_u$ and $2\pi_g$ orbitals have to serve as valence particles to obtain the five lowest states of the molecule. A complete model space would lead to divergence, spanning a broad energy range from $4\sigma_g{}^2$ to $2\pi_g{}^2$, with Q states such as $4\sigma_g 5\sigma_g$ falling within this range turning into intruder states. An incomplete model space was therefore used, comprising the determinants $4\sigma_g{}^2$, $4\sigma_g 4\sigma_u$, $4\sigma_g 2\pi_u$, and $4\sigma_g 2\pi_g$. Two sets of calculations were carried out, one correlating the valence (3s) electrons only, and the other correlating the 2s and 2p electrons as well.

Figure 2 shows the CCSD potential functions of Na_2. The molecular properties are presented and compared with experiment[64] in Table 2. Included are the dissociation energy D_e, the equilibrium separation R_e, the adiabatic excitation energy T_e, and the vibrational constant ω_e. A cubic spline fit of the potential served to extract these constants, and ω_e was fitted from the lowest two or three vibrational levels.

Correlation of the valence 3s orbitals alone does not give accurate parameters for the potential curves, with R_e too long by 0.1-0.15 A, T_e and D_e values 0.1 eV too low. The inclusion of the 2s and 2p correlation greatly improves the results, giving internuclear separations accurate to 0.03 A, excitation and dissociation energies correct to 0.04 eV, and ω_e within 5 cm^{-1} of experiment. The exception is the $B\,^1\Pi_u$ state, which

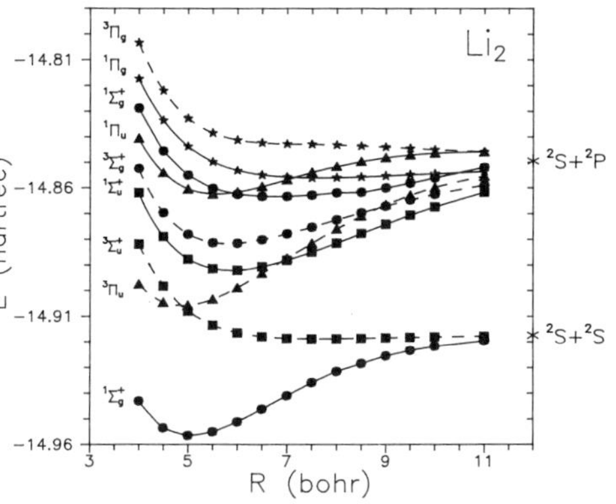

Figure 1. Potential functions of Li_2. Σ_g states are denoted by circles, Σ_u states by squares, Π_u states by triangles, and Π_g - by stars. Solid lines describe singlets, and dashed lines are triplets.

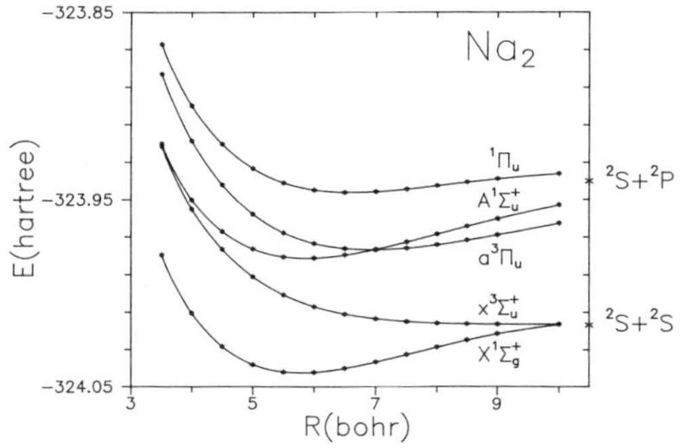

Figure 2. Potential functions of Na_2.

probably needs a better basis for accurate representation. It should be noted that all core (2s and 2p) excitations were included in the calculation, without any adverse effects. This is to be contrasted with the size inextensive CI method, where the inclusion of double excitations from the core[61] gave poor results.

The Triatomic Molecules NO_2, N_3, and NS_2

The spectroscopy of the NO_2 radical is complex enough to have generated two spectral atlases.[65,66] Detailed analysis of the vibrational spectrum of NO_2, giving harmonic frequencies as well as anharmonic corrections, have been reported in recent years by several groups.[67-69] The geometry of the radical has been determined by Morino et al.[70] Ab initio computations of the NO_2 radical have also been reported, the most comprehensive being those of Hirsch and Buenker[71] and of Xie et al.[72] The latter includes a detailed study of the geometry and harmonic vibrational frequencies at different levels of theory and with different basis sets. An interesting observation emerging from the study was that valence-orbital complete-active-space self-consistent field (CASSCF) gave reasonably good prediction of spectroscopic constants, but the configuration interaction method with single and double excitations (CISD), which accounts for a substantially larger part of the total correlation energy, did not do nearly as well. CASSCF includes intra-valence correlation to the limit of the basis set, but completely ignores the large external correlation effects. CISD, on the other hand, takes in a major portion of both internal and external correlation, but neglects triple and higher excitations. The results of Xie et al.[72] indicate that high excitations are important for the NO_2 spectroscopic constants. The coupled-cluster method with singles and doubles (CCSD) includes all possible products of single and double excitations, so that a major part of the contribution of triple, quadruple and higher excitations is taken into account.

The spectroscopy of NO_2^- is less complicated, but most of the experimental work has been done on ions in matrices, crystals, or aqueous solution.[73-76] Only recently has there been a determination of the spectroscopic constants of the anion in gas phase,[69] together with the thermodynamic electron affinity of NO_2. It should be noted that the information available for the anion is not as detailed as for the radical. Thus, no complete analysis of the vibrational spectrum in terms of harmonic frequencies ω_i and anharmonic corrections x_{ij} has been reported, and only fundamental frequencies ($\nu_i=\omega_i+x_{ii}$) are known. Several theoretical determinations of the geometry and electron affinity appear in the literature,[77-79] but we are not aware of a calculation of the vibrational frequencies.

The results reported here for the NO_2 radical were obtained in the triple-zeta plus polarization (TZP) basis, composed of the (9s5p/4s2p) Huzinaga-Dunning[80] orbitals with one set of d functions on the N (ζ=0.80) and O (ζ=0.85) atoms. The 3s combinations of the d orbitals were eliminated, giving a total of 57 basis functions. Diffuse s and p orbitals with exponents[78] 0.048 on N and 0.059 on O were added for the anion, giving the TZP$^+$ basis. The Hartree-Fock function for the ground state of NO_2^-,

$$1a_1^2 1b_2^2 2a_1^2 3a_1^2 2b_2^2 4a_1^2 5a_1^2 1b_1^2 3b_2^2 4b_2^2 1a_2^2 6a_1^2 \qquad {}^1A_1 ,$$

served as a starting point, and the coupled-cluster equations were solved for this state. The $4b_2$, $1a_2$ and $6a_1$ orbitals were then designated as valence holes, and the coupled-cluster energies for the three resulting doublet states of the NO_2 radical obtained. The three 1s orbitals were kept frozen, and the three highest virtual orbitals correlating them were disregarded. The coupled-cluster equations were solved iteratively until all excitation amplitudes converged to 10^{-5} or better.

The NO_2 minimum-energy configuration was found by simple search at the CCSD level. It is precise to 0.001A and 0.3°. Once the minimum geometry has been established, the harmonic vibrational constants were calculated by numerical differentiation. The symmetry coordinates given by Cyvin[81] were used, and the harmonic force constant along symmetry coordinate S_i obtained by

$$F_{ii} = \frac{E(S_e + \delta S_i) + E(S_e - \delta S_i) - 2E(S_e)}{\delta S_i{}^2}$$

where S_e is the equilibrium geometry. The off-diagonal force constant was calculated, when needed, by

$$F_{ij} = \frac{E(S_e + \delta S_i + \delta S_j) + E(S_e - \delta S_i - \delta S_j) - 2E(S_e) - F_{ii}\delta S_i{}^2 - F_{jj}\delta S_j{}^2}{2\delta S_i \delta S_j} .$$

Harmonic frequencies were then obtained by diagonalizing the FG matrix.[82] A similar procedure was followed for NO_2^-, using the TZP+ basis.

The CCSD geometries and harmonic frequencies are listed and compared with experiment and previous calculations in Table 3. Very good agreement with measured quantities is obtained. In particular, the NO_2 frequencies are much more accurate than the CISD values. The spectroscopic constants of the anion, calculated in the TZP+ basis, also show good agreement with experiment. It should be noted that the calculated vibrational frequencies are the harmonic ω_i, whereas the reported experimental values are the fundamentals ν_i. Inasmuch as the latter are generally lower than the former, the agreement is actually better than apparent from the table. To calculate the thermodynamic electron affinity of NO_2, its CCSD energy was computed in the TZP+ basis, using the TZP minimum-energy geometry. The resulting energy, -204.64002 hartree, is 2.13 eV higher than the corresponding energy of the anion. Correcting for zero-point vibrational energy gives 2.15 eV, which may be compared with the experimental[69] 2.27 and the equation-of-motion value[77] of 2.25 eV. The remaining error in the electron affinity is ascribed primarily to the description of the diffuse part of the orbitals by a single set of basis functions. The vertical ionization potentials of NO_2^- are calculated at 2.71 ($6a_1$), 3.68 ($4b_2$), and 4.14 eV ($1a_2$), comparable to the CISD values[78] 2.53, 3.89 and 4.00 eV. The vertical electron affinity of NO_2 at its equilibrium geometry is 1.33 eV.

The NS_2 radical, which is isovalent with NO_2, has recently been detected[83] and studied by the CI method.[84] Preliminary results of CCSD calculations are presented here. The application followed the same lines as for NO_2. The DZP basis was used,

Table 3. Spectroscopic constants of NO_2 and NO_2^-.

Method		Ref	Energy (hartree)	r(N-O) (A)	θ(ONO) (deg)	frequencies(cm^{-1})^a ω_1	ω_2	ω_3
				NO_2				
TZP	CASSCF	72	-204.26976	1.216	133.4	1299	735	1596
	CISD	72	-204.57723	1.189	134.9	1480	783	1812
	CCSD	51	-204.63540	1.206	133.5	1373	748	1679
Exp		70		1.194	133.9			
		67		1.1946	133.83	1353	760	1671
		68				1325	750	1634
		69				1316	748	
				NO_2^-				
TZP+	SCF	78	-204.16044	1.224	117.1			
	CCSD	51	-204.71829	1.265	116.5	1364	796	1296
Exp		69		1.25±0.02	117.5±2	1284±30	776±30	
in Ar		73						1244
in KBr		74				1316	798	1275
$NaNO_2$		75				1330	829	1242
aqueous		76				1332	821	1240

^{a}Harmonic frequencies ω_i are shown, except for the experimental values for NO_2^-, which are the fundamentals ν_i.

which is (9s5p/4s2p) for nitrogen and (11s7p/7s5p) for sulfur. The d function exponents were[84] 0.8 for N and 0.5 for S. The results are summarized and compared with other theoretical predictions in Table 4. The CCSD results fall in between CASSCF and CISD values, and are closer to the former. A similar trend is evident in NO_2, where CCSD gives the best geometry and vibrational frequencies. Since the experimental parameters are not known for NS_2, we cannot state with certainty that this is the case here as well.

The last triatomic species to be discussed is N_3. Both the anion and radical have been of interest to spectroscopists for a long time.[85,86] High-resolution spectroscopic information with attendant determination of vibrational and structural constants became available only recently.[87-93] The investigations were carried out on isolated molecules in the gas phase,[87-90] in N_2 matrices,[91] in KN_3 crystals,[92] or on gas-phase metal azides.[93] The electron affinity of N_3 has been measured by Illenberger et al.[94]

Many *ab initio* calculations have been published for N_3^-. The most extensive investigation is that of Botschwina,[95] who employed the coupled electron pair approximation

Table 4. Spectroscopic constants of NS_2.

Method		Ref	Energy (hartree)	r(S-O) (A)	θ(OSO) (deg)	frequencies(cm^{-1}) ω_1	ω_2	ω_3
DZP	SCF	84	-849.37554	1.543	155.3	742	272	1228
	CASSCF	84	-849.52310	1.602	146.0	656	318	1162
	CISD	84	-849.77707	1.562	150.8	730	300	1299
	CCSD	54	-849.82977	1.581	148.0	675	304	1210

(CEPA) in a large (99 contracted Gaussian-type orbitals) basis. References to earlier literature may also be found in Botschwina's paper. The azido radical has received less attention from theoreticians. An early Hatree-Fock calculation[96] predicted a linear, asymmetric structure. More recently, Tian et al.[91] calculated the geometry and vibrational frequencies of the two species by the Hartree-Fock method, and got a linear, symmetric configuration of the radical. The same geometry is predicted by the coupled-cluster work of Adamowicz.[97] We are not aware of any post-HF computation of vibrational frequencies. The electron affinity of the radical was calculated by Baker et al.[98]

CCSD calculations were carried out in the (9s5p/5s3p) Huzinaga-Dunning set augmented by one set of d orbitals (ζ=0.80, 3s combination excluded) and one set of diffuse s and p orbitals (ζ=0.048) on each atom. The basis, with a total of 69 contracted Gaussian type orbitals, is of triple-zeta plus polarization quality in the valence space, and includes diffuse functions to describe the anion. The Hartree-Fock function for the ${}^1\Sigma_g^+$ ground state of N_3^- in $D_{\infty h}$ symmetry, with the orbital configuration

$$1\sigma_g^2 1\sigma_u^2 2\sigma_g^2 2\sigma_u^2 3\sigma_g^2 3\sigma_u^2 4\sigma_g^2 1\pi_u^4 1\pi_g^4 ,$$

served as starting point, and the coupled-cluster equations were solved for this state. The $1\pi_g$ orbitals are then designated as valence holes, and the coupled-cluster energies for the ${}^2\Pi_g$ ground state of the N_3 radical obtained. The three 1s orbitals are kept frozen, and the three highest virtual orbitals correlating them are disregarded. Corresponding computations were carried out for the asymmetric ($C_{\infty v}$) and bent (C_{2v}) species.

The CCSD energies of N_3 and N_3^- were calculated at twelve geometries, and the results fitted to a potential surface harmonic in the symmetry coordinates given by Cyvin.[81] The rms error of the fit was less than 2×10^{-5} hartree for the anion and 4×10^{-5} hartree for the radical. The minimum-energy configuration and quadratic force constants were determined from the fitted potential, and harmonic vibrational frequencies were calculated from the force constants by the usual FG matrix method.[82]

Table 5. Spectroscopic constants of N_3 and N_3^-.

Method or medium	Ref	Energy (hartree)	r(N-N) (A)	vibrations(cm^{-1}) σ_g	π_u	σ_u
			N_3^-			
HF	96	-163.13262	1.175			
HF	91	-163.31968	1.151	1553	706	2190
CEPA	95	-163.90467	1.1911	1340		1950
CCSD	53	-163.84198	1.190	1352	637	2013
Exp						
KNO_3 crystal	92			1344	642.4	2036.4
N_2 matrix	91					2003.5
gas phase	88		1.1884			1986.4672
			N_3			
HF	96	-163.09704	1.177[a]			
HF	91	-163.28468	1.1538	1495.2	600.92	1674.9
CCSD	97		1.1815			
CCSD	53	-163.75022	1.183	1348	498	1684
Exp						
N_2 matrix	91			1287	472.7	1657.5
gas phase	87		1.1815			
gas phase	90		1.18115			1644.6784
gas phase	89			1320	457	

[a]($D_{\infty h}$ minimum. Lower minimum obtained for $C_{\infty v}$ configuration, r_{12}=1.241A, r_{23}=1.143A).

The geometries and frequencies are listed and compared with experiment and previous calculations in Table 5. Good agreement with measured quantities is obtained for both species. The equilibrium geometry is of $D_{\infty h}$ symmetry, with R_e values accurate to 0.002A. No other minima were found in the vicinity. Vibrational frequencies agree with experimental gas-phase values to 20-40 cm^{-1}, with our results higher than experiment in most cases. It should be noted that values quoted here as experimental are the fundamental frequencies ν_i, whereas the calculated numbers are the harmonic constants ω_i, so that the real disagreement is even smaller. The calculated electron affinity is 2.50 eV, to be compared with the experimental 2.76 eV.[94]

The ground state geometry of NO_3

The geometry of the NO_3 radical in its ground state has been the subject of several recent experimental[99-101] and theoretical[102-106] studies. The observed vibrational bands strongly indicate D_{3h} symmetry,[99,100] although not all the features of the IR spectrum could be explained. A recent study of the photoelectron spectrum of NO_3^-[101] also gave a D_{3h} ground state of the radical. Most theoretical treatments predict a lower C_{2v} symmetry, although the energy difference relative to the D_{3h} configuration is small (~2 kcal/mole). These include Siegbahn's multiconfigurational self-consistent field (MCSCF) and multireference configuration interaction (MRCI) calculations[102] and Davy and Schaefer's full-valence complete-active-space (CASSCF) work.[104] Kim et al.[105] even predict C_s symmetry, with three different N-O bonds. Boehm and Lohr[103] used perturbation theory with unrestricted Hartree-Fock (UHF) orbitals, and found that the minimum geometry depended on the perturbation order, with second order results favoring D_{3h} configuration, third order giving C_{2v}, and full fourth order going back to D_{3h} (geometries were optimized at the second-order level only). The use of UHF orbitals, as well as the low-level geometry optimization, may introduce uncertainties affecting significantly the small D_{3h}-C_{2v} energy difference. Most recently, Stanton et al.[106] performed a single determinant CCSD calculation for the NO_3 radical using Hartree-Fock orbitals for the closed shell NO_3^-, thus avoiding the strong bias of UHF orbitals toward symmetry breaking. They also obtain a C_{2v} minimum, and they suggest that the IR spectrum may be explained by pseudorotation about the D_{3h} maximum. A similar model was tried by Weaver et al.,[101] but they were unable to get parameters corresponding to broken symmetry and concluded the radical had a D_{3h} configuration.

The DZP basis set described above is used, including 60 functions. The Hartree-Fock function for the ground state of NO_3^-

$$\ldots 1a_2''^2 3e'^4 4e'^4 1e''^4 1a_2'^2 \qquad {}^1A_1'$$

served as starting point, with the $1a_2'$ orbital designated as valence-hole. A similar procedure was followed in C_{2v} symmetry, where the NO_3^- ground state is

$$\ldots 1b_1^2 7a_1^2 3b_2^2 8a_1^2 4b_2^2 1a_2^2 5b_2^2 2b_1^2 \qquad {}^2B_1$$

and $2b_1$ is the valence hole. The four 1s orbitals were kept frozen, and the four highest virtual orbitals correlating them were disregarded.

The lowest-energy geometry of the NO_3 ground state was obtained by a simple search, since no energy gradients are coded in our MRCC programs. It was found to have D_{3h} symmetry, with R(N-O)=1.2462A. This minimum of the CCSD potential surface in the DZP basis is precise to 0.001A. The neighborhood of this geometry was searched thoroughly, but no lower energy was found. Several geometries near the C_{2v} minimum reported by Davy and Schaefer[104] were also calculated, but all were higher in energy (by 3 millihartree or more) than the D_{3h} configuration. Once the minimum geometry has been established, the vibrational constants were calculated by numerical differentiation, as described above.

The CCSD molecular constants are compared in Table 6 with experiment[99] and with Davy and Schaefer.[104] Good agreement with experiment is obtained for the minimum-energy configurations and for the a_1' and a_2'' vibrations. Larger errors (100-300 cm^{-1}) occur in the e' frequencies; both these frequencies are, however, real, contrary to the

Table 6. Geometry and vibrational frequencies of NO_3.

	Exp[99,100]	CCSD	CASSCF[104]	
			D_{3h}	C_{2v}
R_1 (A)	1.24	1.246	1.256	1.351
R_2=R_3 (A)	1.24	1.246	1.256	1.224
$\theta_2=\theta_3$	120°	120°	120°	114.7°
$\nu_1(e')$	1492	1163	963	
$\nu_2(a_1')$	1060	1133	1068	
$\nu_3(a_2'')$	762	776	---	
$\nu_4(e')$	380	277	596i[a]	
Energy(hartrees)		-279.56226	-279.09434	-279.09660

[a] Imaginary frequency, as the D_{3h} minimum is a saddle point in CASSCF.

CASSCF predictions.[104] This is, of course, another manifestation of the CCSD minimum being D_{3h}, while CASSCF gives a C_{2v} minimum.

The final word on the ground state of NO_3 has not yet been said. It should be remembered that while the question is qualitative, C_{2v} vs. D_{3h} symmetry, the quantitative differences between the two configurations are small (1-2 kcal/mole) by all calculations, and the one thing we can say with certainty is that the potential surface of the radical ground state is flat near its equilibrium geometry. Further investigation of the NO_3 system, involving larger bases and fuller summation of correlation effects, may be required to resolve the problem.

CONCLUSION

This chapter describes the application of the Fock-space multireference coupled cluster method to the calculation of molecular spectroscopic properties, mostly molecular geometry and vibrational frequencies. Many other applications have been reported in the past.[24-31] In general, highly satisfactory results are obtained, due to the inclusion of most of the important correlation effects. These include summation of one- and two-electron virtual excitations and all their products and powers, as well as the ability to handle multireference situations. Future developments of the method should include three-electron excitations and analytic evaluation of gradients.

REFERENCES

1. J. Hubbard, *Proc. Roy. Soc.* A240:539 (1957); *ibid.* A243:336 (1958).
2. F. Coester, *Nucl. Phys.* 7:421 (1958); F. Coester and H. Kümmel, *Nucl. Phys.* 17:477 (1960); H. Kümmel, K. H. Lührmann and J. G. Zabolitzky, *Phys. Rept.* 36:1 (1978).
3. J. Cizek, *J. Chem. Phys.* 45:4256 (1966); *Adv. Chem. Phys.* 14:35 (1969).
4. J. Paldus, J. Cizek and I. Shavitt, *Phys. Rev. A* 5:50 (1972); J. Paldus, *J. Chem. Phys.* 67:303 (1977); B. G. Adams and J. Paldus, *Phys. 7Rev. A* 20:1 (1979).
5. F. E. Harris, *Intern. J. Quantum Chem.* S11:403 (1977).
6. H. J. Monkhorst, *Intern. J. Quantum Chem.* S11:421 (1977).
7. J. Paldus, J. Cizek, M. Saute and A. Laforgue, *Phys. Rev. A* 17:805 (1978); M. Saute, J. Paldus and J. Cizek, *Intern. J. Quantum Chem.* 15:463 (1979).
8. D. Mukherjee, R. K. Moitra and A. Mukhopadhyay, *Pramana* 4:247 (1975); *Mol. Phys.* 30:1861 (1975); A. Mukhopadhyay, R. K. Moitra and D. Mukherjee, *J. Phys. B* 12:1 (1979); D. Mukherjee and P. K. Mukherjee, *Chem. Phys.* 39:325 (1979); S. S. Adnan, S. Bhattacharyya and D. Mukherjee, *Mol. Phys.* 39:519 (1980); *Chem. Phys. Lett.* 85:204 (1981).
9. R. Offerman, W. Ey and H. Kümmel, *Nucl. Phys.* A273:349 (1976); R. Offerman, *Nucl. Phys.* A273:368 (1976); W. Ey, *Nucl. Phys.* A296:189 (1978).
10. I. Lindgren, *Intern. J. Quantum Chem.* S12:33 (1978); S. Salomonson, I. Lindgren and A. M. Martensson, *Phys. Scr.* 21:351 (1980); I. Lindgren and J. Morrison, "Atomic Many-Body Theory", Springer, Berlin, (1982).
11. I. Lindgren, *Phys. Scr.* 32:291, 32:611 (1985).
12. H. Nakatsuji, *Chem. Phys. Lett.* 59:362 (1978); *ibid.* 67:329 (1979); *Chem. Phys.* 75:425 (1983); *ibid.* 76:283 (1983); *J. Chem. Phys.* 80:3703 (1984).
13. H. Reitz and W. Kutzelnigg, *Chem. Phys. Lett.* 66:111 (1979); W. Kutzelnigg, *J. Chem. Phys.* 77:3081 (1981); *ibid.* 80:822 (1984).
14. B. Jeziorski and H. J. Monkhorst, *Phys. Rev. A* 24:1668 (1981); L. Z. Stolarczyk and H. J. Monkhorst, *Phys. Rev. A* 32:725, 32:743 (1985).
15. A. Banerjee and J. Simons, *Intern. J. Quantum Chem.* 19:207 (1981).
16. V. Kvasnicka, *Chem. Phys. Lett.* 79:89 (1981).
17. A. Haque and D. Mukherjee, *J. Chem. Phys.* 80:5058 (1984); *Pramana* 23:651 (1984).
18. P. Westhaus, *Int. J. Quantum Chem.* S7:463 (1973); P. Westhaus, E. G. Bradford, and D. Hall, *J. Chem. Phys.* 62:1607 (1975).
19. I. Shavitt and L. T. Redmon, *J. Chem. Phys.* 73:5711 (1980).
20. L. T. Redmon and R. J. Bartlett, *J. Chem. Phys.* 76:1938 (1972).
21. J. Arponen, *Ann. Phys. (NY)* 151:311 (1983).
22. K. Tanaka and H. Terashima, *Chem. Phys. Lett.* 106:558 (1984).
23. D. Mukherjee and S. Pal, *Adv. Quantum Chem.* 20:292 (1989).
24. U. Kaldor, *J. Comput. Chem.* 8:448 (1987).
25. U. Kaldor, *J. Chem. Phys.* 87:4693 (1987).
26. A. Haque and U. Kaldor, *Chem. Phys. Lett.* 117:347 (1985).
27. A. Haque and U. Kaldor, *Intern. J. Quantum Chem.* 29:425 (1986).
28. U. Kaldor and A. Haque, *Chem. Phys. Lett.* 128:45 (1986).
29. U. Kaldor, *Intern. J. Quantum Chem.* S20:445 (1986).
30. U. Kaldor, *J. Chem. Phys.* 87:467 (1987).
31. S. Pal, M. Rittby, R. J. Bartlett, D. Sinha, and D. Mukherjee, *Chem. Phys. Lett.* 137:273 (1987); *J. Chem. Phys.* 88:4357 (1988); M. Rittby, S. Pal, and R. J. Bartlett, *J. Chem. Phys.* 90:3214 (1989).
32. B. H. Brandow, *Rev. Mod. Phys.* 39:771 (1967).
33. G. Hose and U. Kaldor, *J. Phys. B* 12:3827 (1979).
34. T. H. Schucan and H. A. Weidenmuller, *Ann. Phys. (NY)* 73:108 (1972); *ibid.* 76:483 (1973).
35. G. Hose and U. Kaldor, *Phys. Scr.* 21:357 (1980); *Chem. Phys.* 63:165 (1981); *J. Phys. Chem.* 86:2133 (1982); *Phys. Rev. A* 30:2932 (1984); U. Kaldor, *J. Chem. Phys.* 81:2406 (1984).
36. D. Mukherjee, *Chem. Phys. Lett.* 125:207 (1986); *Intern. J. Quantum Chem.* S20:409 (1986).
37. I. Lindgren and D. Mukherjee, *Phys. Rep.* 151:93 (1987); W. Kutzelnigg, D. Mukherjee, and S. Koch, *J. Chem. Phys.* 87:5902 (1987); D. Mukherjee, W. Kutzelnigg, and S. Koch, *J. Chem. Phys.* 87:5911 (1987).

38. L. Meissner, S. A. Kucharski, and R. J. Bartlett, *J. Chem. Phys.* 91:6187 (1989); L. Meissner and R. J. Bartlett, *J. Chem. Phys.* 92:561 (1990).
39. D. Mukhopadhyay and D. Mukherjee, *Chem. Phys. Lett.* 163:171 (1989); *ibid.* 177:441 (1991).
40. D. Sinha, S. Mukhopadhyay, and D. Mukherjee, *Chem. Phys. Lett.* 129:369 (1986).
41. U. Kaldor, *Phys. Rev.* 38:6013 (1988).
42. U. Kaldor and S. Ben-Shlomo, *in*: "The Structure of Small Molecules and Ions", R. Naaman and Z. Vager Z, eds., Plenum, New York (1988), p. 199.
43. U. Kaldor, *in*: "Condensed Matter Theories", vol. 3, J. Arponen, R. F. Bishop, and M. Manninen, eds., Plenum, New York (1988), p. 83.
44. U. Kaldor, *in*: "Condensed Matter Theories", vol. 4, J. Keller, ed., Plenum, New York (1989), p. 67.
45. U. Kaldor, *in*: "Aspects of Many-Body Effects in Molecules and Extended systems", D. Mukherjee, ed., Springer Verlag, Berlin (1989), p. 155.
46. U. Kaldor, *in*: "Many-Body Methods in Quantum Chemistry", U. Kaldor, ed., Springer Verlag, Berlin (1989), p. 199.
47. U. Kaldor, *in*: "Condensed Matter Theories", Vol. 5, V. C. Aguillera-Navarro, ed., Plenum, New York (1990), p. 283.
48. S. Berkovic Ben-Shlomo and U. Kaldor, *J. Chem. Phys.* 92:3680 (1990).
49. U. Kaldor, *Chem. Phys.* 140:1 (1990).
50. U. Kaldor, *Isr. J. Chem.*, in press.
51. U. Kaldor, *Chem. Phys. Lett.* 170:17 (1990).
52. U. Kaldor, *Chem. Phys. Lett.* 166:599 (1990).
53. U. Kaldor, *Intern. J. Quantum Chem.* S24:291 (1990).
54. U. Kaldor, unpublished.
55. I. Schmidt-Mink, W. Müller, and W. Meyer, *Chem. Phys.* 92:263 (1985).
56. P. Kusch and M. M. Hessel, *J. Chem. Phys.* 67:586 (1977); M. M. Hessel and C. R. Vidal, *J. Chem. Phys.* 70:4439 (1979); R. A. Bernheim, L. P. Gold, P. B. Kelly, T. Tipton, and D. K. Veirs, *J. Chem. Phys.* 76:57 (1982); J. Verges, R. Bacis, B. Barakat, P. Carrot, S. Churassy, and P. Crozet, *Chem. Phys. Lett.* 98:203 (1983).
57. M L. Olson and D. D. Konowalow, *Chem. Phys. Lett.* 39:281 (1976); *Chem. Phys.* 21:393 (1977); *Chem. Phys.* 22:29 (1977); D. D. Konowalow and M. L. Olson, *J. Chem. Phys.* 67:590 (1977); *J. Chem. Phys.* 71:450 (1979); D. D. Konowalow, M. E. Rosenkrantz, and D. S. Hochhauser, *J. Mol. Spectr.* 99:321 (1983); D. D. Konowalow and P. S. Julienne, *J. Chem. Phys.* 72:5817 (1980).
58. L. R. Kahn, T. H. Dunning, N. W. Winter, and W. A. Goddard, *J. Chem. Phys.* 66:1135 (1977).
59. H. Partridge, C. W. Bauschlicher, and P. E. M. Siegbahn, *Chem. Phys. Lett.* 97:198 (1983).
60. D. D. Konowalow and J. L. Fish, *Chem. Phys.* 77:483 (1983) ; *Chem. Phys.* 84:463 (1984).
61. H. Partridge, C.W. Bauschlicher, S. P. Walch, and B. Liu, *J. Chem. Phys.* 79:1866 (1983).
62. G. Jeung, *J. Phys. B* 16:4289 (1983); *Phys. Rev. A* 35:26 (1987).
63. E. Reinsch and W. Meyer, *Phys. Rev. A* 14:915 (1976).
64. P. Kusch and M.M. Hessel, *J. Chem. Phys.* 68:2591 (1975); M.E. Kaminsky, *J. Chem. Phys.* 66:4951 (1977); J.B. Atkinson, J. Becker, and W. Demtroder, *Chem. Phys. Lett.* 87:92 (1982); J. Verges, C. Effantin, J. D'Incan, A. Topouzkhaian, and R.F. Barrow, *Chem. Phys. Lett.* 94:1 (1983); K.K. Verma, J.T. Bahns, A.R. Rajaei-Rizi, W.C. Stwalley, and W.T. Zemke, *J. Chem. Phys.* 78:3599 (1983); K.P. Huber and G. Herzberg, "Molecular Spectra and Molecular Structure. IV. Constants of Diatomic Molecules", Van Nostrand Reinhold, New York, (1979).
65. D. K. Hsu, D.L. Monts and R.N. Zare, "Spectral Atlas of Nitrogen Dioxide 5530 to 6480 A", Academic Press, New York (1978).
66. K. Uehara and H. Sasada, "High Resolution Spectral Atlas of Nitrogen Dioxide 559-597 nm", Springer, Berlin (1985).
67. J. L. Hardwick and J.C.D. Brand, *Can. J. Phys.* 54:80 (1976).
68. W. J. Lafferty and R.L. Sams, *J. Molec. Spectroscopy* 66:478 (1977).
69. K. M. Ervin, J. Ho and W.C. Lineberger, *J. Phys. Chem.* 92:5405 (1988).
70. Y. Morino, M. Tanimoto, S. Saito, E. Hirota, R. Awata and T. Tanaka, *J. Mol. Spectroscopy* 98:331 (1983).
71. G. Hirsch and R.J. Buenker, *Can. J. Chem.* 63:1542 (1985).
72. Y. Xie, R.D. Davy, B.F. Yates, C.P. Blahous, Y. Yamaguchi and H.F. Schaefer, *Chem. Phys.* 135:179 (1989).
73. M. E. Jacox, *J. Phys. Chem. Ref. Data* 18:945 (1984).

74. R. Kato and J. Rolfe, *J. Chem. Phys.* 47:1901 (1967).
75. K. E. Gotberg and D.S. Tinti, *Chem. Phys.* 96:109 (1985).
76. R. E. Watson and T.F. Brodasky, *J. Chem. Phys.* 27:683 (1957).
77. E. Andersen and J. Simons, *J. Chem. Phys.* 66:2427 (1977).
78. N. C. Handy, J.D. Goddard and H.F. Schaefer, *J. Chem. Phys.* 71:426 (1979); R.J. Harrison and N.C. Handy, *Chem. Phys. Lett.* 97:410 (1983).
79. J. Baker, R.H. Nobes and L. Radom, *J. Comp. Chem.* 7:349 (1986).
80. S. Huzinaga, *J. Chem. Phys.* 42:1293 (1965); T. H. Dunning, *J. Chem. Phys.* 53:2823 (1970).
81. S. J. Cyvin, "Molecular Vibrations and Mean Square Amplitudes", Elsevier, Amsterdam (1968).
82. E. B. Wilson, J. C. Decius, and P. C. Cross, "Molecular Vibrations", Dover, New York, (1981).
83. T. Amano and T. Amano, quoted in ref. 84.
84. Y. Yamagouchi, Y. Xie, R. S. Grev, and H. F. Schaefer, *J. Chem. Phys.* 92:3683 (1990).
85. A. Langseth, J. R. Nielsen, and J. O. Sorenson, *Z. Phys. Chem.* B 27:100 (1934).
86. B. A. Thrush, *Proc. Roy. Soc (London)* A 235:143 (1956).
87. A. E. Douglas and W.J. Jones, Can. J. Phys. **43**, 2216 (1965).
88. M Polak, M. Gruebele, and R.J. Saykally, *J. Am. Chem. Soc.* 109:2884 (1987); M. Polak, M. Gruebele, G.S. Peng, and R.J. Saykally, *J. Chem. Phys.* 89:110 (1988).
89. R. A. Beaman, T. Nelson, D.S. Richards, and D.W. Setser, *J. Phys. Chem.* 91:6090 (1987).
90. C. R. Brazier, P.F.Bernath, J.B.Burkholder, and C.J. Howard, *J. Chem. Phys.* 89:1762 (1988).
91. R. Tian, J.C. Facelli, and J. Michl, *J. Phys. Chem.* 92:4073 (1988); R. Tian, V. Balaji, and J. Michl, *J. Am. Chem. Soc.* 110:7225 (1988).
92. R. T. Lamoureux and D.A. Dows, *Spectrochim. Acta* A 31:1945 (1975).
93. C. R. Brazier and P.F. Bernath, *J. Chem. Phys.* 88:2112 (1988).
94. E. Illenberger, P.B. Comita, J.I. Brauman, H.P. Fenzlaff, M. Heni, N. Heinrich, W. Koch, and G. Frenking, *Ber. Bunsen-Ges. Phys. Chem.* 89:1026 (1985).
95. P. Botschwina, *J. Chem. Phys.* 85:4591 (1986).
96. T. W. Archibald and J.R. Sabin, *J. Chem. Phys.* 55:1821 (1971).
97. L. Adamowicz, quoted in reference 90.
98. J. Baker, R.H. Nobes, and L. Radom, *J. Comp. Chem.* 7:349 (1986).
99. T. Ishiwara, I. Tanaka, K. Kawaguchi, and E. Hirota, *J. Chem. Phys.* 82:2196 (1985); K. Kawagouchi, E. Hirota, T. Ishiwata, and I. Tanaka, *J. Chem. Phys.* 93:951 (1990).
100. R. R. Friedel and S. P. Sander, *J. Phys. Chem.* 91:2721 (1987).
101. A. Weaver, D. W. Arnold, S. E. Bradforth, and D. M. Neumark, *J. Chem. Phys.* 94:1740 (1991).
102. P. E. M. Siegbahn, *J. Comput. Chem.* 6:182 (1985).
103. R. C. Boehm and L. L. Lohr, *J. Phys. Chem.* 93:3430 (1989).
104. R. D. Davy and H. F. Schaefer, *J. Chem. Phys.* 91:4410 (1989).
105. B. Kim, B. L. Hammond, W. A. Lester, and H. S. Johnston, *Chem. Phys. Lett.* 168:131 (1990).
106. J. F. Stanton, J. Gauss, and R. J. Bartlett, *J. Chem. Phys.*, in press.

THEORY AND COMPUTATION OF NONSTATIONARY STATES OF POLYELECTRONIC ATOMS AND MOLECULES

Cleanthes A. Nicolaides

Theoretical and Physical Chemistry Institute
National Hellenic Research Foundation
48 Vas. Constantinou Ave., Athens 116 35
Greece

ABSTRACT

I present a theory of polyelectronic atomic and molecular nonstationary states decaying into free particle continua, which shows how to define and compute efficiently correlated wavefunctions, energies, energy shifts and transition rates for phenomena such as autoionization, multiphoton ionization, static and dynamic polarization, predissociation etc. The problem is formulated in a unified manner as a complex eigenvalue Schrodinger equation (CESE) which is derived rigorously for each state of interest starting from Fano's basic equation of discrete -continuous spectra mixing. The form of the resulting resonance wavefunction is $\Psi = a\Psi_o + bX_{as}$, where Ψ_o is the maximum square-integrable projection of Ψ, excluding the open channels. Knowledge of the exact asymptotic form of the resonance wavefunctions thus derived, allows the imposition of suitable perturbations of the asymptotic boundary conditions in the form of coordinate transformations, in order to render the resonance wavefunction normalizable. Thus, for short-range or Coulomb potentials, the corresponding coordinate transformation is $r \rightarrow \rho = re^{i\theta}$, first introduced for short-range potentials by Dykhne and Chaplik in 1961. For the LoSurdo-Stark potential, the herein derived resonance asymptotic form and subsequent analysis justifies previous choices of the coordinate rotation and translation transformations as well as of

Applied Many-Body Methods in Spectroscopy and Electronic Structure
Edited by D. Mukherjee, Plenum Press, New York, 1992

their combination $r \rightarrow \rho = re^{i\theta} - z_0/F$, where z_0 is the complex eigenvalue and F is the field strength. Now, Ψ_0 and X_{as} are represented by different function spaces which are optimized separately, the first on the real axis once, yielding a real energy, E_0, and the second partly on the real axis for its bound part and partly in the complex energy plane. The second part of the computation depends only on the structure of the continuous spectrum and yields the energy shift, Δ, and the width, Γ. For Coulomb autoionization where the interaction operator is nonseparable, the state-specific calculation of Ψ_0 is based on the existence of a localized, self-consistent multiconfigurational zeroth order function satisfying the virial theorem and the physically correct orbital nodal structure. The remaining localized electron correlation is obtained variationally. Both the zeroth order and the localized correlation part exclude the open channels by construction and by satisfying orthogonality constraints to electronic structure dependent core orbitals. It is shown how this scheme can be used for the construction of diabatic or quasidiabatic states. X_{as} incorporates the interchannel couplings and a simple expression yields the partial widths to all orders. When the nonstationary state is caused by an external ac-field, the present theory is adapted to solving the ensuing many-electron, many-photon (MEMP) problem as a time-independent CESE where the computed multiphoton ionization rates and frequency-dependent hyperpolarizabilities constitute the Floquet averages over the field cycle. The following, field-free, examples are discussed: The inner-hole Auger state Ne^+ $1s2s^22p^6\ ^2S$; the triply excited He^- $2s2p^2\ ^2D$ and ^4P resonances; the He_2^+ $^2\Sigma_g^+$ diabatic spectrum. The formulation of the MEMP theory and its applications are relegated to the references.

I. PREFACE

The scope of this article is to present the essentials of a self-consistent theory of nonstationary states of atoms and molecules, whose orientation is computational, i.e. how to obtain from first principles reliable numbers for real polyelectronic systems.

In the following sections only the significant theoretical steps which explain and justify the present theory are presented. In particular, it is shown how advanced electronic structure theory and methods suitable for excited states can be integrated in a practical way into selected elements of the rigorous theory of the continuous spectrum in order to compute properties such as positions and widths of multiply excited states, multiphoton ionization rates, multichannel predissociation lifetimes etc.

In order to keep the article short, certain aspects of the computational methodology, such as the solution of the Floquet matrices in the many-electron, many-photon (MEMP) problem, and most of the applications which have been made thus far are left out .

II. THE CONCEPT OF THE NONSTATIONARY STATE: THE TIME-DEPENDENCE OF A NONSTATIONARY STATE IS DRIVEN BY A COMPLEX POLE OF THE RESOLVENT OF THE TOTAL HAMILTONIAN

In the continuous spectrum of multiparticle Hamiltonians, there exist nonstationary states whose temporary formation and decay have a direct effect on a variety of spectroscopies. These range from resonances in nuclear reactions, to autoionizing states in atoms, to predissociating states in molecules. To these, we add the nonstationary states which are created when a free atomic (molecular) state is "dressed" with an external dc- or ac-field which ionizes or breaks the molecule or both.

The basic concept underlying these phenomena is the existence of an initially (t=0) localized wavefunction which mixes via the interactions present in the total Hamiltonian with the adjacent continuum of scattering states and decays into it. In the case of an isolated atom (molecule), the continuous spectrum which results from the free motion of an electron or a nucleus is an intrinsic characteristic of the Hamiltonian with all the interparticle interactions present. In the case of a "dressed" nonstationary state, the continuous spectrum becomes accessible because of the interaction with the external field (E.g. photoionization due to an ac-field or tunneling due to a dc-field).

The word nonstationary implies a time evolution which is different than the one characterizing the stationary state. In general, if the quantum system is found at t=0 in a state $|\Psi_0(t=0)>$, then its evolution is given by :

$$|\Psi(t)> = e^{-\frac{i}{\hbar}Ht}|\Psi_0(t=0)> \qquad (1)$$

If the state at t=0 is stationary, $|\Psi(t)|^2$ is time independent and Ψ_0 satisfies the standard Schrödinger equation :

$$H|\Psi_0(t=0)> \;=\; E_0|\Psi_0(t=0)> \qquad (2)$$

H is Hermitian , E_0 is real

On the other hand, if Ψ_0 is nonstationary, eq.2 is not satisfied, i.e. $\Psi_0(t=0)$ is not an eigenstate of H. In this case, the time evolution of $\Psi(t)$ is determined by [1,2]

$$|\Psi(t)> = \frac{1}{2\pi i}\oint R(z)e^{-\frac{i}{\hbar}zt}dz|\Psi_0> \qquad R(z) \equiv \frac{1}{z-H} \qquad (3)$$

Nonstationary state

where z is a complex variable and $\oint$ is an appropriate contour over the spectrum of H.

In order for eq.3 to be integrated exactly for an arbitrary spectrum, the mathematical properties of the resolvent R(z) must be known. For an isolated nonstationary state, the physically meaningful result in lowest order is [1-3]

$$R(z) \approx \frac{1}{z - E_o - A} \tag{4}$$

where E_o is the energy of Ψ_o (t=0) and A is a small complex number compared to E_o, which is obtained to a very good approximation by computing the "self-energy" function A(z) at E_o, i.e.

$$A(z) \approx A(E_o) = \Delta(E_o) - \tfrac{i}{2}\Gamma(E_o) \tag{5}$$

Eqs. 3-5 show how, by assuming the existence of the localized Ψ_o, the time-dependent treatment leads to the result that the observable intrinsic properties, energy and lifetime, are determined by a complex pole of R(z) on the second Riemann sheet just below the real energy axis, at

$$z_o = E - \tfrac{i}{2}\Gamma \approx E_o + A(E_o) \tag{6}$$

When eq.4 is used in eq.3 and the contour is taken from $-\infty$ to ∞, the result is

$$\Psi(t) = e^{-\frac{i}{\hbar}z_o t}\Psi_o, \qquad |\Psi(t)|^2 = e^{-\frac{\Gamma}{\hbar}t} \tag{7}$$

the well-known exponential decay, for which the lifetime $\tau = \frac{\hbar}{\Gamma}$

The approximate results (4) and (7) are expected to be reliable for a large number of nonstationary states, especially when their widths are narrow relative to E_o. The same holds for "dressed" nonstationary states when the field is not very strong (The strength of the field is relative to the position and character of the state under consideration). However, for nonstationary states caused by extraordinary, effective real or model potentials, eqs. 4-7 need not be valid with high accuracy.

III. COMPLEX EIGENVALUE SCHRODINGER EQUATION (CESE)

A. The Gamow Complex Energy as an Imposed Requirement on the Asymptotic Behavior of the Resonance Wavefunction

Nonstationary states manifest themselves in spectroscopy as resonances i.e. peaks over a backround of a smooth variation of the energy-dependent probability (cross-section) of a particular reaction.

The original advances in the formalism of resonances occured in connection with the theory of nuclear reactions and of matter-radiation interaction in the period 1928-1960. Reviews and references can be found in the books of Heitler [1], of Goldberger and Watson [2], of Newton [4] and of Mahaux and Weidenmüller [5]. Concepts and methods such as complex energies, complex poles of the S-matrix, time delay, optical potential, coupled scattering equations, R-matrix, etc, emerged from this extensive research and were later carried over to atomic and molecular physics and chemical reactive scattering.

In many cases, these theories aim at providing information on- and off-resonance. On the other hand, in most of the spectroscopic situations which are physically interesting, it is desirable to obtain just the intrinsic resonance properties, represented by z_0 of eq. 6. This problem can be formulated as a time-independent, *complex eigenvalue Schrödinger equation (CESE).*

The original introduction of a complex energy for the characterization of a nonstationary state is due to Gamow [6]. He studied the problem of α -particle decay in terms of a simple finite barrier model and imposed on the resulting free-particle solution a complex energy in order to obtain the physics depicted by eq.7. In other words, using a short-range potential, the simplest Schrodinger equation yields for the case of no incoming wave the asymptotic solution

$$\Phi^{res} \underset{r\to\infty}{\sim} b(u_0)e^{i\sqrt{2u_0}r} \tag{8}$$

where u_0 is the energy of the emitted particle. If u_0 is real, $\Phi^{res}(r,t)$ cannot represent a decaying state. Thus, Gamow [6] required that u_0 be complex, in which case the physics of decay is satisfied.

A few years later, Siegert [7] showed that eq. (8) with u_0 complex corresponds to a pole of the S-matrix representing a resonance. However, neither Gamow's ad hoc treatment nor Siegert's rigorous one yield the explicit form of the coefficient $b(u_0)$. This is done in the next section in a way which allows the derivation of the asymptotic form for resonances of potentials which are not short-range, such as the Coulomb or the linear one induced by an external electric field.

B. Exact Asymptotic Form of the Resonance Function

Given the shell structure of atoms and molecules and the conceptual simplicity of the notion of configuration-interaction (CI) , Fano's CI in the continuum (CIC) formalism of the early 60's [8,9] constitutes a rigorous, yet easily visualized treatment of multichannel scattering phenomena on- and off- resonance. This formalism examines the physics in terms of wave-functions and reaction (K-) matrices of real energies. These are defined or constructed from the mixing of one or more discrete solutions of zeroth order Hamiltonians with their orthogonal prediagonalized scattering solutions belonging to the continuous spectra into which the discrete eigenvalues are embedded [10].

The phenomenology of mixing of discrete states with the continuous spectrum allows the explicit derivation of the asymptotic form of the resonance function for each particular problem of interest and leads naturally to the CESE satisfied by nonstationary states. Below I repeat the derivation [11] and apply it to two cases. One is that which corresponds to the short- range potential situation treated by the nuclear physicists, i.e. the case of "compound" states of negative ions, resulting in resonances on the cross-section of electron-atom scattering. The other is that of the LoSurdo-Stark effect [12].

The two examples demonstrate that by applying this general method it is possible to obtain explicit asymptotic forms for a variety of resonance wavefunctions, if the pure scattering basis functions corresponding to the problem of interest are known analytically, possibly including JWKB solutions. In turn, this result can be utilized in calculational studies of state-specific, optimized square-integrable basis sets for the representation of the asymptotic part of the resonance function particular to each nonstationary process .

In the energy dependent picture, the scattering function in the vicinity of an isolated decaying state is expressed as a linear combination of a bound function, Ψ_o, representing a discrete state of a zeroth order Hamiltonian H_o, and of an orthogonal complete set of scattering functions of H_o, $U_n(E)$, spanning the continua, n, into which Ψ_o is embedded. If the interactions causing the decay are internal to the system (e.g. Coulomb autoionization) then H_o should be taken as an effective operator whose utility , at this stage, is formal. The same holds for the corresponding effective perturbation operator, V [13].

According to Fano [8], the total scattering wavefunction, Ψ (E) , satisfying the Schrödinger equation on the real axis and expressing the above situation, is written as

$$\Psi(r;E) = a(E)\Psi_o(r) + \int dE' b_{E'}(E) U(r;E') \tag{9a}$$

$$= a(E)\left[\Psi_o + P.V. \int dE' \frac{V_{OE'}}{E-E'} U(r;E') + \lambda(E) V_{OE} U(r;E)\right] \tag{9b}$$

where

$$V_{oE} \equiv \langle \Psi_o | H | U(r;E) \rangle \tag{10}$$

$$E = E_o + P.V. \int dE' \frac{|V_{oE'}|^2}{E - E'} + \lambda(E)|V_{oE}|^2 \tag{11}$$

$\lambda(E)$ is fixed by the asymptotic boundary conditions [15].

Eq. 9 contains all the information on- and off-resonance. $\Psi(E)$ being a scattering state, E is a real parameter and not an eigenvalue. Is there a time-independent eigenvalue equation that is satisfied on resonance? In order to answer this question we must impose the boundary conditions which a nonstationary state must satisfy. Thus, we start from eq. 9b and seek to determine its asymptotic form pertaining to each nonstationary situation under study. To achieve this, we must know the corresponding asymptotic expression for the basis functions U(r;E). For the case of the atomic negative ion "compound" resonance, where the potential is of short-range,

$$U(r,k) \underset{r\to\infty}{\sim} \sqrt{\frac{2}{\pi k}} \sin(kr + \delta) \tag{12}$$

$$k^2 = 2E \tag{13}$$

In the case of the LoSurdo-Stark problem, the potential is -Fx and the corresponding differential equation can be written as

$$\frac{d^2U}{d\xi^2} + \xi U = 0 \tag{14}$$

where

$$\xi = \left(x + \frac{E}{F}\right)(2F)^{1/3} \tag{15}$$

F is the field strength

The well-known solution of eq.14 is the Airy function, whose asymptotic form is

$$U(k) \underset{\xi\to\infty}{\sim} \sqrt{\frac{2}{\pi k}} \sin\left(\frac{k^3}{3F} + \frac{\pi}{4} + \delta\right) \tag{16}$$

with

$$k^2 = 2(E + Fx) = (2F)^{2/3}\xi \tag{17}$$

Let $N \equiv kr + \delta$, $\qquad L \equiv \frac{k^3}{3F} + \frac{\pi}{4} + \delta \qquad (18)$

Substituting eq. 12 or eq. 16 in 9b we obtain [11]

$$\Psi(r) \underset{r\to\infty}{\sim} -\pi V a \sqrt{\frac{2}{\pi k}} \left\{ \frac{1}{2}\left(1 - \frac{\lambda(E)}{i\pi}\right) e^{iN} + \frac{1}{2}\left(1 + \frac{\lambda(E)}{i\pi}\right) e^{-iN} \right\} \qquad (19)$$

for the U(r) of eq. 12, and the same expression for the U(r) of eq.16, where N is replaced by L.

On resonance, the coefficient of the incoming wave must be zero, so that, for both problems,

$$\lambda(E) = -i\pi \qquad \text{on resonance} \qquad (20)$$

Therefore, the energy of eq. 11 becomes complex

$$E \to z_o = E_o + \text{P.V.} \int dE' \frac{|V_{oE'}|^2}{E - E'} - i\pi |V_{oE}|^2 \qquad (21a)$$

$$= E_o + \Delta - \frac{i}{2}\Gamma \qquad (21b)$$

In lowest order, $\Delta \approx \Delta(E_o)$ and $\Gamma \approx \Gamma(E_o)$, so that eq. 21 is the same as eqs. 5 and 6. In general, the extent of energy dependence of Δ and Γ depends on the characteristics of the nonstationary state, on the strength of the effective perturbation and on the functions used in actual computations to represent the localized and the asymptotic components.

According to eqs. 13 and 17, the momenta also become complex

$$k^2 = 2z_o \qquad \text{on resonance} \qquad (13)'$$

$$k^2 = 2(z_o + Fx) \qquad \text{on resonance} \qquad (17)'$$

Given eqs. 19-21, the exact asymptotic form of the resonance functions for the two nonstationary states is obtained

$$\Psi^{res}(r) \underset{r\to\infty}{\sim} -\pi V a \sqrt{\frac{2}{\pi k}} e^{i(kr+\delta)} \qquad (22)$$

Negative ion compound resonance

$$\Psi^{res}(k) \underset{\xi\to\infty}{\sim} -\pi V a \sqrt{\frac{2}{\pi k}}\, e^{i\left(\frac{k^3}{3F}+\frac{\pi}{4}+\delta\right)} \tag{23}$$

LoSurdo-Stark resonance

where the two k are given by eqs. (13)' and (17)' . V and a are evaluated at E_o or in its neighborhood.

Now, if we call the energy-dependent coefficient of the outgoing wave $b(k_o)$, (this is $b(u_o)$ of eq.8), eqs. 22 and 23 show that the flux of emitted particles, represented by $|b|^2$, is given by (ref. 11, p.363, ref.3, p.479)

$$|b|^2 = |a|^2 \frac{\Gamma}{|k|} \tag{24}$$

In conclusion, it follows from the above development that there is a *complex eigenvalue Schrödinger equation*

$$(H - z_o)\Psi = 0 \tag{25}$$

which is satisfied by the eigenfunctions of nonstationary states. The solution Ψ is not square integrable and its explicit asymptotic form depends on the phenomenon it describes . Eqs. 22 and 23 are two such examples.

IV. COMPLEX SCALING FOR THE REGULARIZATION OF RESONANCE WAVEFUNCTIONS WITH ASYMPTOTIC FORMS GIVEN BY EQS. 22 AND 23

The Gamow resonance function 8, (the form 22 has the same coordinate dependence), is not square integrable. This fact has negative consequences as regards the evaluation of matrix elements and the solution of eq.25 by standard methods of quantum mechanics. In order to circumvent the norm problem, Kemble [16, p.195] suggested the insertion in the integrals involving the Gamow Φ^{res} of a "modulating factor", e^{-ar^n}, in order to render them finite. Years later, this recipe was followed by Zeldovich [17] in the simple case of an s-wave resonance from a short-range potential. Soon afterwards, Dykhne and Chaplik [18] showed that the same effect is achieved, i.e. regularization of the resonance function for a short-range potential, if "the integral over the coordinate r is computed along some contour C which lies in the upper half-plane and which makes an angle $\alpha>\Gamma/2E$ with the real axis". The Dykhne-Chaplik proposal marks the beginning of the use of the transformation

$$r \to \rho = r e^{i\theta} \tag{26}$$

for the normalization of resonance functions of the Gamow form [19]. This property allows the formal transformation of the CESE, eq. 25, with the asymptotic boundary condition 8 or 22, into the equivalent

$$(H(\rho) - z_0)\Psi = 0 \tag{27}$$

where Ψ, which by construction corresponds to a particular Ψ_0 (eqs. 3,9), can be a function of real or complex coordinates with the requirement that it is square integrable.

The use of complex scaling (26) as a normalization technique is rigorously applicable to resonance functions of the asymptotic form 8 (or 22). Because of the way that (22) is derived, it is clear that a similar form holds for the Coulomb potential [11]. Thus, the transformation (26) and the ensuing CESE (27) are applicable in the Coulomb case as well.

It can also be shown [28] that the rotation (26) as well as the translation $x \rightarrow x+iq$ are the transformations which render the resonance function of eq.23 square integrable. Furthermore, the combined form below is also regularizing

$$x \rightarrow xe^{i\theta} - \frac{z_0}{F} \tag{28}$$

$$0 < \theta < \frac{2\pi}{3}$$

Both rotation and translation transformations have been used in previous studies of the LoSurdo-Stark problem using complex scaling as a means of studying the spectrum [24]. First came the numerical results of Reinhardt and coworkers [24] that the rotation yields stabilized results although the unbounded linear potential is not dilatationally analytic. Elaborate mathematical analysis by Herbst and coworkers (see the papers quoted in [24]) justified a posteriori the validity of these transformations. The present theory shows how these transformations emerge from a different, and physically transparent point of view. Application of the present theory (see below) has been made [28] to a model potential of a shape resonance in a dc-field where it is demonstrated that a "field-dressed" Ψ_0 (see eq.29) secures very good convergence with small expansions for the asymptotic X_{as} (eqs.29,36).

V. STATE-SPECIFIC THEORY FOR THE AB INITIO COMPUTATION OF NONSTATIONARY STATES

The results of sections II-IV show how the subject of computing from first principles the energy and the lifetime of a nonstationary state can be formulated in terms of a CESE, (eq. 27), with a square integrable solution and a nonHermitian Hamiltonian, both obtained as a result of a coordinate transformation appropriate for the problem under study. The steps leading to eq.27 are rigorous and the approach offers an alternative picture to the one presented by the mathematics of the CCR work [21].

In this section I present the essentials of the theory for solving eq. 27 in the case of polyelectronic atoms and molecules in the absence of an ac-field. (The CESE many-electron, many-photon (MEMP) theory for dressed nonstationary states uses the same concepts but more complicated computational methods. In order to keep the article short, its formulation is not presented here and the reader is referred to [35-40]). In general, applications of the theory have been made to many-electron systems for phenomena such as

- Multiply and inner hole excited states [29,3,11,30-34]
- Multichannel autoionization [11,30-34]
- Multiphoton ionization and ATI without and with a DC-field [35-38]
- Nonlinear static and dynamic polarization [39,40]
- DC-field-induced tunneling [36,37]
- Single channel and multichannel predissociation [27,41]

Given that Ψ of eq. 27 is square-integrable, solution of eq. 27 for a particular nonstationary state means that, in principle, one is able to diagonalize to a high degree of accuracy the corresponding nonHermitian Hamiltonian matrix, $\underline{\underline{E}}(\theta)$, with or without the presence of an external ac- or dc-field. The magnitude of the conceptual, formal and computational complexity of such a desideratum can be appreciated immediately if one considers the amount of work-useful or not- that has been done over the decades on the much simpler many-electron problem of diagonalizing to a good approximation the Hermitian energy matrix of an atomic or molecular ground state.

Even though the standard CCR calculations have constructed and solved $\underline{\underline{E}}(\theta)$ by brute-force diagonalization with and without an external field, [21-24], this is neither efficient nor possible for arbitrary N-electron systems [42]. For it is clear that the standard CCR approaches face a computational bottleneck by the required size of the N-electron function spaces. As was pointed out in ref.3, p. 506, "an arbitrarily chosen set of square-integrable functions may be a good representation on the real axis, but in the θ-plane they need not". Furthermore, the physically significant partial widths cannot be obtained in this way.

The theory and methods developed in this institute since 1977 reduce significantly this complexity to the level where small yet reliable calculations can be done, taking into account the specific electronic structure characteristics of each nonstationary state under examination. The fundamental steps in simplifying the solution of eq.27 involve:

i) The choice of a form of the trial wavefunction which is in accordance with the physics of the particular nonstationary state, without or with the presence of an external field.

ii) The state-specific choice and efficient computation of the relevant one-particle (electron or nuclear) and multielectron function spaces, which are appropriate for the phenomenon under investigation. (E.g., if the laser field intensity for the multiphoton ionization of a negative ion is strong, the discrete representation of the continuous spectrum must include a much higher number of angular momenta with accurate radial representation than the one dictated by the lowest-order perurbation theory applicable to the weak-field case [35-38]).

iii) The ensuing invariance properties of the Hamiltonian diagonal and off-diagonal matrix elements under coordinate complex scaling and the concomitant drastic reduction of the size of $\tilde{\tilde{E}}(\theta)$.

The related implications allow the separation of the overall calculation into two parts. In the first, state-specific calculations are done on the real axis, yielding Ψ_0 and E_0. In the second part, the calculations are done in the complex energy plane, using different optimization procedures in conjunction with a different one-particle (electron or nuclear) and multielectron function space containing the information -through its interaction with Ψ_0 - on the energy shift, Δ, and width Γ. Regardless of the size of the system, given that Ψ_0 is already computed for a given electronic or vibrational state, the magnitude of the second part of the overall computation is small, even for the multichannel case.

More specifically. The very derivation of the CESE, eq.25 , shows that the resonance N-particle wavefunction should have the form

$$\Psi = a\Psi_o + bX_{as} \tag{29}$$

Being a bound function, Ψ_0 satisfies for all particle coordinates

$$\Psi_o(0) = 0, \qquad \lim \Psi_o(r)_{r \to \infty} = 0 \tag{30}$$

On the other hand, for real coordinates, X_{as} behaves asymptotically as an outgoing wave of one coordinate, call it r_N, with forms such as those of eqs. 22 and 23 . Considering for simplicity electronic wavefunctions only, I write

$$X_{as} = A[X_b(r_1, \ldots r_{N-1}) \otimes g_N(r_N)] \tag{31}$$

where A is the antisymmetrizer, X_b is the bound core function of (N-1) electrons and g_N is the Gamow orbital satisfying

$$g_N(0) = 0, \qquad g_N(r_N) \underset{r \to \infty}{\approx} \text{outgoing wave (eqs. 22 and 23)} \tag{32}$$

Now, consider a perturbation of the asymptotic boundary conditions which is a function of the coordinates, f(r), and which renders Ψ square integrable [3, p.481-483]. Such a transformation, $\rho \equiv f(r)$, changes the Hamiltonian into $H(\rho)$ and the multiparticle Ψ into $\Psi(\rho)$, for all coordinates. Let $\rho = re^{i\theta}$. Eq. 29 becomes

$$\Psi(\rho) = a(\theta)\Psi_0(\rho) + b(\theta)X_{as}(\rho), \qquad a^2 + b^2 = 1 \tag{33}$$

The corresponding nonHermitian energy matrix, $\tilde{\tilde{E}}(\theta)$, contains the matrix elements $<\Psi_o(\rho), H(\rho), \Psi_o(\rho)>, <\Psi_o(\rho), H(\rho), X_{as}(\rho)>, <X_{as}(\rho), H(\rho), X_{as}(\rho)>$.

The corresponding nonHermitian energy matrix, $\underset{\approx}{E}(\theta)$, contains the matrix elements $<\Psi_o(\rho),H(\rho),\Psi_o(\rho)>,<\Psi_o(\rho),H(\rho),X_{as}(\rho)>,<X_{as}(\rho),H(\rho),X_{as}(\rho)>$. Application of mathematical analysis applicable to integrals of analytic functions [26] as well as explicit demonstration on solvable bound state systems such as the hydrogen atom and the harmonic oscillator, and on the usual matrix elements with STOs and GTOs, led to the conclusion [3,26] that for the bound function Ψ_o the following invariance property holds

$$<\Psi_o(\rho),H(\rho),\Psi_o(\rho)> = <\Psi_o(r),H(r),\Psi_o(r)> = E_o \tag{34}$$

In order to appreciate the role that the back-rotated eq. 34 plays in the present theory, consider the ground state of the hydrogen atom, |1s>. The standard CCR theory and computation [21-24] says that if $\underset{\approx}{E}(\theta)$ is constructed from the rotated hydrogenic Hamiltonian and a complete basis set, e.g. Sturmians, then its diagonalization will yield the exact, real E_{1s}. How good is the convergence in actual such calculations? Nicolaides and Beck [3] diagonalized a 10x10 $\underset{\approx}{E}(\theta)$ over hydrogenic functions and found that only complex eigenvalues appeared, i.e. unphysical solutions. On the other hand, since for hydrogen <1s(ρ), H(ρ), 1s(ρ)> $\equiv E_{1s}$, it follows that only one appropriate state-specific function, 1s(ρ), does the job of a complete set of functions of real coordinates.

This example shows that rather than thinking in terms of constructing the nonHermitian matrix, $\underset{\approx}{E}(\theta)$, from a large basis set with real or complex functions and search for the proper complex eigenvalue directly, it is much more efficient, physically meaningful and suitable for polyelectronic multichannel nonstationary states, to first construct the projection of Ψ onto the localized part Ψ_o (ref.29, eqs. 1-4), in terms of which the expectation value of H on the real axis is E_o, and then, knowing E_o, to obtain the correction to it, $\Delta - \frac{i}{2}\Gamma$,using a different function space.

Specifically, eqs. 33 and 34 imply that $\underset{\approx}{E}(\theta)$ can be written as (field-free case for simplicity)

$$\underset{\approx}{E}(\theta) = \begin{bmatrix} E_o & <\Psi_o|H|X_{as}> \\ <X_{as}|H|\Psi_o> & <X_{as}|H|X_{as}> \end{bmatrix} \tag{35}$$

whose size depends only on the size of the expansions of the Gamow orbitals $g_N(\rho_N)$ for each channel, since both Ψ_o and $X_b(N\text{-}1)$ and their expectation values are fixed from previous calculations on the real axis. The actual computation of the matrix elements $<\Psi_o(\rho), H(\rho), X_{as}(\rho)>$ and $<X_{as}(\rho), H(\rho), X_{as}(\rho)>$ can be carried out over complex coordinates or over real and complex coordi-

boundary conditions of eq.28 -first employed in a general basis set expansion by Bardsley and Junker [20]- has been used in conjunction with the consistency criteria associated with the satisfaction of the virial theorem and of eq. 24 [3,11]. An alternative choice, which is convenient for use with numerical zeroth order bound orbitals belonging to Ψ_0 or to $X_b(N-1)$ or with calculations involving the vibrational continuum [41], is to expand the square-integrable $g_N(\rho_N)$ for each channel i in terms of Sturmians, of STOs or of GTOs of real coordinates. Thus,

$$g^i(\rho) = \sum_n a_n \phi^i_n(r) \tag{36}$$

where the g^i are coupled to the symmetry adapted core $X_b(N-1)$ to form angular momentum states. When the back-rotation (eq. 34) is applied to the two matrix elements, for the bound functions Ψ_0 and X_b as well as for the Hamiltonian, ρ is replaced by r while for the basis set of eq. 36 r is replaced by $\rho^*=re^{-i\theta}$.

How is the optimization of the various g^i achieved when the basis set expansion, eq. 36, is used and the Φ_n are made functions of $\rho^*=re^{-i\theta}$? Two criteria have proven sufficient and practical. First, because of the overall construction of the trial Ψ, the corresponding solution has to be closest to the pair (Ψ_0, E_0) with

$$\tilde{z}_0 = E_0 + \Delta - \frac{i}{2}\Gamma \tag{37}$$

$$< \tilde{\Psi}, \Psi_0 > \approx \max \tag{38}$$

Second, given that z_0 is an eigenvalue, it is independent of θ, since θ is just the parameter which renders the resonance wavefunction square integrable via transformations such as 26 or 28. For a trial Ψ, this independence is not perfect. Therefore, in a calculation which follows the present theory, where E_0 is already very close to $\mathrm{Re}z_0$, the g^i can be optimized as to the size of expansion (36) or the parameter in Φ_n, by searching for that range of values of θ where, after the diagonalization of the small matrix $E(\theta)$ $(\mathrm{Re}z_0 - E_0)$ and $\mathrm{Im}z_0$ are stable. The range of θ values for which stability is observed depends, of course, on the choice of g^i while it varies from system to system. For example, for dressed non-stationary states it is a function of frequency and field strength [35,37].

It should finally be added that since this approach allows the efficient analysis of the localized electron correlation in Ψ_0 (see next section), further simplification in the calculation of $E(\theta)$ is possible -if desired- by reducing the size of the computation of the off-diagonal matrix element $<\Psi_0|H|X_{as}>$ without seriously compromizing accuracy. This is possible since, having obtained E_0 accurately, the expansion of Ψ_0 can be truncated, keeping only the terms with coefficients larger than a reasonable threshold value and neglecting the large number of correlation vectors with smaller coefficients. A case in point is the calculation of the energy and partial and total widths of the 1s-hole Ne^+ ^2S Auger state [30]. This is a 9-electron system which decays into a 5-channel electronic continuum

correlation vectors with smaller coefficients. A case in point is the calculation of the energy and partial and total widths of the 1s-hole Ne^+ ^2S Auger state [30]. This is a 9-electron system which decays into a 5-channel electronic continuum and involves all electrons. The standard CCR prescription [21-24], (diagonalize $H(re^{i\theta})$ in a 9-electron basis set), is obviously useless. On the contrary, according to the state-specific CESE theory the problem is broken down into a series of electronic structure- dependent steps and accurate solutions have been obtained [30]. In particular, first Ψ_o was obtained on the real axis as (symbolic expression)

$$\Psi_o \approx \Phi_{HF} + \Phi_{HF}^{-1}\sigma(r_1) + \Phi_{HF}^{-2}\pi_{loc}(r_1, r_2) \tag{39}$$

where Φ_{HF} is the numerical Hartree-Fock function for the Ne^+ $1s2s^22p^6$ ^2S state, $\sigma(r_1)$ signifies the single orbital excitations and $\pi_{loc}(r_1,r_2)$ the symmetry-adapted localized pair correlation functions. Given the smallness of the coefficients of all the correlation vectors, in the calculation of $E(\theta)$ Ψ_o was replaced by Φ_{HF} while for the final ionic bound states only the most important correlations were included [30]. The result for Γ was 0.844 eV with experiment being 0.99 ± 0.1 eV [30].When the energy shift $\Delta = -0.09$ eV was added to E_o=870.4 eV, (including relativistic corrections), the agreement with experiment was perfect, suggesting a not surprising cancellation of the higher order correlation effects.

As regards the partial widths to the five channels, these were obtained according to the formulae of the next subsection. For the two strongest transitions, the 1s-$2p^2$ ^1D and 1s-2s2p $^1P^o$, we obtained $\gamma(^1D)$= 0.560 eV, $\gamma(^1P^o)$=0.154 eV while the experimental values are $\gamma_{exp}(^1D)=0.604 \pm 0.06$ eV and $\gamma_{exp}(^1P^o)=0.174 \pm 0.017$ eV [30].

Partial Widths to All-Orders

In the case of multichannel decay, X_{as} can be written as

$$X_{as} = \sum_i X^i_{as} \equiv \sum_i A\left[X^i_b(N-1) \otimes g^i_N\right] \tag{40}$$

where i signifies each available open channel and the type of coordinates used for the calculation of the matrix elements follows from the previous discussion. The bound X^i_{as} are computed by a convenient bound state method. Using each X^i_{as} separately, we can optimize the corresponding g^i_N from the solution of their $E^i(\theta)$. We have called this approach the Independent Asymptotic Pair Approximation (IAPA) [11, 30]. After the g^i_N have been obtained, we construct the total $E(\theta)$, where the open channels are mixed in the presence of Ψ_o, or of many Ψ_os if we generalize the theory to include many interacting nonstationary states . The solution after diagonalization is given by

$$Q = \Psi_o > < \Psi_o \quad (45)$$

then the projection of H onto Q, QHQ, indeed satisfies

$$QHQ|\Psi_o> = E_o|\Psi_o> \quad (46)$$

but eq. 46 can be thought of only as a self-consistent equation subject to the constraints of eq. 44. The combination of 44 with 46 expresses a concept of dynamic localization, where the major role is played by the self-consistent choice and optimization of the function space representing Ψ_o.

Of course, conceptual simplifications are allowed without significant loss of accuracy. Thus, in the case of a dressed nonstationary state, QHQ is written explicitly as the exact field-free atomic or molecular Hamiltonian and Ψ_o is the exact square integrable solution for the state of interest [43]. A similar convenient assumption can be made in certain cases where the decay is due to relativistic operators or to nuclear motion coupling in the Born-Oppenheimer approximation. In both situations QHQ may be taken to correspond to an exact nonrelativistic Hamiltonian form. On the other hand, in the problems of Coulomb autoionization or of construction of molecular diabatic states [44, and section VII] such a correspondence is impossible since the coupling operator cannot be separated explicitly.

The MCHF Zeroth Order Approximation to Ψ_o

How then should Ψ_o be computed including the effects of electron correlation? The theory of ref. 29, unlike other approaches to the calculation of resonances, advocated the advantages of starting the calculation of Ψ_o by first obtaining a state-specific, self-consistent, single- or multiconfigurational wavefunction. (For details and applications see refs. 29, 3,11 and 46-49).

As is well-known, in ground state many-body theory the Hartree-Fock zeroth order approximation, (and MCSCF in recent years), has constituted the standard starting point for a systematic incorporation of correlation effects. However, unlike the ground state wavefunction where convergence to the absolute energy minimum is guaranteed as the multiparameter space is increased, in the case of the Ψ_o of an effective Hamiltonian QHQ the search is for a minimum embedded in the continuous spectrum, whose existence can be determined only if the physically correct input of functions and careful execution of the computation subject to eqs. 30 and 44 is achieved. In this respect, the MCHF zeroth order wavefunction is the optimal choice for highly excited nonstationary states since it is in harmony with the criterion of dynamic localization: Firstly, it satisfies the *virial theorem* which is a fundamental feature of localized systems. Secondly, it can automatically take into account square-integrability boundary conditions by using -at least for atoms and perhaps diatomics- the orbital occupancy and the number of orbital nodes of the zeroth order configuration(s) as a constraint. Both of the above allow the interpretation that *a proper N-electron*

$$\Psi = a\Psi_o + \sum_i b_i X^i_{as} \qquad (41)$$

$$a^2 + \sum_i b_i^2 = 1 \qquad (41a)$$

so that the partial energy shifts and widths, with interchannel coupling to all orders, are given by

$$z^i_o - E_o = \frac{b_i}{a} < \Psi_o, H, X^i_{as} > = \delta_i - \frac{i}{2}\gamma_i \qquad (42)$$

and

$$\sum_i (\delta_i - \frac{i}{2}\gamma_i) = \Delta - \frac{i}{2}\Gamma \qquad (43)$$

VI. DYNAMIC LOCALIZATION AS A FUNDAMENTAL CONCEPT IN THE COMPUTATION OF Ψ_0

When dealing with model potentials supporting nonstationary states, it is possible to carry out experimental computations (e.g. vary parameters of interaction strength or of box size) in order to see some of the wavefunction characteristics. However, for real, polyelectronic atoms and molecules this luxury is prohibitive. In this respect, the practical definition and computation of Ψ_0 of eqs. 25, 27 is of paramount importance. This was emphasized in [29, 3, 11] where, in the study of the conceptually most difficult case, that of Coulomb autoionization, it was argued that an a priori defined systematic approach must be based on general guiding principles such as orbital nodal behaviour of zeroth order functions as an index of boundary conditions, the satisfaction of the virial theorem on the real axis as a criterion of localization, distinction between localized correlations contributing to the stability of the nonstationary state and asymptotic correlations contributing to its decay, use of electronic structure dependent one-electron orthogonality conditions in variational calculations, etc. More specifically: (see ref. 29, eqs. 1-4, 31 and ref. 3 sections 2 and 5).

We seek a Ψ_0 which represents in an optimal way the part of the exact resonance function that, through a minimal perturbation of the asymptotic boundary condition characterizing eq.21, is square integrable. Such a Ψ_0 would correspond to a projection of Ψ satisfying

$$< \Psi, \Psi_o > = \text{max but finite} \qquad (44)$$

Since the existence of a nonstationary state is the result of the interactions present in the total Hamiltonian, it is impossible to write explicitly the zeroth order Hamiltonian of which Ψ_o is an eigenfunction [13]. Thus, if we define

bound solution of the MCHF equations in the continuous spectrum implies the existence of a localized state in the neighborhood of the MCHF energy and, therefore, if symmetry allows interaction with the continuum, of a resonance.

Furthermore, as regards the fidelity of the results obtained at the MCHF level, by proper choice of configurations they incorporate the most significant radial and angular correlation effects which are usually very important in multiply excited states [46-48]. Finally, if the computation is carried out numerically (atoms and diatomics), the radial details of the major features of the resonance function (which are, most often, so extended that commonly used analytic basis sets, including r_{ij}-dependent functions, are not necessarily advantageous) are computed with high accuracy [40, 46-50].

Given the above, for each nonstationary state of interest, Ψ_o is written formally as

$$\Psi_o = \Phi_o + X_{loc} \tag{47}$$

where Φ_o is the bound HF or MCHF function and X_{loc} is the remaining correction, i.e. the localized correlation contributing to the stability of the nonstationary state and not to its decay. X_{loc} is developed in terms of one, two, three etc. symmetry adapted correlation functions, each expanded in terms of its own virtual orbitals which are optimized variationally, first separately and then in a total CI [49,50] (see eq.39).

By construction and by imposed orthogonality to state-dependent core orbitals, the aim is for both Φ_o and X_{loc} to contain only those configurations and correlation vectors contributing to the stability of E_o and to exclude the possibility of following a different root, whether physical or unphysical [29,3,44-48]. Since Φ_o and X_{loc} are obtained from equations which are variational and based on the total Hamiltonian, the significance of a converged solution for the square-integrable Ψ_o is that a discrete N-electron solution is found in the continuous spectrum of H which is eigenfunction (in the limit) of QHQ of eq 46 .

VII. THEORY AND COMPUTATION OF Ψ_o AND OF Ψ_{core}

For the problems already tackled by the present CESE theory of nonstationary states, methods for the computation of correlated electronic as well as of vibrational resonance wavefunctions have been developed and applied. Below I will outline our approach to the calculation of isolated autoionizing and of diabatic states. The final core state bound wavefunction, Ψ_{core} , is calculated in terms of the same theory and methods [51].

As regards the treatment of nonstationary states which fragment into the diatomic vibrational continuum, the reader is referred to [27,41]. The problems which are discussed and solved there are: i) Calculation of tunneling rates via numerical integration in the complex coordinate plane [27]. ii) Ab initio calculation of energy curves and off-diagonal matrix elements for the determination of predissociation rates of excimer states in HeH, NeH and HeF, using analytic methods which involve real and complex distributed Gaussians for the representation of the vibrational Ψ_o and g^i (eq.36) [41]. It is interesting to note that in HeF, calculation of the nonadiabatic partial width to all orders for the decay of the first excited $^2\Sigma$ repulsive ground state shows that is is an order of magnitude larger than that predicted by the golden rule formula.

Autoionizing States

The calculation of Ψ_o starts with eq. 47. Φ_o is the Fermi-Sea [50,53], state-specific, multiconfigurational Hartree-Fock description which for atoms and diatomics is obtained numerically and therefore, no basis set errors intrude in the important zeroth order description. The Fermi-Sea Φ_o contains configurations that are expected to be most important - excluding those which contribute to the decay of Ψ_o. As regards ordinary ground and excited states, guidelines for the choice of the FS orbitals (also called "active orbitals" in quantum chemical methods developed more recently) have been given as a combination of empirical with a priori arguments [53]. If needed, some of the orbitals of Φ_o are made orthogonal to core orbitals of lower-lying configurations in order to exclude to open channels. The importance of starting the computation of autoionizing states with such a Φ_o has been repeatedly demonstrated since the introduction of the present theory [29,11,46-50]. The remaining localized correlation, X_{loc}, is developed in terms of correlation functions which are computed variationally subject to orthogonality constraints to all the orbitals of Φ_o as well as to the core orbitals of lower-lying configurations [29, 11,3, 46-50].

In order to demonstrate the above approach with new results, let me consider two *triply excited* negative ion resonances of **He⁻**, the $2s2p^2\ ^4P$ and ^2D.

The He^- $2s2p^2\ ^4P$ state has been computed variationally by Chung and Davis [54] and by Bylicki [55], using large CI wavefunctions which are made orthogonal to lower states through the application of hole-projection operators [29,3]. Following the present theory, the important step is to obtain an MCHF solution. Indeed,

$$\Phi_{MCHF}(^4P) = 0.9902\ (2s2p^2) + 0.1052\ (2p^23d) - 0.9020(2s3d^2) \qquad (48)$$

$$(\text{with } \langle 2s, 1s_{He^+} \rangle = 0)$$

whose energy (E=-0.78521 a.u. = 57.648 eV above He) is only about 0.25 eV above the previous values [54, 55]. Such a property of rapid convergence for the self-consistent field solutions for multiply excited states has been observed in a number of applications of the state-specific theory [46,50] including the well-known case of the He^- $1s2s2p\ ^2P^0$ broad resonance [48], and the He^- $1s2p^2\ ^4P$ core excited shape resonance [56].

Going beyond $\Phi_{MCHF}(^4P)$ in terms of only 7 additional, variationally optimized virtual orbitals, yields a compact Ψ_o with E_o= 57.414 eV, in agreement with the results of refs.54 and 55.

The application of the theory is continued by computing Δ and Γ due to the interaction with the continuum. Only the He 1s2p $^3P^0$ channel is available for autoionization. Then, using the Φ_{MCHF} instead of Ψ_o (for reasons of economy and without great loss of accuracy-see previous section) together with a ten-term square-integrable representation of the Gamow orbital (eq. 36), leads to a small 11x11 complex matrix whose diagonalization yields Δ=0.027 eV and Γ=0.015 eV.

Now I turn to the He^- $2s2p^2$ ^2D resonance whose existence has been disputed by Chung and Davis [54]. I show that with a proper identification of the important zeroth order mixing configurations, the MCHF solution exists and so does the ^2D resonance regardless of whether it is above or below the He 2s2p $^3P^0$ threshold. As with most multiply excited states, there are more than one thresholds below which the resonance may occur. In the present case, the He $2p^2$ ^1D state is definitely above the He^- $2s2p^2$ ^2D state and plays the role of the closed channel.

In the 60's and early 70's, experimentalists [57, 58] and theoreticians [59, 60, 29] alike agreed on the existence of the He^- $2s2p^2$ ^2D resonance around 58.3 eV above the He ground state. Chung [54] and Chung and Davis [55] disputed the correctness of this work and claimed that this resonance does not exist. The arguments were based on their unconverged, saddle-point technique calculations and on a critical appraisal of the previous theoretical work. For example, in ref.54, p.94, they conclude "This result does not corroborate the previous assignment by Fano and Cooper. What has been seen in the experiment could be the result of a postcollision interaction effect". In discussing the particularities of this state, Chung stressed that "one must exclude the [(s,s) ^1Sd] and [(p,p) ^1Sd] angular terms in the variational calculation", otherwise it will either collapse to the He $2s^2$ ^1S energy or it will not be very meaningful. Then he pointed to the characteristics of the Nicolaides wavefunction [29]- as quoted by Schulz [58] , p.406 - which contained 83% of HF $2s2p^2$ and 17% from the correlation vectors, with the main contribution coming from $2s^2nd$ and $2p^2nd$ configurations. Thus, he concluded that these early calculations did not offer "conclusive evidence for the existence of this Feshbach resonance since the inclusion of these terms may lead to erroneous results".

The answer to these remarks is based on the very essence of the nature of the state-specific theory of autoionizing states. As it has already been underlined, this theory puts emphasis on the dynamical localization which must characterize all N-electron autoionizing states regardless of their mode of excitation or decay, and on the recognition that this can best be brought out from an MCHF (or similar) calculation of the important zeroth order near-degeneracy effects. Since MCHF theory is variational, the justification for its implementation to the calculation of states embedded in the continuum is the shell-structure constraint of orthogonality to core orbitals, the proper orbital nodal structure and the satisfaction of the virial theorem- a property which is intrinsic to converged HF or MCHF solutions as well as to localized systems. Thus, the theory associates MCHF zeroth order solutions with physical, quasilocalized states in the continuous spectrum.

A converged MCHF solution means that at the particular energy of the continuous spectrum this solution has the largest coefficient in the full expansion of the wavefunction. The correlating orbitals describe state-specific localized correlations and not open channels. For example, in the case of the $2s2p^2\ ^2D$ state, the MCHF 3d bound orbital of the $2s^23d$ correlating configurations is square integrable and nodeless and does not resemble the εd scattering functions. Yet, since this 3d function is obtained independently of the εd channel and is not made orthogonal to it, it contains contributions from the continuous spectrum -just like the correlating bound orbitals of ground states do. The net overall effect is that the $2s^23d$ eigenvalue is above the $2s2p^2$ solution which corresponds to the lowest root. Thus, in the state-specific variational scheme, whether using an analytic HF $2s2p^2$ solution and radially compact virtual orbitals of d symmetry [29], or the numerical MCHF solution as in this work, the optimized $2s^23d$ and $2p^23d$ configurations simply contribute to the localized Ψ_0.

This can be further understood from the arguments of ref. 3 (pages 460-465) and of ref. 46 (p.109).That is, the correlating virtual orbitals, such as the 3d in the He^- case, pick up just a part of the continuous spectrum which does not destroy the square-integrability of the zeroth order function. Of course,the remaining physically important portion of the continuum must be computed via an appropriate theory, with real or complex functions.

In order to carry out the self-consistent variational calculations for the best zeroth order description of the He^- $2s2p^2\ ^2D$ state, the appropriate correlating configurations much be chosen. In the present case I chose the important configurations representing angular and radial correlations in the n=2 and n=3 shells. In both cases, well-converged numerical MCHF solutions were computed subject to the orthogonality condition $<2s \text{ or } 3s_{MCHF}|1s_{ion}>=0$. The most important configurations in the expansion are (in parentheses are the absolute values of the coefficients): $2s2p^2$(.927), $2s^23d$(.225), $2p^2(^3P)3d$ (.211), $2s3p^2$(.151), $(3s3p)^3P^02p$(0.073), $2s3d^2$(.105), $2p^2(^1S)3d$(.072).

I conclude that this state exists as a resonance although its exact position and widths will be decided after the remaining interelectronic interactions representing closed and open channels have been added. The energy of the compact, 12-term $He^-\ ^2D$ function is 58.61 eV above the He ground state while experiment [57,58] yielded 58.30±0.04 eV. How much of this difference is caused by X_{loc} and how much by X_{as}, remains to be seen.

Ψ_0 for Molecular Diabatic States

The state specific theory and methods for computing the Ψ_0 of autoionizing states is also applicable to the a priori construction of correlated wavefunctions for *diabatic* or *quasidiabatic* states.

As is well known [61-63] diabatic states do not diagonalize the full electronic Hamiltonian. However, knowing their potential energy curves and their wavefunctions may often fascilitate the interpretaion of collisional processes and of certain spectroscopic phenomena. The question of practical interest is how to compute them from first principles directly, without transformation from the adia-

ic states in diatomic molecules by making the formal correspondence with the then available treatments of autoionizing states. In ref. [66] he applied his idea of "projected atomic orbitals" (PAO) in conjunction with simple valence-bond functions to construct diabatic wavefunctions corresponding to covalent states of H_2, He_2^+, He_2 and HeH^+. In most cases, a valid qualitative picture emerges. However, such simple wavefunctions cannot be considered quantitatively reliable. Furthermore, since the extension of valence-bond type methods to include more electron correlation in the wavefunctions of diabatic states of polyelectronic molecules has not been accomplished, it is desirable to be able to treat such problems in the MO picture. This subject was commented upon in 1972 [29, footnote 73] in the context of the question of including the important zeroth-order near-degeneracy effects in the Ψ_0 of N-electron autoionizing states. The proposal was made that the self-consistent computation of the diabatic states could start in the dissociation region where their zeroth order single or multiconfigurational representation can be defined from the physics. Once such a diabatic state has been computed in the dissociation region, its state-specific wavefunction, Ψ_0, can be computed for smaller distances by maximizing the overlap with the previous solution [67]. This method would secure in most cases the construction of a good Ψ_0 for all geometrical arrangements, by excluding self-consistently the undesirable crossing of discrete or scattering states. An application of this idea was made in ref.45 , were it was shown using state-specific zeroth order functions, that crossing diabatic curves of the triatomic molecules HeH_2, NeH_2 and ArH_2 could indeed be constructed a priori in this way, thereby explaining the previously observed quenching of the radiation from $H_2^*B^1\Sigma_u^+$ by He and predicting similar phenomena for Ne and Ar.

The state-specific theory [49,50] has recently been implemented for the calculation of electron correlation of diatomic molecules [68]. Thus, the correlated Ψ_0 for a ground or an excited state is obtained by computing a zeroth order numerical [69] multiconfigurational Hartree-Fock function, followed by the computation of the localized correlation effects variationally, using orbitals in cylindrical symmetry.

Using this approach, we studied the He_2^+ states of $^2\Sigma_g^+$ and $^2\Sigma_u^+$ symmetry [68, 70, 71]. Fig. 1 shows the well-known diabatic system [63] of the $^2\Sigma_g$ ($1\sigma_g 1\sigma_u^2$, $1\sigma_g^2 n\sigma_g$, n=2,3....) states. The valence diabatic curve crosses the diabatic Rydberg states reaching the $He_2^{++1}\Sigma_g$ curve near its minimum, at R=1.207 a.u. . From then on it represents an autoionizing state. The comparison with O'Malley's calculation [66] for the valence state is as follows:

R(a.u.)	1.0	1.25	1.5	2	3	4	6
-E(a.u.) (Ref.66)	3.084	3.736	4.135	4.553	4.803	4.855	4.872
Present theory (Ref.68)	3.807	3.881	4.175	4.552	4.805	4.856	4.872

Although the two-term valence-bond description with POAs introduced by O'Malley [66] works very well for large R, where the structure of the system $He+He^+$ is simple and the diabatic curve essentially coincides with the adiabatic one [71] , in the inner region (R<2 a.u.) there are additional correlation effects which push the diabatic curve to lower energies.

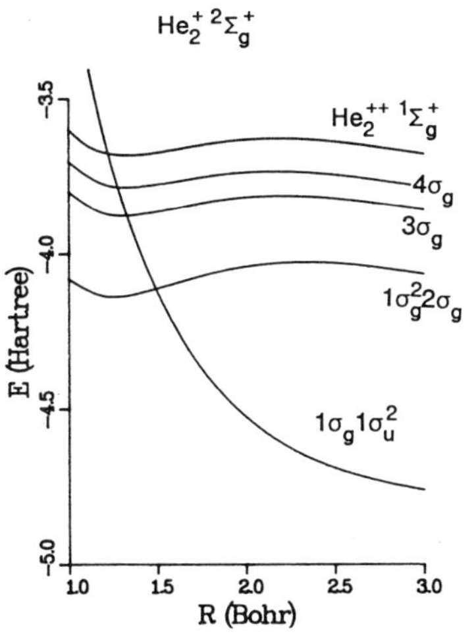

Fig. 1. Diabatic curves for the $He_2^+\ {}^2\Sigma_g^+$ system of states ($1\sigma_g 1\sigma_u^2$, $1\sigma_g^2 n\sigma_g$, n=2,3,4, $He_2^{++}\ 1\sigma_g^2\ {}^1\Sigma_g^+$) computed directly, according to the present theory. The computational method follows from the state-specific theory [50]. Its implementation [68] involves the calculation of Ψ_o (eq. 47), with Φ_o being the numerical [69] MCHF function and X_{loc} the variationally obtained important localized correlation corrections.

REFERENCES

1. W. Heitler, "The Quantum Theory of Radiation" 3rd Ed., Oxford 1954.
2. M.L. Goldberger and K.M. Watson, "Collision Theory", J.Wiley N.Y.(1964).
3. C.A. Nicolaides and D.R. Beck, Int.J.Qu.Chem. 14 475 (1978).
4. R.G.Newton, "Scattering Theory of Waves and Particles" 2nd Ed., Springer-Verlag, N.Y. (1982).
5. C.Mahaux and H.A. Weidenmüller, "Shell Model Approach to Nuclear Reactions", North Holland, Amsterdam (1969).
6. G. Gamow, Z.f.Physik 51, 204 (1928).
7. A.F.J. Siegert, Phys.Rev. 56, 750 (1939).
8. U.Fano, Phys.Rev. 124 1866 (1961).
9. U.Fano and F.Prats, J.Natl.Acad.Sci. (India) A33 533 (1963).

10. The Fano-Prats work (ref.9) is limited to the case of only open channels. The general theory with the inclusion of closed channels, the derivation of multichannel quantum defect formulae independently of the properties of the Coulomb function and the implementation in terms of state-specific numerical and analytic functions for the many-electron computation of photoabsorption cross-sections to perturbed Rydberg states close to threshold and to doubly excited Rydberg series of resonances, was presented recently by Y.Komninos and C.A.Nicolaides, Z.Phys. B4 301 (1987); Phys.Rev. A34 1995 (1986).
11. C.A. Nicolaides, Y. Komninos and Th.Mercouris, Int.J.Qu.Chem. S15, 355 (1981).
12. For a short historical account of the LoSurdo-Stark effect, see the article by H.J.Silverstone in "Atoms in Strong Fields" , eds. C.A.Nicolaides, C.W.Clark and M.H.Nayfeh, Plenum (1990), p. 295.
13. In the theory of resonances by Feshbach (ref.14) and by Fano (ref.8), the projection operators P and Q and the corresponding effective Hamiltonians and projected interaction operators, as well as the prediagonalized zeroth order Hamiltonians, are defined only formally. These theories are fundamental in explaining the phenomenology of resonances. However, for real atomic and molecular nonstationary states, it is equally important to have theories and methods which provide not only the framework for the definition of their properties but also for their systematic, electronic structure-dependent computation. For example, in order to demonstrate the dissolution of a discrete level into the continuous spectrum, the doubly excited states of the He atom have been used as a prototype example in the following way. H_o is taken to be the interactionless hydrogenic operator and $V=1/r_{12}$. However, such a model is conceptually unsatisfactory since the interelectronic interactions are, in fact, never turned-off! Furthermore, it is obviously computationally naive and cannot lead to accurate results with a reasonable amount of effort. This difficulty is accentuated for polyelectronic atoms. The theory of this article aims at justifying those essential formal results which allow the understanding and practical computation of nonstationary states many-electron atoms and molecules.
14. H. Feshbach, Ann.Phys.(N.Y.) 5 357 (1958); 19 287 (1962).
15. P.A.M. Dirac, "The Principles of Quantum Mechanics", Oxford Univ. Pr., 4th Ed., (1957), chapter 8.
16. E.C. Kemble, "The Fundamental Principles of Quantum Mechanics" Dover, (1958).
17. Ya.B.Zeldovich, Sov.Phys. (JETP) 12 542 (1961).
18. A.M.Dykhne and A.V.Chaplik, Sov.Phys.(JETP) 13 1002 (1961).
19. The Dykhne-Chaplik paper (ref.18) was apparently first quoted in the literature of atomic and molecular physics in ref. 11, as soon as it was discovered. In the meantime, their transformation (eq.26) had been used and had become known in the 70's , both as a regularization technique

(ref.20,26,3) and as a formal means of studying the spectral properties of the atomic Coulomb Hamiltonian (ref.21). The results of ref.21 led to the so-called complex coordinate rotation (CCR) method (refs 22-24) whereby $H(re^{i\theta})$ is diagonalized repeatedly in a large real or complex square-integrable basis set and the resonances are identified by the regions of stability observed as a function of the rotation angle θ or of the size of the basis sets. Upon rereading their one-page article for the purpose of writing this review, I saw that Dykhne and Chaplik also proposed the possibility of starting the complex integration beyond a point on the real axis to circumvent problems of nonanalyticity. In recent years this idea has been known as "exterior complex scaling" (refs. 25-27).

20. J.N.Bardsley and B.R.Junker, J.Phys.B5 L178 (1972) .
21. J.Aguilar and J.M.Combes, Commun. Math.Phys. 22 269 (1972); E.Balslev and J.M.Combes, Commun.Math.Phys. 22 280 (1972);B.Simon, Ann.Math.97 247 (1973).
22. G.Doolen, J.Nuttall and R.W.Stagat, Phys.Rev. A10 1612 (1974); G.Doolen, J.Phys.B8 525 (1975); R.A. Bain, J.N.Bardsley, B.R.Junker and C.V.Sukumar, J.Phys. B7, 2189 (1974); B.R.Junker, Int.J.Qu.Chem. 14 371 (1978);. N. Moiseyev, P.R.Certain and F.Weinhold, Phys.Rev. A24 1254 (1981).
23. B.R.Junker, Adv.At.Mol.Phys. 18 207 (1982); Y.K.Ho, Phys.Reports, 99 1 (1983).
24. W.P.Reinhardt, Ann.Rev.Phys.Chem. 33 223 (1982); C.Cerjan, R.Hedges, C.Holt, W.P.Reinhardt, K.Scheibner and J.J.Wendoloski, Int.J.Qu.Chem. 14 393 (1978).
25. B.Simon, Phys.Lett. A71 211 (1979).
26. C.A. Nicolaides and D.R.Beck, Phys.Lett. A65 11 (1978).
27. C.A.Nicolaides, H.J.Gotsis, M.Chrysos and Y.Komninos, Chem.Phys.Lett. 168 570 (1990) and refs. therein.
28. C.A. Nicolaides and S.Themelis, unpublished.
29. C.A. Nicolaides, Phys.Rev. A6, 2078 (1972).
30. C.A.Nicolaides, Th.Mercouris and Y.Komninos, Int.J.Qu.Chem.26 1017 (1984); C.A.Nicolaides and Th.Mercouris, Phys.Rev. A32 3247 (1985).
31. Th. Mercouris and C.A. Nicolaides, J.Phys. B17 4127 (1984).
32. C.A. Nicolaides and Th. Mercouris , Phys.Rev. A36 390 (1987).
33. Th. Mercouris and C.A. Nicolaides, Z. Phys. D5 1 (1987).
34. M. Chrysos, Y. Komninos, Th.Mercouris and C.A. Nicolaides, Phys. Rev. A42 2634 (1990).
35. Th. Mercouris and C.A. Nicolaides, J.Phys. B21 L285 (1988).
36. C.A. Nicolaides and Th. Mercouris, Chem. Phys. Lett. 159 45 (1989).
37. Th. Mercouris and C.A. Nicolaides, J.Phys. B23 2037 (1990).
38. Th. Mercouris and C.A. Nicolaides, J. Phys. B24 L 57 and L165 (1991).
39. C.A. Nicolaides, Th. Mercouris and G.Aspromallis, J.Opt. Soc. Am. B7 494 (1990).

40. C.A. Nicolaides, Th. Mercouris and N. A. Piangos, J. Phys.B23 L669 (1990).
41. I.D. Petsalakis, Th. Mercouris, G.Theodorakopoulos and C.A. Nicolaides, J.Phys. B23 L89 (1990); J.Chem.Phys. 93 6642 (1990); Chem.Phys.Lett. (1991).
42. A systematic perturbative treatment of the rotated atomic Coulomb Hamiltonian, $H(re^{i\theta})$ is also possible, based on the form of eq. 9.3 of ref. 3. Formally, the infinite nonHermitian Hamiltonian matrix is written as $H(\theta)=H(0)+K(\theta)$. $K(\theta)=(e^{-2i\theta}-1)T+(e^{-i\theta}-1)V$, where T and V are the kinetic and potential energy matrices respectively. This formulation is an expression of the idea that the calculation of the complex eigenvalue , z_0, should constitute a continuation from E_0, the expectation value of H(0) on the real axis, and allow the possibly interesting study of the autoionization shift and width, of say a doubly excited state, in the complex plane via CI-based small-or large-order perturbation theory (J.N.Silverman and C.A.Nicolaides, Chem.Phys.Lett. 153 61 (1988); in "Atoms in Strong Fields" eds. C.A.Nicolaides, C.W.Clark and M.H.Nayfeh, Plenum (1990), p. 309.
43. Even this picture breaks down in principle, when the strength of the external field increases to the point that the Ψ_0 cannot represent only the unperturbed, free atomic or molecular state.
44. Although the thrust of the work of ref. 29 was on N-electron autoioinizing states of arbitrary electronic structure, its concepts and methods are applicable to the subject of the a priori construction of correlated *diabatic* molecular states. This was pointed out in footnote 73 of ref. 29 but no such computations were possible at that time. Later on, an application of the idea of starting with the properly projected diabatic solution in the dissociated region and moving into the interaction region while exluding unwanted interacting configurations by maximizing the wavefunction of each state-specific solution at each geometry to the previous one, was applied to the analysis of the potential energy surfaces of HeH_2, NeH_2 and ArH_2 (ref. 45).
45. C.A.Nicolaides and A.Zdetsis, J.Chem.Phys. 80 1900 (1984).
46. Y.Komninos, N.Makri and C.A.Nicolaides, Z.Phys. D2 105 (1986).
47. Y.Komninos and C.A.Nicolaides, J.Phys.B19 1701 (1986).
48. Y.Komninos, G.Aspromallis and C.A.Nicolaides, Phys.Rev. A27 1865 (1983).
49. C.A. Nicolaides in "Advanced Theories and Computational Approaches to the Electronic Structure of Molecules" ed. C.E. Dykstra, Reidel (1984), p. 161.
50. C.A.Nicolaides, in "Quantum Chemistry - Basic Aspects, Actual Trends", ed., R.Carbo, Elservier (1989).
51. It is obvious from the structure of the theory that interacting scattering resonances as well as intermediate dressed bound states can be included in the formalism and the overall computation using the same methods for obtaining their correlated wavefunctions (see refs. 37, 46-48, 52).

52. C. A. Nicolaides and Th. Mercouris, in "Atoms in Strong Fields", eds C.A.Nicolaides, C.W.Clark and M.H. Nayfeh, Plenum (1990), p. 353.
53. D.R. Beck and C.A. Nicolaides, in "Excited States in Quantum Chemistry" eds. C.A.Nicolaides and D.R.Beck, Reidel (1978), p. 105.
54. K.T.Chung and B.F. Davis, in "Autoionization", ed. A.Temkin, Plenum, N.Y. (1985) p. 73; K.T. Chung, Phys.Rev. A22 1341 (1980).
55. M. Bylicki, Phys.Rev. A , in press (1991).
56. C.A. Nicolaides, Y. Komninos and D.R. Beck, Phys.Rev. A24 1103 (1981).
57. C.E. Kuyatt, J.A. Simson and S.R. Mielczarek, Phys.Rev. 138 A385 (1965); P.J. Hicks, C.Cvejanovic, J. Comer, F.H. Read and J.M. Sharp, Vacuum 24, 573(1974).
58. G.J. Schulz, Rev.Mod.Phys. 45 378 (1973).
59. U. Fano and J.W.Cooper, Phys.Rev. 138 A400 (1965).
60. K.Smith, D.E.Golden, S.Ormonde, B.W. Torres and A.R.Davis, Phys. Rev. A8 3001 (1973).
61. W. Lichten, Phys.Rev. 131 229 (1963).
62. F.T. Smith, Phys.Rev. 179 111 (1969).
63. T.F.O'Malley, Adv.At.Mol.Phys. 7 223 (1971).
64. C.A. Mead and D. G. Truhlar, J.Chem.Phys. 77 6090 (1982).
65. T.F.O.Malley, Phys.Rev. 162 98 (1967).
66. T. F.O.Malley, J.Chem.Phys. 51 322 (1969).
67. For a diatomic molecular electronic spectrum, the analogy with the atomic spectra as a function of Z, treated as a continuous parameter, is enlightening. Consider the mixing of a valence configuration (V) with a Rydberg (R) series and the scattering (S) states of the same channel. The V-R-S mixing is Z-dependent . For large Z, the V state is found below the R states which acquire more hydrogenic character and are raised in energy. Call the large Z region, the "dissociation" region . Here, the definition and computation of the Ψ_0 for a V state is straight forward (For example, the $1s^22p^2\ {}^1S$ valence excited state is represented mainly by $a(1s^22p^2)+b(1s^22s^2)$). As Z is decreased, the V state may start "crossing" the R states which start coming down. At the neutral or negative ion end, the V state may lie in the continuous spectrum, mixing with the scattering states of the same symmetry and configuration as those of the R states below the ionization threshold. This is indeed the case with the $1s^22p^2\ {}^1S$ V state. For Z=4 (Be) it lies in the continuous spectrum. For Z=5 (B^+), it lies below the $1s^22sns\ {}^1S$ series and above the ground state $1s^22s^2\ {}^1S$. For $4<Z<5$, it "crosses" the Rydberg states. If its Ψ_0, which is defined unambiguously for Z=5, is optimized for each noninteger value of Z between 5 and 4 with its state-specific numerical zeroth order and analytic correlation functions excluding by construction or orthogonality the R-S ${}^1S$ channel, an "atomic diabatic state" is calculated.
68. N. Bacalis, Y.Komninos and C.A. Nicolaides, unpublished.
69. E.A. McCullough, J.Chem.Phys. 62 3991 (1975).
70. C.A. Nicolaides, Chem.Phys.Lett. 161 547 (1989);
71. A. Metropoulos, C.A.Nicolaides and R.J.Buenker, Chem.Phys. 114 1 (1987).

ON THE CONSTRUCTION OF SIZE EXTENSIVE EFFECTIVE HAMILTONIANS IN GENERAL MODEL SPACES USING QUASI-HILBERT AND QUASI-FOCK STRATEGIES

Debasis Mukhopadhyay (Jr) and Debashis Mukherjee

Theory Group, Department of Physical Chemistry
Indian Association for the Cultivation of Science
Calcutta 700-032, India

I. INTRODUCTION

There has been an ever-growing interest over the last two decades towards developing and implementing the formalism of effective hamiltonians [1] in the framework of many-body perturbation theory (MBPT) and coupled cluster (CC) theory [2, 3] . One major thrust in all these developments has been the desire to obtain size-extensive effective hamiltonians H_{eff} which give size-extensive energies upon diagonalization within the model space. The traditional formulations of both MBPT [4] and CC theory [5-9] for the open-shells were built upon complete model spaces (CMS), and a connected H_{eff} could be obtained in all these formalisms. The energies obtained on diagonalization were automatically size-extensive owing to the completeness of the model space.

It is now well-documented that the theories utilizing the CMS have serious numerical instabilities stemming from the "intruder states" [1-3, 10]. It also does not seem to be physically sensible to start with a large model space, chosen simply because it is complete-without reference to the wide spread in their energies, when we are interested in the low-lying region of the spectrum. As an example, for most of the low-lying excited states relative to a closed shell ground state, we should start with just the required hole-particle (hp) and 2 hole-particle (2h2p) excited determinants warranted due to their proximity in energies and develop analogous MBPT and CC theory. These model spaces are incomplete, necessitating a formulation of size-extensive effective hamiltonians in incomplete model spaces (IMS).

It should be emphasized here that there are two aspects of size-extensivity which should be borne in mind when we use an IMS. One is the generation of a connected H_{eff}. The other is the guarantee that the energies on diagonalization are also size-extensive. The second aspect is specially relevant if we recall that a CI matrix in an incomplete model space generates size-inextensive terms <u>even if</u> the microscopic hamiltonian H is connected. One thus has to impose special structure to an H_{eff}, defined in an IMS, to ensure that such terms do not appear.

All the open-shell many-body theories can be classified into two groups : (a) those using a model space of a given number of electrons, and (b) those involving a "valence universal" wave-operator Ω for not only the parent model space but also the subduced lower valence model space. It is customary now to categorize them as belonging to Hilbert space type and Fock space type of open-shell theories [2, 3] . For a CMS, connected H_{eff} and size-extensive energies can be generated using both these approaches in MBPT [4] and CC theories [5-9].

The first attempt to abandon a CMS in favour of an IMS in MBPT was by Hose and Kaldor [10]. They obtained a disconnected H_{eff} whose disconnected components themselves were however, not legitimate terms of H_{eff} of lower order. They hence concluded that the size-extensivity of energies is not violated in their formulation. Sheppard [11] showed that the size-extensivity is destroyed when the energies are obtained on diagonalization. More extensive theoretical studies made later [12-14] bolstered this conclusion. Hose and Kaldor used a Hilbert space approach. The analogous formulation in Hilbert space for CC theory of Jeziorski and Monkhorst [9] using IMS led also to a disconnected H_{eff} and the associated energies.

In contrast to the Hilbert space approach, the use of a Fock space approach with a valence-universal Ω [2, 3, 5, 7] leads to a clear separation of the working equations for either MBPT or CC theory into various n-body components. Discerning the sources of disconnectedness and their amelioration become rather more transparent in such a strategy. It is thus not surprising that the first rigorous formulations of an extensive H_{eff} for an IMS could be achieved using the Fock space approach [15, 16] . Subsequent developments continued to use a valence-universal Ω [3,17-20]. In these works, Bloch equation is cast in Fock space, and we thus obtain energies for the different valence sectors simultaneously. It turns out (vide infra) that one should pay attention to two distinct but interrelated aspects of size extensivity : (a) choosing a normalization for Ω that is compatible with the connectivity of H_{eff}, and (b) the use of suitable "decoupling conditions" for determining the cluster components of Ω which ensures that the subsequent diagonalization produces size-extensive energies. It is remarkable that the Fock space approach satisfies both these conditions in a natural manner. The traditional choice of intermediate normalization (IN) for Ω, used in all the previous works with IMS, was found to be incompatible with the connectedness of H_{eff} . This normalization has therefore to be generally abandoned in the formulations using IMS [2,15].

Once, however, the conditions leading to extensivity are well-understood, it is possible to adopt the same strategy (i.e. the same choice of normalization and the same decoupling conditions) for a Hilbert space formulation. We are then led to a size-extensive formulation for H_{eff} and the energies in a Hilbert space [21] . We wish to view this scheme as a Fock space theory for IMS that is <u>projected onto</u>

a given n-valence model space [21, 22]. The information regarding all the lower valence model space gets integrated out by this projection. This is why we call this approach as using a quasi-Hilbert space formulation [21, 22]. We may want to go a step still further ; we may , e.g., demand that our formulation generates an H_{eff} for not only the n-valence IMS, but also for some, specific, lower valence IMS sectors as well. Thus we envision here a strategy of projecting the Fock space working equations onto several valence sectors. This strategy has been termed a quasi-Fock space formulation by us [21, 22] and is well-suited for the direct calculation of such energy differences of spectroscopic interest as excitation energies [23] or double ionization potentials [24] . In practice, we never solve the Fock space MBPT or CC equations, and then project them on the valence sectors of interest. This way of viewing the situation is only for conceptual clarity. In actual application, we derive the analogous working equations directly for just the model spaces of interest -either one valence sector as in Quasi-Hilbert space theory or for several valence sectors as in Quasi-Fock space theory.

If one is concerned in generating a size-extensive formulation for just one valence-sector, then one may envisage using another alternative route. One may start with a formulation for CC theory as in the approach of refs [21, 22] . Following the development of Mukherjee [15, 16] , one then abandons the IN for Ω. One may then look for the sources of disconnectivity in the equations for the cluster components of Ω and introduce extra flexibility, i.e. extra decoupling conditions, which cancel these disconnected terms. This is the approach taken by Meissner *et al* [25] and Meissner and Bartlett [26]. The resultant formulation is essentially equivalent to our Quasi-Hilbert space theory. These authors did not consider the Quasi-Fock space extension, however.

We want to present in this article a survey of these two alternative ways of generating an extensive H_{eff}. We shall omit most of the proofs ; our main objective is to present an overview of the developments-delineating the logical interconnections among themselves and with the Fock space approach. In order to fix our terminology and notations, and also to set the scenario, we start out with a brief introduction to the CC theory in Fock space using an IMS.

II. FOCK SPACE CC THEORY WITH AN IMS : A RESUME'

Let us suppose that we have either holes or particles in our parent n-valence IMS. All the crucial aspects of proving the connectivity of H_{eff} can be described in such a situation. The case of IMS involving nh-mp excited determinants with fixed (n, m) is exactly analogous and the more complicated case of several nh-mp determinants with different (n, m) can be transcribed in terms of valence particles only.

The Bloch equation for the n-valence IMS can be written as

$$H \Omega P^{(n)} = \Omega P^{(n)} H_{eff} P^{(n)} \qquad (2.1)$$

where $P^{(n)}$ is the projector onto the model space.

By construction, H_{eff} is a "closed" operator [15-17] i.e. it cannot lead to excitations out of the model space. We call an operator "external" [2, 27] if there is at least one model space function for which the action of the operator leads to an excitation out of the model space. We include in Ω , following ref. [15] , <u>all</u> the possible external operators in Ω. We write Ω in normal order, with respect to the zero-valence determinant $|\Phi_0>$ as the vacuum, as

$$\Omega = \{\exp (S)\} \qquad (2.2)$$

where S contains all the external operators. It may so happen that some external operators lead to scattering within the model space for some model space functions. The powers of S may even be closed. Thus $P^{(n)} \Omega P^{(n)} \neq P^{(n)}$ in general, and IN for Ω is abandoned. The cluster components of S are to be determined from the Fock space decoupling conditions that all the different m-body components of the external part of the transformed hamiltonian $L = \Omega^{-1} H \Omega$ are vanishing :

$$[L_m]_{ext} P^{(n)} = \left[(\Omega^{-1} H \Omega)_m\right]_{ext} P^{(n)} = 0, \forall m \tag{2.3}$$

Clearly, H_{eff} will be the closed part of L.

The essential difference between a naive Hilbert space approach, and the Fock space approach lies in using eq. (2.3) in the former and a condition

$$Q^{(n)} L P^{(n)} = 0 \tag{2.4}$$

imposed in the latter. Eq. (2.3) implies that all the external operators of L are to be eliminated. If for some model space function, this same operator induces a model space → model space scattering, then H_{eff} in that case will have a null entry, owing to the validity of eq. (2.3). In eq. (2.4), one does not individually equate the various m-body external components of L to zero, rather only specific combinations thereof contributing to $Q^{(n)} L P^{(n)}$. Ω in that case does not have all the external operators. The null entries indicated above does not thus arise at all [14].

It is straightforward to demonstrate that eq. (2.3) implies Ω to be a valence universal wave-operator. Let us generate all the lower k-valence model spaces $P^{(k)}$ obtained by deleting (n-k) valence orbitals from all the determinants comprising $P^{(n)}$. Since $H_{eff} = L_{cl}$ is a closed operator by construction (with appropriate null entries in positions implying scattering by an external operator), even for each $P^{(k)}$ it can scatter within that model space. Thus $P^{(k)} L_{cl} P^{(k)}$ is the appropriate connected H_{eff} for the k-valence model space. We thus have two equivalent options of generating the CC equations in Fock space for an IMS. One is the way we have just indicated : i.e. use of eqs (2.1) and the decoupling conditions (2.3). The other would be to use Bloch equations for each k-valence model space :

$$H \, \Omega \, P^{(k)} = \Omega \, P^{(k)} \, H_{eff} \, P^{(k)}, \ \forall \ 0 \leqslant k \leqslant n \qquad (2.5)$$

The ansatz, eq. (2.2) for Ω, remains the same as before.

It still remains to be shown explicitly that the energies on diagonalization of H_{eff} in the IMS are also extensive. Let us assume, following ref [14] that we are inerested in the root E_k , and we want to follow its emergence from a perturbative series via Bloch equation where the starting function is a model function ϕ_m and the other model functions ϕ_l $(l \neq m)$ are the virtual functions. We call the associated wave-operator as W_m .

In close analogy to the corresponding CI-problem using the same space of functions, we expect to generate disconnected contribution from the norm correction terms. Thus, for example, at the fourth order of perturbation, we expect to obtain disconnected terms of the form $\left[-\left(VW_m^{(1)}\right)^a \left(VW_m^{(1)}\right)^b\right]$, having no orbital labels in common between the components $(\ )^a$ and $(\ )^b$. By using the generalized time-ordering [28], these terms can be rewritten as sum over certain "direct" terms with a negative sign where the intermediate functions generated by $W_m^{(3)}$ would contain the multiply excited victual functions generated by the operator product $W_k^{(1)a} \, W_k^{(1)b}$ on ϕ_m. If these same virtual functions are generated by the direct term $VW_k^{(3)}$, then they would cancel each other. Thus the only source of disconnectedness in E_k can arise if the functions generated by $W_k^{(1)a} \, W_k^{(1)b} \, \phi_k$ are not present in $W_k^{(3)}$. For this to be true, $W_k^{(1)a} \, W_k^{(1)b} \, \phi_k$ must not belong to the set of functions ϕ_l in the IMS. But this is impossible, since $W_k^{(1)}$ is generated by the operator $H_{eff} = L_{cl}$ which being closed by construction can never generate any other functions besides the ϕ_l . This feature is quite generally valid at all orders, and hence there are no disconnected terms in the energies obtained by diagonalizing H_{eff} in the IMS. We emphasize here that in the analogous CI problem in the IMS, there are <u>no null entries</u> in <u>the posi-</u>

tions of the external operators (since H, unlike L_{cl}, also contains H_{ext}). Thus, it is possible in that case to generate a function outside the model space. This will happen whenever $W_k^{(1)a}\ \phi_k$ and $W_k^{(1)b}\ \phi_k$ are model functions but $W_k^{(1)a}\ W_k^{(1)b}\ \phi_k$ is not. For a more elaborate discussion of this point, we refer to ref [14].

It is instructive to go back again to the Bloch equation (2.1) for $P^{(n)}$ where we now explicity take a specific function ϕ_l of $P^{(n)}$. Then eq. (2.1) leads to

$$H\,\Omega\,\phi_l = \Omega\,P^{(n)}\,H_{eff}\,\phi_l \tag{2.6}$$

The decoupling conditions, eq. (2.3), also can be written as

$$[L_m]_{ext}\,\phi_l = 0 \tag{2.7}$$

If we now confine ourselves to only the specific n-valence IMS, and project $L = \Omega^{-1}\,H\,\Omega$ onto the virtual functions characterized by $Q^{(h)}$ orthogonal to $P^{(n)}$, then we have

$$Q^{(n)}\,[L]\,\phi_l = 0 \tag{2.8}$$

Only certain specific combinations of various m-body external operators can contribute to eq. (2.8), depending upon which function in $Q^{(n)}$ is used in the projection, and the eq. (2.8) will hold good as a consequence of eq. (2.7). But eq. (2.7) tells us more. We know that an external operator in an IMS can in general lead to scattering within the model space depending on the function ϕ_l it acts upon. Such cases are not covered by eq. (2.8), and this is the reason we do not have null entries in H_{eff} in a naive Hilbert space version. But with the hindsight of our Fock space decoupling conditions, we may now demand that projection of $L\,\phi_l$ on all the model functions generated by the external operators in Ω on ϕ_l should also be vanishing. Calling this part of model space as $P_l^{(k)}$, we may write these additional conditions as

$$P_l^{(k)} \; L \; \Phi_l \; = 0 \qquad (2.9)$$

Ω should contain all such external operators leading to scattering within the IMS by their action on Φ_l. They should be determined from eq. (2.9).

Since eqs. (2.8) and (2.9) are compatible with the Fock space decoupling conditions, eq. (2.7), it should be possible to formulate an extensive CC theory for just the model space $P^{(n)}$. Of course, Ω in the Fock space contains much more cluster amplitudes than that in a (would be) Hilbert space Ω. Thus the sets of equations generated by eqs (2.8) and (2.9) are much less numerous than the set (2.7). However, by taking each Φ_l as the vacuum, we can combine the redundant cluster amplitudes of the lower valence problems suitably to have precisely the right number of unknowns in the Hilbert space Ω. We demonstrate the emergence of this scheme in the next section.

III. QUASI-HILBERT SPACE CC THEORY FOR AN IMS

Let us assume that we have an IMS, with the associated projector P. To be compatible with the Fock space decoupling conditions, eq (2.3) or (2.7), we call an operator as "external" if there is at least one model function which gets excited by it to a virtual function. Operators leading to excitations within the IMS are called "closed". We want H_{eff} to be closed.

We choose a cluster ansatz for Ω in the IMS is analogous to the Hilbert space ansatz for the CMS [9] :

$$\Psi_k = \Omega \; \Psi_k^0 = \Sigma_\mu \; C_{\mu k} \; \Omega^\mu \; \phi^\mu \qquad (3.1)$$

where $\{\phi^\mu\}$ spans the IMS, and Ω^μ is written as

$$\Omega^\mu = \exp \; (S^\mu) \qquad (3.2)$$

S^μ involves all the external operators corresponding to the chosen model space. We write S^μ in terms of holes and particles taking ϕ^μ as the vacuum. There are thus no orbital labels in common in each S^μ before and after the scattering. Since some external operators in S^μ can connect ϕ^μ to another model function ϕ^ν, and since the powers of S^μ may even be closed, we do not have IN for Ω in general. Let us note, however, that Ω^μ has no term (except unity) that causes a scattering from ϕ^μ onto itself.

To determine Ω, we start from the Bloch equation :

$$H\,\Omega^\mu\,\phi^\mu = \Omega\,P\,H_{eff}\,\phi^\mu\ \forall\mu \tag{3.3}$$

For each equation for a particular ϕ^μ, it is convenient to rewrite all the operators in normal order with respect to ϕ^μ as the vacuum. Thus, we introduce the following operators :

$$H = \langle H\rangle^\mu + H_0^\mu + V^\mu \tag{3.4}$$

$$H_{eff} = \langle H_{eff}\rangle^\mu + H_0^\mu + G_{eff}^\mu \tag{3.5}$$

where $\langle A\rangle^\mu = \langle\phi^\mu|A|\phi^\mu\rangle$. The operator H_0^μ is a one-body operator diagonal in the orbital basis, and the superscript μ implies that the normal ordering is with respect to ϕ^μ. In general V^μ is a sum of one body and two body operators :

$$V^\mu = -\Sigma_{ij}\langle i|u^\mu|\rangle\left\{a_i^+ a_j\right\}^\mu + \frac{1}{2}\,\Sigma_{ijkl}\langle ij|g|kl\rangle\left\{a_i^+ a_j^+ a_l\, a_k\right\}^\mu \tag{3.6}$$

G_{eff}^μ is the operator part of V_{eff}, written in normal order with respect to ϕ^μ.

Premultiplying eq. (3.3) by $\exp(-S^\mu)$, and using eqs. (3.1) and (3.2), we have

$$\left[S^\mu,\ H_0^\mu\right]\phi^\mu = Z^\mu\,\phi^\mu - \Sigma_{\nu\neq\mu}\,\Upsilon^{\nu\mu} \tag{3.7}$$

where

$$Z^\mu = \overline{\{V^\mu\ \exp\ (S^\mu)\}^\mu} - \overline{\langle V^\mu\ \exp\ (S^\mu)\rangle^\mu} \tag{3.8}$$

and

$$Y^{\nu\mu} = \left\{e^{-S^{\mu}} e^{S^{\nu}} G^{\nu\mu}_{eff}\right\}^{\mu} - \langle e^{-S^{\mu}} e^{S^{\nu}} G^{\nu\mu}_{eff} \rangle \qquad (3.9)$$

The product of exponentials in eq. (3.7) may be expanded using Baker-Campbell-Hausdorff formula [21] as :

$$\exp(-S^{\mu}) \exp(S^{\nu}) = \exp \left[(S^{\nu}-S^{\mu}) + \frac{1}{2} [S^{\nu},S^{\mu}] + \frac{1}{12} [[S^{\nu},S^{\mu}],S^{\mu}] - \frac{1}{12} [[S^{\nu},S^{\mu}],S^{\nu}] + \ldots\right] \qquad (3.10)$$

From eq. (3.8), the equations for S^{μ} are given by

$$\left[S^{\mu}_{m} , H^{\mu}_{0}\right]_{ext} \phi^{\mu} = \left\{Z^{\mu}_{m}\right\}_{ext} \phi^{\mu} - \Sigma_{\nu\neq\mu} \left[Y^{\nu\mu}_{m}\right]_{ext} , \forall m,\mu \qquad (3.11)$$

where m stands for various particle ranks. Clearly, to solve for eq. (3.11) we have to project it not only onto Q, the orthogonal complement of P, but also onto $\overline{P}_{\mu}$, the space reached by the external operators S^{μ} acting on ϕ^{μ} -as indicated at the end of the previous section.

Although the proof of the connectivity of S^{μ} is somewhat involved, we indicate here rather schematically the essential steps, since this will help us later to see the connection of this formalism with the ones proposed by Meissner _et al_ [25] and Meissner and Bartlett [26]. For more detailed discussions, we refer to our original papers [21, 23].

Since Z_m is manifestly connected if S^{μ} is connected, it is enough to show that $[Y^{\nu\mu}_{m}]$ is connected. To show this, we also need the closed components of eq. (3.7), which essentially defines G_{eff} :

$$Z_{cl} \phi^{m} = \Sigma_{\nu\neq\mu} Y^{\nu\mu}_{cl} = \Sigma_{\nu\neq\mu} \left\{\exp(-S^{\mu}) \exp(S^{\nu}) G^{\nu\mu}_{eff}\right\}^{\mu}_{cl} \qquad (3.12)$$

Since $Y^{\nu\mu}_{cl}$ involves the _differences_ $(S^{\nu} - S^{\mu})$ apart from commutators as follows from eq. (3.10), we have to show that $(S^{\nu}-S^{\mu}) G^{\nu\mu}_{eff}$ is a connected operator. The rest of the expansion in eq. (3.10) involve commutators and the connectivity can be discerned easily. Since each eq. (3.11) is of exactly the same structure for each $\Upsilon^{\nu\mu}$, the solutions S^{μ} for each ϕ^{μ} should have <u>the same functional dependance</u> on μ. Quite generally, then, $(S^{\nu}-S^{\mu})$ may be written as

$$S^{\nu} - S^{\mu} = \Delta S^{\nu} - \Delta S^{\mu} \tag{3.13}$$

where ΔS^{ν} and ΔS^{μ} are those components of S^{ν} and S^{μ} which have explicitly the labels from the group of orbitals distinguishing ϕ^{ν} from ϕ^{μ}. Since $G_{eff}^{\nu\mu}$ induces a transition from ϕ^{μ} to ϕ^{ν}, it must be labelled by <u>all</u> the orbitals needed to differentiate ϕ^{μ} and ϕ^{ν}. The term $(S^{\nu}-S^{\mu})\ G_{eff}^{\nu\mu}$ is thus connected in the sense of having orbitals common to $(S^{\nu}-S^{\mu})$ and $G_{eff}^{\nu\mu}$. Since the rest of expansion of the eq. (3.10) involve commutators, the connectivity of $\Upsilon^{\nu\mu}$ is established, if $G_{eff}^{\nu\mu}$ is connected. $G_{eff}^{\nu\mu}$ can also be proved to be connected, since we find that

$$\left[G_{eff}^{\nu\mu}\right] = \langle\phi^{\nu}|Z_{CI}|\phi^{\nu}\rangle - \Sigma_{\lambda\neq\mu}\langle\phi^{\nu}|\{[S^{\lambda}-S^{\mu}]+ \ldots\}|\phi^{\lambda}\rangle G_{eff}^{\lambda\mu} \tag{3.14}$$

Using again the argument that $(S^{\lambda}-S^{\mu})\ G_{eff}^{\lambda\mu}$ have always orbital labels in common, the connectness of $G_{eff}^{\nu\mu}$ is established. Moreover $\langle H_{eff}\rangle^{\mu}$ is also connected, and is given by the expression :

$$\langle H_{eff}\rangle^{\mu} = \langle H\rangle^{\mu} + \langle V^{\mu} \exp(S^{\mu})\rangle^{\mu}$$

$$- \Sigma_{\nu\neq\mu}\langle\phi^{\mu}|\exp(-S^{\mu})\exp(S^{\nu})|\phi^{\nu}\rangle G_{eff}^{\nu\mu} \tag{3.15}$$

Eqs. (3.17) to (3.15) are the principal working equations of the Quasi-Hilbert space CC theory for an IMS, derived by us [21, 23]. Once S^{μ}, s are obtained, the matrix of H_{eff} in the IMS can be constructed as

$$\langle\phi^{\nu}|H_{eff}|\phi^{\mu}\rangle = \langle H_{eff}\rangle^{\mu}\ \delta_{\mu\nu} + G_{eff}^{\nu\mu} \tag{3.16}$$

For all the pairs (ϕ^{ν}, ϕ^{μ}), connected by an external operator, $\langle\phi^{\nu}|H_{eff}|\phi^{\mu}\rangle = 0$ by construction. We recall that in a Hilbert space formulation, it is not a priori clear that we need this decoupling at all, since <u>there is no com-</u>

pelling reason in a Hilbert space formulation to decouple the components of H_{eff} within the model space prior to a diagonalization. We see the emergence of these extra decouplings as a consequence of eq. (2.9) (also, see the discussion after eq. (3.11)), since they have to conform to the Fock space decoupling conditions. For a more detailed discussion, we refer to our earlier papers [14,21-23].

IV. SIZE-EXTENSIVITY OF ENERGIES IN QUASI-HILBERT SPACE CC THEORY

We now demonstrate the connectivity of the roots E^{μ} obtained by diagonalizing H_{eff} in the IMS. From now on, for a particular μ, we rename ϕ^{μ} and E^{μ} as simply ϕ^0 and E^0. Any other ϕ^{ν} hence forth will imply $\nu \neq \mu$. We may follow the diagonalization via Bloch equation defined in the model space, with ϕ^0 as the starting function and ϕ^{ν}'s as the virtuals. However since we have only operators of the form $G^{\nu\mu}_{eff}$ which depend on the ϕ^{μ} it acts upon, the generalized time ordering argument used in the Fock space formulation cannot be exploited. For this, we need a single operator, independent of the function it acts upon. Again a special proof is needed, which we sketch below to give a flavour of the type of arguments one has to give for Hilbert space developments [21-23].

For our proof, we need to rewrite H_{eff} in normal order with respect to ϕ° as the vacuum. Since the definition of the external operators depends on the IMS and not on the choice of the vacuum, the null entries in the matrix of H_{eff} are not affected by this charge of the vacuum, and moreover, H_{eff} remains connected.

For proving the connectivity of E°, it is unnecessary to distinguish between numbers and operators in H_{eff}. Thus, we repartition H_{eff} as

$$H_{eff}\ \phi^0 = \langle H_0 \rangle\ \phi^0 + H_0^{\mu}\ \phi^0 + \Sigma^{\mu}\ \phi^0 \qquad (4.1)$$

where Σ^{μ} is given by

$$\Sigma^{\mu} = \langle H_{eff} - H_0 \rangle^{\mu} + G^{\mu}_{eff} = \langle V_{eff} \rangle^{\mu} + G^{\mu}_{eff} \quad (4.2)$$

Also, from now on, $\Sigma^{\mu} = \Sigma^{0}$.
To follow the diagonalization via Bloch equation, we follow an order by order expansion of the following expression :

$$H_{eff}\ W\ \phi^0 = W\ \phi^0\ E^0 = (I + X)\ \phi^0\ E^0 \quad (4.3)$$

and use $\langle \phi^0 | W | \phi^0 \rangle = 1$ to fix the normalization of the associated wave operator W. We have proved [21] that W can be written as an operator exp (T) where T contains only connected operators including transitions out of ϕ^0 to all other ϕ^{ν}'s. T contains only "closed" operators as introduced for the IMS in that T can never cause transitions outside the IMS in which H_{eff} is diagonalized. T has operators T_{λ} which produce ϕ_{λ} from ϕ^0 :

$$T = \Sigma_{\lambda}\ T_{X}\ ;\ T_{\lambda}\ \phi^0 = t_{\lambda}\ \phi^{\lambda} \quad (4.4)$$

The products of T similarly produce multiple excitations, though these functions are still in the IMS :

$$T_{\lambda}\ T_{\nu}\ \phi^0 = t_{\lambda}\ t_{\nu}\ \phi^{\lambda+\nu} \quad (4.5)$$

We illustrate our proof by explicitly considering the terms of W upto third order. To show this, we note that $\Sigma^{\nu} - \Sigma^{0}$ must involve operators which are <u>explicitly labelled</u> by the orbitals from the group distinguishing ϕ^0, using arguments entirely similar to the one leading to eq. (3.14). We may write this difference as

$$\Sigma^{\nu} - \Sigma^{0} = \Delta^{(\nu,o)} \quad (4.6)$$

where $\Delta^{(\nu,o)}$ has orbital labels from the group of orbitals distinguishing ϕ^{ν} and ϕ^{o}. We also generalize eq. (4.6) as

$$\Sigma^{\lambda+\nu} - \Sigma^{o} = \Delta^{(\lambda+\nu,o)} = \Delta^{(\lambda,o)} + \Delta^{(\mu,o)} + \Delta^{(\lambda,\nu,o)} \quad (4.7)$$

where $\Delta^{(\lambda,\nu,o)}$ has some labels from each group of orbitals that distinguish ϕ^{λ} and ϕ^{v} . This notation is an obvious extension involving more superscripts. To avoid multiple counting, $\Delta^{(\lambda,o)}$ and $\Delta^{(\mu,o)}$ exclude terms included in $\Delta^{(\lambda,\nu,o)}$.

It may be verified by iteration in order of perturbation that eq. (4.3) leads in first order to the following expression for X_v, inducing excitation from ϕ^0 to ϕ^v :

$$X_v^{(1)} = R_o \left[G^0_{eff}\right]_v = T_v^{(1)} \qquad (4.8)$$

with R_o as the reduced resolvent with respect to ϕ^0. We also have,

$$E^{(1)0} = \langle V_{eff}\rangle^0 \qquad (4.9)$$

At second order, using eq. (4.6), we have

$$[X_v^{(2)}, H_o^0] = \Sigma_\lambda \left\{G^0_{eff}\ T_\lambda^{(1)}\right\}_v + \overline{\left\{G^0_{eff}\ T_\lambda^{(1)}\right\}}_v$$
$$+ \Sigma_\lambda \left\{\Delta^{(\lambda,0)}\ T_\lambda^{(1)}\right\}_v + \Sigma_\lambda \overline{\left\{\Delta^{(\lambda,0)}\ T_\lambda^{(1)}\right\}}_v \qquad (4.10)$$

The terms containing $\Delta^{(\lambda,o)}$ and $T_\lambda^{(1)}$ are all connected -whether they have explicit contractions between them or not, since $T_\lambda^{(1)}$ is labelled by all the orbitals distinguishing ϕ^λ and ϕ^0, and $\Delta^{(\lambda,o)}$ has some subsets from this group of labels. Introducing a <u>connected entity</u> $((\Delta A))$ as

$$((\Delta A)) = \Sigma_\lambda \left[\left\{\Delta^{(\lambda,o)}\ A_\lambda\right\} + \overline{\left\{\Delta^{(\lambda,o)}\ A_\lambda\right\}}\right] \qquad (4.11)$$

we have

$$X^{(2)} = \frac{1}{2} T^{(1)^2} + T^{(2)} \qquad (4.12)$$

with

$$T_\nu^{(2)} = R_o \left[\overline{\left\{G^0_{eff}\ T^{(1)}\right\}}_v + ((\Delta T^{(1)}))_v\right] \qquad (4.13)$$

$T_v^{(2)}$ is thus a connected operator. $E^{(2)0}$ is given by

$$E^{(2)0} = \langle\phi^0|\overline{G^0_{eff}\; T^{(1)}}|\phi^0\rangle + \langle\phi^0|((\Delta T^{(1)}))|\phi^0\rangle \tag{4.14}$$

and is connected.

To proceed at the third order, we introduce the following generalized connected terms :

$$((\Delta A))^{(v,o)} = \Sigma_\lambda \left[\left\{\Delta^{(\lambda,v,o)}\; A_\lambda\right\} + \overline{\left\{\Delta^{(\lambda,v,o)}\; A_\lambda\right\}}\right] \tag{4.15}$$

$$((\Delta AB)) = \Sigma_v \left[((\Delta A))^{(v,o)}\; B_v + \overline{((\Delta A))^{(v,o)}\; B_v}\right] \tag{4.16}$$

$((\Delta A))^{(v,o)}$ and $((\Delta AB))$ are, by construction, connected entities. The above definitions allow of obvious extention to more superscripts. Defining $X^{(3)}$ as

$$X^{(3)} = \frac{1}{3}\, T^{(1)^3} + T^{(1)}\; T^{(2)} + T^{(3)}, \tag{4.17}$$

we have

$$T_3^{(2)} = R_o \left[\frac{1}{2}\left\{\overline{\overline{G^0_{eff}\, T^{(1)}}\, T^{(1)}}\right\}_v + \left\{\overline{G^0_{eff}\, T^{(2)}}\right\}_v + \frac{1}{2}((\Delta T^{(1)}\, T^{(1)}))_v\right] \tag{4.18}$$

Thus X correct to the order γ, $X^{[\gamma]}$, emerges from an exponential structure of W : $W^{[\gamma]} \sim \exp\,(T^{[\gamma]})$, with $T^{[\gamma]}$ a connected operator. $T^{[\gamma]}$ has been obtained for $\gamma = 1, 2, 3$. It follows by induction [21] that the exponential structure for W and the connectedness of T as well as of E° are valid to all orders. Since any ϕ^μ can serve as vacuum for an appropriate E^μ_o the extensivity of the roots follows.

V. CONNECTED HILBERT SPACE FORMULATION BY INSPECTION : AN APPROACH DE NOVO

One may envisage an alternative approach where one generates a connected S^μ by inspection. This is the approach used by Meissner <u>et al</u> [25] and Meissner and Bartlett [26] in their development of Hilbert space type of extensive CC theory.

Suppose we take a cue from the Jeziorski and Monkhorst formulation [9] and write Ω in the same manner as in eqs. (3.1) and (3.2). Unlike in the Quasi-Hilbert formulation of Sec. III, where we were guided by the Fock space decoupling conditions to intercept that S^{μ} has to include all the external operators, Meissner et al [25] proceeded by inspection and in steps. Thus, as befitting a Hilbert space formulation, they first included in S^{μ} only those operators leading to excitations onto Q. With such a choice of S^{μ}, they were then led to consider the expression analogous to eq. (3.7). Now, in the expression for Z, they should again encounter a term of the form $(S^{\nu} - S^{\mu})\ G_{eff}^{\nu\mu}$. When the type of operator in consideration in S^{μ} is such it leads to excitation when acting on ϕ^{μ} as well as ϕ^{ν}, S^{ν} contains the same operator also, and one can adduce the same arguments as the ones given in Sec. III to show that $(S^{\nu} - S^{\mu})\ G_{eff}^{\nu\mu}$ is connected. The trouble arises when this operator in S^{μ} is such that its action on ϕ^{ν} leads to a scattering within the model space. The corresponding operator is then absent from S^{ν} by choice, and $(S^{\mu} - S^{\nu})G_{eff}^{\nu\mu}$ cannot remain connected. One way out of this difficulty is to consider special IMS's where external operators can never lead to scattering within the model space. Such model spaces were first introduced by Lindgren [29] and studied further by others [17] and are called quasi-complete model spaces. For such choices of model space, whenever there is an external operator in S^{μ}, there is always the same external operator in any other S^{ν}, and $(S^{\nu} - S^{\mu})$ remains connected. In their first formulation, Meissner et al [25] considered only the quasi-complete IMS. In a later paper, Meissner and Bartlett [26] showed that if one includes in S^{ν} also all the operators taken in S^{μ}, then $(S^{\nu} - S^{\mu})$ can be made to be connected. To be conistent, it essentially implies that one uses the same ansatz for Ω as the one used by us [21-23], since S^{μ} now contains all the external operators in the same sense as used by us. The two approaches are thus essentially equivalent. In our opinion, our way of viewing the solution as a Fock space theory projected onto a specific n-electron Hilbert space offers the additional insight in that we know beforehand which kind of operators to include in S^{μ} without vitiating the Fock space decoupling conditions. It also motivates us to look for extensive formulations which are more general in their information content-viz the Quasi-Fock space theories [21-23].

VI. QUASI-FOCK SPACE CC THEORY FOR AN IMS

We now describe a quasi-Fock space theory where information of some subduced k-valence model spaces is deliberately retained in Ω. For example, in computing the excitation energies of a closed shell ground state ψ_{gr}, S_{gr} is computed first taking ϕ_o for the Hartree-Fock ground state as the vacuum. In the Fock space strategy, one then proceeeds hierarchically upwards by first solving the one valence (one hole and one particle) cluster amplitudes first and then going to the desired one hole-one particle functions $\phi^{\alpha p} = a_p^+ a_\alpha \phi_o$ as forming an IMS. Although, in this formulation one obtains the ionization potential and electron affinity of the ground state as bonus, we may not always want to compute them. In the quasi-Fock space theory, we side-step this procedure and go over directly to the one hole-one particle IMS after solving the ground state problem. This leads to a drastic reduction of the numbers of S amplitudes without losing the information of interest.

Having solved for S_{gr}, we introduce the "dressed" hamiltonian $\tilde{H}$ via

$$\tilde{H} = \exp(-S_{gr})\, H \exp(S_{gr}) = E_{gr} + \bar{H} \tag{6.1}$$

with

$$\bar{H} = H_o + \bar{V} \tag{6.2}$$

where H_o and $\bar{V}$ are written in normal order with respect to ϕ_o.

For Ω defined in the h-p IMS, we use an ansatz of the form

$$\Omega = \exp\,(S_{gr})\;\overline{\Omega} \qquad (6.3)$$

From the Bloch equation for the h-p IMS :

$$H\,\Omega\,\phi^{\alpha p} = \Omega\,P\,H_{eff}\,\phi^{\alpha p} \qquad (6.4)$$

it then follows that

$$\overline{H}\,\overline{\Omega}\,\phi^{\alpha p} = \overline{\Omega}\,P\,\overline{H}_{eff}\,\phi^{\alpha p} \qquad (6.5)$$

where $\overline{H}_{eff} = H_{eff} - E_{gr}$ generates the excitation energies directly on diagonalization in the h-p IMS. We now write $\overline{\Omega}$ as

$$\overline{\Omega} = \Sigma_{\mu}\,\exp\,(T^{\mu})\;|\phi_{\mu}><\phi_{\mu}| \qquad (6.6)$$

where μ stands for all the collective indices (αp). Just as in the Quasi-Hilbert space theory, T^{μ} should include in it all the external operators. We are then led to CC equations for the quasi-Fock space theory :

$$\left[T^{\mu},\,H_{o}^{\mu}\right]\,\phi^{\mu} = Z^{\mu}\,\phi^{\mu} - \Sigma_{\nu\neq\mu}Y^{\nu\mu} \qquad (6.7)$$

where $\overline{Z}$ and $\overline{Y}$ have expressions similar to those in eqs. (3.8) and (3.9), with $\overline{V}$ replacing V. Since $\overline{H}$ is a connected operator by construction, it is clear that T^{μ} and $\overline{H}_{eff}^{\nu\mu}$ are all connected -using arguments entirely similar to those used in the quasi-Hilbert space theory. The special IMS considered here (the h-p IMS) is quasi-complete, but the formalism sketched above is clearly valid for any general IMS. It is also possible to compute the double ionization potentials in a similar manner by solving for the ground state problem first and then directly going over to the two-hole determinants forming a model space. For the first few double ionization potentials, one needs to include in the model space only the outer-outer, and the outer-inner ionized determinants, which is an IMS [24]. For a recent application of both quasi-Hilbert and quasi-Fock theory we refer to one of our recent papers [23].

VII. CONSTRAINTS IN THE TRUNCATION SCHEMES IN QUASI-HILBERT AND QUASI-FOCK THEORIES

Unlike in the Fock-space developments, there are essential constraints in truncating the operators in S^{μ}'s and also in the expansion of Ω^{μ}'s in the CC equations in the quasi-Hilbert or quasi-Fock theories. In a Fock space formulation, each term in the CC equation is manifestly connected so that S can be truncated after any particle rank. Also, due to normal ordering the highest power of $S^{(n)}$ - i.e. the operator in S with n valence destruction operators - for the CC equation for the n-valence problem is two. The situation is completely different for the quasi-Hilbert and quasi-Fock space formulations. The powers of S^{μ}'s in the CC equations for these theories can be much greate than two. Moreover, in these theories $(S^{\nu} - S^{\mu})\, G^{\nu\mu}_{eff}$ is a connected operator although individually $S^{\mu}\, G^{\nu\mu}_{eff}$ etc. are not. The crucial argument leading to the connectivity had been to assume that S^{μ} and S^{ν} have the same functional dependence on the superscript μ and ν. This implies that any operator included in S^{μ} has to be in S^{ν} and, moreover, they must appear <u>up to the same power</u> in the CC equation. Clearly this imposes quite nontrivial restrictions on the truncation schemes. Any external operator included in a S^{μ} must appear in the S^{ν}'s $(\nu \neq \mu)$, and each must be expanded in the same manner in all the Ω^{ν}'s as in Ω^{μ} in the CC equations. This has to be done even if a particular operator has rather small excitation amplitude for a particular Φ^{ν}.

In thus appears that the advantages in using the quasi-Hilbert and quasi-Fock theories (i.e. working in specific n-electron Hilbert spaces) are offset somewhat by not only the more involved book-keeping processes because of multiple vacua and diagrammatic transcription of an operator due to a change in the vacuum but also due to the restriction in the truncation schemes that one should carefully conform to.

VIII. CONCLUDING REMARKS

We have described quasi-Hilbert and quasi-Fock space CC theories using arbitrary IMS which provide size-extensive H_{eff} and the associated energies. For this to hold good, it is essential to choose a size-extensive normalization for Ω and the use of decoupling conditions for Ω that are compatible with the Fock space decoupling conditions. One may, in principle, retain in the formalism the explicit information of the subduced k-valence model spaces upto some k. Let us assume that k=n corresponds to the parent model space of interest. When information upto k=n-1 are suppressed, we have a theory for only one model space, which has been called by us as a quasi-Hilbert space theory. When information upto some k<n-1 are retained, we have a quasi-Fock space theory. Two alternative ways of achieving a connected formulation are described if only one IMS is of our interest. In one, we start from the Bloch equation in that IMS, and impose normalization and decoupling conditions appropriately, by choosing S^{μ} to contain all the external operators. This is the approach chosen by us [21-23], which may be viewed as a Fock space CC theory projected on a given IMS. In another approach, one first includes in S^{μ} only those operators needed to excite ϕ^{μ} to virtual space. Connectivity then is shown to demand that the same operators must also excite ϕ^{v}'s ($v \neq \mu$) to the virtual space. This is possible only for some special IMS, viz quasi-complete model space. For more general model spaces, one is forced to include all the external operators in all the S^{μ}'s. This is the approach followed by Meissner et al [25] for quasi-complete model space and by Meissner and Bartlett [26] for the general IMS. Our approach seems to be more general, since we can envision the quasi-Fock strategy as well.

Although the problem of intruders are to a large extent avoided in the formulations with IMS, they are not altogether absent when computing potential surfaces. The trouble with the potential surfaces is that, states which are nicely separated in energy in one nuclear configuration come precariously close to the model space energies in another. It is thus very difficult to separate the model space and the virtuals uniquely, unambiguously and naturally over all the nuclear configurations of interest |C| see,e.g. ref. [34] for a concrete case involving potential curves of Li_2 molecule). A way out of this dilemma is to use the concept of intermediate hamiltonians [1, 30-32]. Although it appears that a Fock space formulation leading to a size-extensive intermediate hamiltonian is possible [35], much more work is necessary to formulate a quasi-Hilbert or quasi-Fock size-extensive theory for the intermediate hamiltonians.

IX. ACKNOWLEDGEMENTS

The authors thank F. Courte and G. Dedieu for their kind help in preparing the manuscript in photoready form. The work was completed in Laboratoire de Physique Quantique of Université Paul Sabatier. D.M. thanks all its members for their warm hospitality. D.M. also gratefully acknowledges stimulating discussions with J.P. Malrieu, J.P. Daudey, P. Durand, J.L. Heully, A.V. Zaitsevski, W. Kutzelnigg and S. Koch on effective hamiltonians and the related formalisms. Thanks are also due to the CSIR and DST (India) and the CNRS (France) for their financial support.

REFERENCES

1. See, e.g. P. Durand and J.P. Malrieu, Adv Chem Phys 57, 321 (1987) for a general Survey of the effective hamiltonian formalisms.

2. See, e.g. D. Mukherjee and S. Pal, Adv Quantum Chem, 20, 291, (1989) for an extensive and critical survey of both MBPT and CC theory, both for complete and incomplete model spaces.

3. For an earlier exposition, see, e.g. I. Lindgren and D. Mukherjee, Phys Rep., 151, 93 (1987)

4. B. Brandow, Rev. Mod. Phys. 39, 771 (1967) ; I. Lindgren, J. Phys B 7, 2441 (1974)

5. D. Mukherjee, R.K. Moitra and A. Mukhopadhyay, Mol Phys. 33, 955 (1977). D. Mukherjee, Pramana, 12, 203 (1979) A. Haque and D. Mukherjee, J Chem Phys. 80, 5058 (1984)

6. R. Offerman, W. Ey and H. Kümmel, Nucl Phys. A273, 349 (1976). R. Offerman, Nucl Phys. A273, 368 (1976). W. Ey, Nucl Phys., 296, 189 (1978)

7. I. Lindgren, Int. J. Quantum Chem Symp. 12, 33 (1978)

8. A. Haque and U. Kaldor, Chem. Phys. Lett. 117, 347 (1985) ; 120, 261 (1985). U. Kaldor, Int. J. Quantum Chem. Symp. 20, 445 (1986)

9. B. Jeziorski and H.J. Monkhorst, Phys Rev. A24, 1668 (1981).

10. G. Hose and U. Kaldor, J Phys B12, 3827 (1979) ; Phys. Scripta. 21, 357 (1980) ; Chem Phys. 62, 469 (1981) ; J. Phys. Chem. 86, 2133 (1982)

11. H.G. Sheppard, J. Chem Phys. 80, 1225 (1984)

12. L. Meissner, K. Jankowski and J. Wasilewski, Int. J. Quantum Chem. 34, 535 (1988)

13. L. Meissner and R.J. Bartlett, J. Chem. Phys. 91, 4800 (1989)

14. R. Chaudhuri, D. Sinha and D. Mukherjee, Chem. Phys Lett. 163, 165 (1989)

15. D. Mukherjee, Proc. Ind. Acad. Sci. 96, 145 (1986) ; Chem Phys Lett. 125, 207 (1986) ; Int J Quantum Chem Symp. 20, 409 (1986)

16. D. Sinha, S.K. Mukhopadhyay and D. Mukherjee, Chem Phys Lett. 129, 369 (1986)

17. W. Kutzelnigg, D. Mukherjee and S. Koch, J. Chem Phys.

87, 5902 (1987). D. Mukherjee, W. Kutzelnigg and S. Koch, J. Chem Phys. 87, 5911 (1987)

18. R. Chaudhuri, D. Mukhopadhyay and D. Mukherjee, in Lecture Notes in Chemistry, Vol 50 (Ed. D. Mukherjee, Springer Verlag, Heidelberg, 1989)

19. S. Pal, M. Rittby, R.J. Bartlett, D. Sinha and D. Mukherjee, Chem Phys. Lett. 137, 273 (1987) ; J. Chem. Phys. 88, 4357 (1988)

20. M. Rittby, S. Pal and R.J Bartlett, J. Chem Phys. 90, 3214 (1989) ; R. Mattie, M. Rittby, R.J. Bartlett and S. Pal, in Lecture Notes in Chemistry, Vol 50 (Ed : D. Mukherjee, Springer Verlag, Heidelberg, 1989)

21. D. Mukhopadhyay and D. Mukherjee, Chem Phys Lett. 163, 171 (1989)

22. D. Mukhopadhyay, R. Chaudhuri, S.K. Mukhopadhyay and D. Mukherjee, Theoretica Chimica Acta (Special Issue on CC Theory, in press)

23. D. Mukhopadhyay and D. Mukherjee, Chem Phys Lett. 177, 441 (1991)

24. D. Mukhopadhyay and D. Mukherjee, to be published

25. L. Meissner, S.A. Kucharski and R.J. Bartlett, J. Chem. Phys. 91, 6187 (1989)

26. L. Meissner and R.J. Bartlett, J. Chem Phys. 92, 561 (1990)

27. D. Mukherjee, in Condensed Matter Theories, Vol 3 (Ed : J. Arponen, R.F. Bishop and M. Manninen, Plenum Press, N.Y., 1988)

28. L.M. Frantz and R.L. Mills, Nucl Phys. 15, 16 (1960)

29. I. Lindgren, Phys. Scripta. 32, 291, 611 (1985)

30. J.P. Malrieu, P. Durand and J.P. Daudey, J. Phys. A18, 809 (1985)

31. J.L. Heully and J.P. Daudey, J. Chem. Phys. 88, 1048 (1988)

32. A. V. Zaitsevski and A.I. Dement'ev, J. Phys. B23, L517 (1990)

33. S. Koch, submitted to Theoretica Chim. Acta

34. U. Kaldor, Chem. Phys. 140, 1 (1990)

35. The last entry of ref. [15], and D. Mukhopadhyay and D. Mukherjee, to be published.

CONTRIBUTORS

1) Steven A. Blundell
University of California
Lawrence Livermore National Laboratory
Livermore, CA 9455(U.S.A.

2) I. Cacelli
Scuola Normale Superiore., Pisa, Italy

3) V. Carravetta
I.C.Q.E.M. del C.N.R., Pisa Italy

4) L.S. Cederbaum
Theoretische Chemie
Universitat Heildelberg
D-6900 Heidelberg, Germany

5) Shih-I. Chu
Department of Chemistry
University of Kansas
Lawrence, KS 66045, U.S.A.

6) Uzi Kaldor
School of Chemistry, Tel Aviv University
69978 Tel Aviv, Israel

7) Werner Ketzelnigg
Lehrstuhl fur Theoretische Chemie
Ruhr-Universitat Bochum
D-4630 Bochum, Germany

8) Jean Paul Malrieu
Laboratoire de Physique Quantique
(U.R.A. 505 du C.N.R.S.)
Universite Paul Sabatier
118, Route de Narbonne
31062 Toulouse Cedex, France

9) R. Moccia
Dipartimento di Chimica e Chimica Industriale
Pisa Italy

10) Debashis Mukherjee
Theory Group, Depatment of
Physical Chemistry
Indian Association for the Cultivation
of Science
Calcutta 700-032, India

11) Debasis Mukhopadhyay, Jr.
Theory Group, Department of
Physical Chemistry
Indian Association for the Cultivation
of Science
Calcutta 700-032, India

12) Cleanthes A. Nicolaides
Theoretical and Physical Chemistry Institute
National Hellenic Research Foundation
48 Vas. Constantinou Ave.
Athens 11635, Greece

13) A. Rizzo
I.C.Q.E.M. Del C.N.R., Pisa, Italy

14) A. Sgamellotti
Dipartimento di Chimica
Universita di Perugia, I-06100
Perugia, Italy

15) F. Tarantelli
Dipartimento di Chimica
Universita di Perugia, I-06100
Perugia, Italy

16) Danny. L. Yeager
Chemistry Department
Texas A & M University
College Station, Texas 77843-3255, U.S.A.

INDEX